Global Perspectives on Climate Change, Social Resilience, and Social Inclusion

Aly Abdel Razek Galaby
Alexandria University, Egypt

Mohammed Abo ElEnein
American University, UAE

Hassan Mohamed
Alexandria University, Egypt

A volume in the Practice, Progress, and
Proficiency in Sustainability (PPPS) Book Series

Published in the United States of America by
    IGI Global
    Engineering Science Reference (an imprint of IGI Global)
    701 E. Chocolate Avenue
    Hershey PA, USA 17033
    Tel: 717-533-8845
    Fax: 717-533-8661
    E-mail: cust@igi-global.com
    Web site: http://www.igi-global.com

Copyright © 2024 by IGI Global. All rights reserved. No part of this publication may be reproduced, stored or distributed in any form or by any means, electronic or mechanical, including photocopying, without written permission from the publisher. Product or company names used in this set are for identification purposes only. Inclusion of the names of the products or companies does not indicate a claim of ownership by IGI Global of the trademark or registered trademark.

Library of Congress Cataloging-in-Publication Data

Names: Galaby, Aly Abdel Razek, 1941- editor. | Aboelenein, Mohammed
    editor. | Mohamed, Hassan, 1945- editor.
Title: Global perspectives on climate change, social resilience, and social
    inclusion / edited by Aly Galaby, Mohammed Aboelenein, Hassan Mohamed.
Description: Hershey, PA : Engineering Science Reference, [2024] | Includes
    bibliographical references and index. | Summary: "The objective of this
    book is to study the important social, cultural, and political factors
    that affect the living conditions of the people in their societies. This
    study can provide the decision-makers with solutions to the problems
    that may face their societies and destabilize their structure"--
    Provided by publisher.
Identifiers: LCCN 2023045047 (print) | LCCN 2023045048 (ebook) | ISBN
    9781668489635 (hardcover) | ISBN 9781668489642 (paperback) | ISBN
    9781668489659 (ebook)
Subjects: LCSH: Climatic changes--Social aspects.
Classification: LCC QC903 .G576 2024 (print) | LCC QC903 (ebook) | DDC
    304.2/8--dc23/eng/20231102
LC record available at https://lccn.loc.gov/2023045047
LC ebook record available at https://lccn.loc.gov/2023045048

This book is published in the IGI Global book series Practice, Progress, and Proficiency in Sustainability (PPPS) (ISSN: 2330-3271; eISSN: 2330-328X)

British Cataloguing in Publication Data
A Cataloguing in Publication record for this book is available from the British Library.

All work contributed to this book is new, previously-unpublished material. The views expressed in this book are those of the authors, but not necessarily of the publisher.

For electronic access to this publication, please contact: eresources@igi-global.com.

Practice, Progress, and Proficiency in Sustainability (PPPS) Book Series

Ayman Batisha
International Sustainability Institute, Egypt

ISSN:2330-3271
EISSN:2330-328X

Mission

In a world where traditional business practices are reconsidered and economic activity is performed in a global context, new areas of economic developments are recognized as the key enablers of wealth and income production. This knowledge of information technologies provides infrastructures, systems, and services towards sustainable development.

The **Practices, Progress, and Proficiency in Sustainability (PPPS) Book Series** focuses on the local and global challenges, business opportunities, and societal needs surrounding international collaboration and sustainable development of technology. This series brings together academics, researchers, entrepreneurs, policy makers and government officers aiming to contribute to the progress and proficiency in sustainability.

Coverage

- Sustainable Development
- Outsourcing
- Strategic Management of IT
- Technological learning
- Global Content and Knowledge Repositories
- Green Technology
- Global Business
- E-Development
- Environmental informatics
- ICT and knowledge for development

IGI Global is currently accepting manuscripts for publication within this series. To submit a proposal for a volume in this series, please contact our Acquisition Editors at Acquisitions@igi-global.com or visit: http://www.igi-global.com/publish/.

The Practice, Progress, and Proficiency in Sustainability (PPPS) Book Series (ISSN 2330-3271) is published by IGI Global, 701 E. Chocolate Avenue, Hershey, PA 17033-1240, USA, www.igi-global.com. This series is composed of titles available for purchase individually; each title is edited to be contextually exclusive from any other title within the series. For pricing and ordering information please visit www.igi-global.com/book-series/practice-progress-proficiency-sustainability/73810. Postmaster: Send all address changes to above address. Copyright © 2024 IGI Global. All rights, including translation in other languages reserved by the publisher. No part of this series may be reproduced or used in any form or by any means – graphics, electronic, or mechanical, including photocopying, recording, taping, or information and retrieval systems – without written permission from the publisher, except for non commercial, educational use, including classroom teaching purposes. The views expressed in this series are those of the authors, but not necessarily of IGI Global.

Titles in this Series

For a list of additional titles in this series, please visit: www.igi-global.com/book-series/practice-progress-proficiency-sustainability/73810

Opportunities and Challenges for Women Leaders in Environmental Management
Perfecto Gatbonton Aquino, Jr. (The University of Cambodia, Cambodia) Revenio Cabanilla Jalagat, Jr. (Al-Zahra College for Women, Oman) and Mercia Selva Malar Justin (Xavier Institute of Management and Entrepreneurship, India)
Engineering Science Reference • copyright 2024 • 259pp • H/C (ISBN: 9781668459867) • US $250.00 (our price)

The Role of Women in Cultivating Sustainable Societies Through Millets
Raghvendra Kumar (GIET University, India) and Ishaani Priyadarshini (University of California, Berkeley, USA)
Engineering Science Reference • copyright 2024 • 350pp • H/C (ISBN: 9781668498194) • US $265.00 (our price)

Water-Soil-Plant-Animal Nexus in the Era of Climate Change
Ahmed Karmaoui (Moulay Ismail University, Meknès, Morocco & Moroccan Center for Culture and Science, Morocco)
Engineering Science Reference • copyright 2024 • 330pp • H/C (ISBN: 9781668498385) • US $240.00 (our price)

Quality of Life and Climate Change Impacts, Sustainable Adaptation, and Social-Ecological Resilience
Kasturi Shukla (Symbiosis International University) Yogesh B. Patil (Symbiosis International University, India) Ronald C. Estoque (Forestry and Forest Products Research Institute (FFPRI), Japan) and Pedro Antonio López de Haro (Universidad Autónoma Indígena de México, Mexico)
Engineering Science Reference • copyright 2024 • 300pp • H/C (ISBN: 9781668498637) • US $250.00 (our price)

Sustainable Development in AI, Blockchain, and E-Governance Applications
Rajeev Kumar (Moradabad Institute of Technology, India) Abu Bakar Abdul Hamid (Infrastructure University, Kuala Lumpur, Malaysia) Noor Inayah Binti Ya'akub (Infrastructure University, Kuala Lumpur, Malaysia) Hari Om Sharan (Amity University, Kolkata, India) and Sandeep Kumar (Rightzone Technologies Pvt. Ltd., India)
Engineering Science Reference • copyright 2024 • 350pp • H/C (ISBN: 9798369317228) • US $285.00 (our price)

Achieving the Sustainable Development Goals Through Infrastructure Development
Cristina Raluca Gh. Popescu (University of Bucharest, Romania & The Bucharest University of Economic Studies, Romania) Poshan Yu (Xi'an Jiaotong-Liverpool University, China & European Business University of Luxembourg, Luxembourg) and Yue Wei (Institution of Public Private Partnerships, Hong Kong)
Engineering Science Reference • copyright 2023 • 303pp • H/C (ISBN: 9798369307946) • US $230.00 (our price)

701 East Chocolate Avenue, Hershey, PA 17033, USA
Tel: 717-533-8845 x100 • Fax: 717-533-8661
E-Mail: cust@igi-global.com • www.igi-global.com

Table of Contents

Preface xii

Chapter 1

Building Community Resilience and Mitigating the Impacts of Climate Change Risks on Social
Inclusion: Exploratory Study on Alexandria, Egypt 1
    Aly Abdel Razek Galaby, Faculty of Arts, Alexandria University, Egypt

Chapter 2

Climate Change and Its Impact on Sustainable Development Goals in the Mediterranean
Countries: An Evaluation Study Through Geographic Information Systems (GIS) 18
    Bdor Osama, Alexandria University, Egypt

Chapter 3

Climate Change in World and Egyptian Cinema: A Sociocultural Study of Highlighted Values 52
    Arij Elbadrawy Zahran, Independent Researcher, Australia

Chapter 4

Climate-Resilient Crops and Tribal Women Empowerment: A Model of Odisha Millets Mission
in India 70
    Prageetha G. Raju, Rainbow Management Research and Consultants, India

Chapter 5

ENGO Communication Management Towards Climate Change: Solutions in Port City 90
    Valentina Burkšienė, Klaipeda University, Lithuania
    Jaroslav Dvorak, Klaipeda University, Lithuania

Chapter 6

People-Centered Urban Governance in Latin America and the Caribbean: Sociocybernetics,
Climate Justice, and Adaptation 108
    Shar-Lee E. Amori, McGill University, Canada

Chapter 7

Self-Defeat and Psychological Fragility as Predictors of Psychosomatic Disorders: In Light of
Climate Changes Among University Students in Egypt and the Emirates 131
    Islam Hassan Abdel Wareth, Faculty of Education, Alexandria University, Egypt
    Marwa Abdel Hamid Tawfiq, Ain Shams University, UAE

Chapter 8
Smart Sustainable Cities to Treat the Economic and Climate Repercussions in of Global Climate .. 153
Ehab Atalah, Independent Researcher, Egypt

Chapter 9
Social Resilience's Significance in Managing Extreme Climatic Events in the Cities of the Eastern
Area: A Field Study 174
Yasmin Alaa Ali Youssef, Ain Shams University, Egypt
Ahmed Zayed Abdalla Zayed, Assuit University, Egypt
Mahmoud Zayed Abdalla Zayed, Cairo University, Egypt

Chapter 10
The Relationship Between Consumer Culture and Food Security in the Context of Climate
Change: A Prospective Study in Sulaymaniya, Iraq 190
Nieaz Mohammed Fatah, Halabja University, Iraq

Chapter 11
The Social Responsibility of Third Sector Institutions in Planning to Face Climate Change in the
Kingdom of Saudi Arabia 210
Yasmin Alaa Ali, Ain Shams University, Egypt
Asmaa Hassan Omran Hassan, Helwan University, Egypt
Yasmin Alaa Ali, Ain Shams University, Egypt

Compilation of References 231

Related References 255

About the Contributors 275

Index 278

Detailed Table of Contents

Preface xii

Chapter 1
Building Community Resilience and Mitigating the Impacts of Climate Change Risks on Social
Inclusion: Exploratory Study on Alexandria, Egypt 1
    Aly Abdel Razek Galaby, Faculty of Arts, Alexandria University, Egypt

The current research addressed the issue of building community resilience and the mitigation of climate
change risks on social inclusion of Alexandria city. The research developed a theoretical and methodological
background to guide the exploratory study of this city. Starting with the re-analysis of the results of the
literature, analyzing the available secondary data and statistics, beside other tools. The research came
out with several results about the multiple manifestations of climate change risks and varied losses in
the city. The repercussions of climate risks on social inclusion, and building community resilience,
the mechanisms of mitigating, the effects of climate change, and features of strength and weakness.
Recommendations and issues that need more study in the future are discussed.

Chapter 2
Climate Change and Its Impact on Sustainable Development Goals in the Mediterranean
Countries: An Evaluation Study Through Geographic Information Systems (GIS) 18
    Bdor Osama, Alexandria University, Egypt

The study was divided into three main sections. The first section is terminology related to climate change.
The second is to introduce the sustainable development goals in relation to climate change. The author
inquired about the positive or negative impacts of climate change in relation to achieving them in 2030.
The last section contains the results of impacts through GIS software. The assessment processes include
(1) the data preparation stage, (2) the data entry stage, and (3) the geographical processing of the data.
This section also sheds light on sustainable development with regard to the social and economic aspect,
as this aspect is reflected in the economy, politics and the environment. The researcher reveals (1) how
economics and politics protect the environment and (2) how economics and politics help society protect
the environment for future generations. She concluded by identifying the relationship between climate
change and sustainable development as a cycle.

Chapter 3

Climate Change in World and Egyptian Cinema: A Sociocultural Study of Highlighted Values 52
Arij Elbadrawy Zahran, Independent Researcher, Australia

The chapter focuses on the climate changes and environmental transformations that swept the world through two aspects: a realistic aspect and an artistic one. The first section of the chapter identifies the problem of climate change and its causes, most of which are attributed to human action, in addition to natural disasters. The second section presents the international and global action towards these environmental problems, the global conferences that were held and the agreements that were concluded in this regard, and the importance of regional institutions and constitutions was also highlighted. The artistic aspect was dealt through the third and fourth sections of the chapter. The third section presents cinema art, film sociology, and the term environmental cinema. The fourth and final section deals with climate changes as presented by international cinema films - especially American cinema - and Egyptian cinema, and tried to explain the reason for the difference from a sociological point of view.

Chapter 4

Climate-Resilient Crops and Tribal Women Empowerment: A Model of Odisha Millets Mission
in India 70
Prageetha G. Raju, Rainbow Management Research and Consultants, India

The state of Odisha in India has a grim malnutrition problem. The growing risk of climate change is rendering rice-wheat cropping unsustainable given the resource intensiveness. The present chapter narrates the potential of nutrient-rich, resource-efficient, and climate-resilient millets to address the malnutrition among tribal populations. The dual challenge of malnutrition coupled with growing food insecurity, environmentally unsustainable agricultural practices on account of climate change vis-à-vis empowerment of tribal women was solved by the government initiative called the Odisha Millets Mission (OMM). OMM aimed to promote the cultivation of millets in the state to boost the livelihood concerns, and nutritional deficiencies. Women empowerment in tribal areas was taken up because vulnerability to climate change is exacerbated by inequalities and marginalization related to gender, low income, and other socio-economic factors. Solutions to combat climate change will be more effective if they can address this reality.

Chapter 5

ENGO Communication Management Towards Climate Change: Solutions in Port City 90
Valentina Burkšienė, Klaipeda University, Lithuania
Jaroslav Dvorak, Klaipeda University, Lithuania

This chapter explores environmental NGOs' (ENGOs) communication management towards climate change issues. E-communication was researched in the Lithuanian port city of Klaipeda. ENGOs residing in the city unsuccessfully fight against the polluting port business. The authors argue that communication management could bring success. A case study to research the e-communication of Klaipeda ENGOs was used. Content analysis of Facebook (FB) profiles of ENGOs and semi-structured interviews were chosen as approaches. The research revealed a weak understanding of communication management that impacts common e-communication practices and misunderstanding of this instrument for effective networking and joint actions.

Chapter 6
People-Centered Urban Governance in Latin America and the Caribbean: Sociocybernetics,
Climate Justice, and Adaptation.. 108
Shar-Lee E. Amori, McGill University, Canada

This chapter delves into the complexity of sustainable urbanization, climate justice, social inclusion, and participatory governance. Grounded in a one-year descriptive ethnographic study and meta-synthesis, the analysis deconstructs the disparities between the urban rich and poor in Jamaica, Panama, Trinidad and Tobago, and Columbia, across five key development domains- wellbeing, education, security, infrastructure, and governance. Through sociocybernetics, the decision-making processes in urban ecosystems are interrogated, revealing unique challenges faced by the urban poor, trapped in a cycle of recovery, versus the mitigation-oriented urban rich. The analysis extends to the role of urban citizens, designers and integrators, governance structures, levels of social inclusion, resource allocation, and their amalgamated implications for socio-climate justice. It evaluates the international policy arena, translation of global mandates into local development plans, and the need for hyperlocal strategies that encourage a more people-centred planning approach for sustainable urbanization.

Chapter 7
Self-Defeat and Psychological Fragility as Predictors of Psychosomatic Disorders: In Light of
Climate Changes Among University Students in Egypt and the Emirates 131
Islam Hassan Abdel Wareth, Faculty of Education, Alexandria University, Egypt
Marwa Abdel Hamid Tawfiq, Ain Shams University, UAE

The current study aimed to reveal the relationship of self-defeat and psychological fragility with psychosomatic disorders among samples of university students in Egypt and the UAE. The study sample consisted of (200) university students in Egypt and the Emirates, and the results revealed a correlation between each of self-defeat and psychological fragility, and between psychosomatic disorders, and also found statistically significant differences between the mean scores of university students on measures of self-defeat and psychological fragility, and psychosomatic disorders according to gender in favor of females, as well as found statistically significant differences between Average scores of university students on measures of self-defeat, psychological fragility, and psychosomatic disorders according to the difference of nationality in favor of the Egyptian nationality.

Chapter 8
Smart Sustainable Cities to Treat the Economic and Climate Repercussions in of Global Climate .. 153
Ehab Atalah, Independent Researcher, Egypt

Sustainable smart cities aim to provide a number of opportunities to achieve sustainable development and face challenges related to climate change, that is the great challenge facing the future of life on the planet. Information and communication technology is one of the basics in the field of monitoring climate change, mitigating its effects and adapting to it, such as the field of early warning system and smart applications. That transforms the way services are provided in areas including energy, waste, and water management to reduce carbon footprint and address environmental challenges, the role of modern technology, the internet of things (IOT) and artificial intelligence (AI) applications.

Chapter 9
Social Resilience's Significance in Managing Extreme Climatic Events in the Cities of the Eastern
Area: A Field Study 174
> *Yasmin Alaa Ali Youssef, Ain Shams University, Egypt*
> *Ahmed Zayed Abdalla Zayed, Assuit University, Egypt*
> *Mahmoud Zayed Abdalla Zayed, Cairo University, Egypt*

This current chapter starts from a main question: What is the extent of social resilience as a function of managing extreme climatic events in the Kingdom of Saudi Arabia? The study will present three basic concepts: social resilience, crisis management, and extreme climatic phenomena. As for the approach used for the study, it is the descriptive approach, and an electronic questionnaire is relied upon as a tool for collecting data within cities in the eastern region. The study reached a number of results, the most important of which are: dehydration is one of the many manifestations of the impact of climatic phenomena on cities, including: increasing rates of economic pressures, giving priority to smart cities, and restructuring development plans for infrastructure within cities.

Chapter 10
The Relationship Between Consumer Culture and Food Security in the Context of Climate
Change: A Prospective Study in Sulaymaniya, Iraq 190
> *Nieaz Mohammed Fatah, Halabja University, Iraq*

The current study is one of the future studies which attempt to evaluate what societies may go through under the influence of global and trade openness; it also evaluates the industrial development and the merger between agriculture and the food industry. The majority aspires to make greater profit through a higher level of production, unaware of the impacts of their greed on consumers' diet and food security, nor the negative impact they have on climate change. The main conclusions are the citizens do not have enough knowledge about what is offered in the markets they make their choices blindly under the urge of practicing their freedom to the fullest. At the same time, due the large number of offers presented to the people they became daily purchasers who are careless about reading the written labels on the products. Consequently, the markets became a place for the liquidation of all products which led to three major issues: the change of the culture of consumption, a high level of consumer purchases, and the misuse of nature in many aspects which has caused climate change.

Chapter 11
The Social Responsibility of Third Sector Institutions in Planning to Face Climate Change in the
Kingdom of Saudi Arabia 210
> *Yasmin Alaa Ali, Ain Shams University, Egypt*
> *Asmaa Hassan Omran Hassan, Helwan University, Egypt*
> *Yasmin Alaa Ali, Ain Shams University, Egypt*

This current chapter stems from a main question: "What is the reality of the social responsibility of third-sector institutions in planning to face climate change in the Kingdom of Saudi Arabia?" The study will present three basic concepts: social responsibility, third-sector institutions, and climate change. It has a theoretical approach consisting of four theoretical approaches: the theory of social system, the theory of social action, the theory of risk society, and finally the theory of social capital. The current study is one of the descriptive analytical studies, and an electronic questionnaire is relied upon as a tool for collecting data. The study reached a number of results, the most important of which are: that there is a social responsibility for third sector institutions in planning to face climate changes in the eastern region.

Compilation of References 231

Related References 255

About the Contributors 275

Index 278

Preface

As editors of *Global Perspectives on Climate Change, Social Resilience, and Social Inclusion*, we find ourselves at the intersection of urgent global challenges and the imperative for comprehensive, interdisciplinary solutions. Aly Abd ElRazek Galaby, Mohamed Abo Elenein, and Hassan Mohamed, our esteemed colleagues and co-editors, join us in presenting this compilation as a timely response to the complex and interconnected issues posed by climate change.

In our pursuit, we are guided by the recognition that climate change is not merely an environmental crisis; it is a multifaceted challenge woven into the fabric of biophysical, economic, political, and socio-cultural factors. Sociologists, as keen observers of societal dynamics, have highlighted the imperative to approach climate change through an interdisciplinary lens. It has become evident that the success of climate mitigation plans is contingent upon considering Diversity, Inclusion, and Equity as fundamental considerations in their design and execution.

Our contributors delve into the profound understanding that addressing the hazardous effects of climate change necessitates cooperation among all members of societies at national and international scales, transcending racial, cultural, and economic differences. The impacts of climate change exacerbate existing social and economic inequalities within countries, making it imperative to view sustainable development as a framework that bestows resilience and opportunities upon societies to cope with structural inequalities.

In this volume, we explore the vital role of voluntary associations and activists in championing the interests of those victimized by climate change. We contend that reducing the risks associated with climate change must be integral to a broader resilient development framework, one that seeks to uplift the economic positions of disadvantaged groups, enhance their quality of life, and foster equity among all members of society.

Chapter 1 lays the foundation for our exploration by addressing the pressing issue of building community resilience and mitigating climate change risks in Alexandria City, Egypt. The research employs a theoretical and methodological background to guide an exploratory study, re-analyzing literature results and utilizing secondary data and statistics. The findings encompass multiple manifestations of climate change risks, examining their repercussions on social inclusion and community resilience. Mechanisms for mitigation, the effects of climate change, and both strengths and weaknesses are scrutinized, concluding with recommendations and identifying areas for future research

Divided into three sections, Chapter 2 begins by examining terminology related to climate change and introduces the Sustainable Development Goals. It explores the positive and negative impacts of climate change on achieving these goals by 2030. The final section delves into GIS-based assessment processes, shedding light on the social and economic aspects of sustainable development. The researcher elucidates

Preface

how economics and politics intersect to protect the environment and help societies safeguard it for future generations, emphasizing the cyclical relationship between climate change and sustainable development.

Focusing on climate changes and environmental transformations, Chapter 3 unfolds in two aspects: a realistic examination and an exploration of artistic expressions. It identifies the causes of climate change, attributing many of them to human actions and natural disasters. The chapter also delves into international efforts, conferences, agreements, and the role of regional institutions. The artistic aspect is explored through cinema, film sociology, and the representation of climate changes in international and Egyptian cinema, providing a sociological perspective on the differences portrayed.

Chapter 4 discusses the malnutrition problem in Odisha, India, exacerbated by the growing risk of climate change. The Odisha Millets Mission (OMM) is presented as a government initiative addressing malnutrition, food insecurity, and environmentally unsustainable agricultural practices through the cultivation of millets. The chapter highlights the connection between vulnerability to climate change and gender-related inequalities, emphasizing the need for solutions that address this complex reality.

Exploring the communication management of environmental NGOs in Klaipeda, Chapter 5 focuses on e-communication efforts to combat climate change. It investigates the struggles of ENGOs in fighting against polluting port businesses, emphasizing the importance of effective communication management. Utilizing a case study approach and content analysis of ENGOs' Facebook profiles, the research uncovers a weak understanding of communication management hindering effective networking and joint actions.

Chapter 6 delves into the complexity of sustainable urbanization, climate justice, social inclusion, and participatory governance. Grounded in ethnographic study and meta-synthesis, it analyzes disparities between urban rich and poor in Jamaica, Panama, Trinidad & Tobago, and Colombia. The focus extends to decision-making processes, social inclusion levels, resource allocation, and implications for socio-climate justice. The need for hyperlocal strategies in sustainable urban planning is emphasized.

Chapter 7 explores the relationship between self-defeat, psychological fragility, and psychosomatic disorders among university students in Egypt and the UAE. Findings reveal correlations and statistically significant differences based on gender and nationality. The chapter contributes to understanding the psychological aspects of climate change, emphasizing the need for tailored interventions and support for students facing psychological health challenges.

Focusing on sustainable smart cities, Chapter 8 underscores the role of information and communication technology in monitoring climate change, mitigating its effects, and adapting to it. It explores the applications of modern technology, the Internet of Things (IoT), and artificial intelligence (AI) in areas such as energy, waste, and water management. The chapter emphasizes the transformative potential of technology in addressing environmental challenges and reducing carbon footprints.

Examining social resilience as a function of managing extreme climatic events in the Kingdom of Saudi Arabia, Chapter 9 utilizes a descriptive approach and an electronic questionnaire. Results highlight manifestations of climatic impacts on cities, economic pressures, and the need for prioritizing smart cities and restructuring development plans for infrastructure. The chapter contributes to understanding the challenges faced by communities in the region and suggests strategies for resilience.

Chapter 10 delves into the impacts of global and trade openness, industrial development, and the merger between agriculture and the food industry on societies. It highlights the lack of consumer awareness, leading to issues such as a change in consumption culture, high levels of purchases, and the misuse of nature contributing to climate change. The chapter emphasizes the need for informed consumer choices to address these challenges.

Finally, exploring the social responsibility of third-sector institutions in facing climate change in Saudi Arabia, Chapter 11 employs a theoretical approach encompassing social systems, social action, risk society, and social capital theories. Utilizing a descriptive analytical approach and an electronic questionnaire, the research identifies the social responsibility of third-sector institutions in planning for climate change in the eastern region. The findings contribute to understanding the role of non-profit organizations in climate resilience.

OBJECTIVES, IMPACT, AND VALUE

This book responds to the imperative that, despite the perilous consequences of climate change, the natural sciences have disproportionately dominated the discourse. Our objective is to redress this imbalance by focusing on the crucial social, cultural, and political factors shaping the living conditions of people in their societies. By examining the profound impact of climate change on food sources, water, social stability, and occupational patterns, we aim to equip decision-makers with the insights necessary to navigate these challenges and fortify societal structures.

TARGET AUDIENCE

Designed for university students, researchers, policy makers, voluntary associations and for all those passionate about sustainable development in the context of diversity, inclusion, and equity, this book spans a wide array of topics. Its coverage includes the sociological perspective on climate change as a burgeoning field of research, causes and impacts of climate change on populations, coping mechanisms employed by societies, and the intricate interplay of climate change with health, food security, poverty, migration, water scarcity, and societal consciousness.

The book also addresses the social responsibilities of governments and institutions in achieving equality, explores the role of social institutions and movements in advocating for change, and delves into the intersection of sustainable development with diversity, inclusion, and equity.

As editors, we believe that this volume will serve as a valuable resource, fostering a deeper understanding of the sociological dimensions of climate change and inspiring informed action towards a more resilient, inclusive, and equitable future for all.

Aly Abdel Razek Galaby
Alexandria University, Egypt

Mohammed Abo ElEnein
American University, UAE

Hassan Mohamed
Alexandria University, Egypt

Chapter 1

Building Community Resilience and Mitigating the Impacts of Climate Change Risks on Social Inclusion:
Exploratory Study on Alexandria, Egypt

Aly Abdel Razek Galaby
Faculty of Arts, Alexandria University, Egypt

ABSTRACT

The current research addressed the issue of building community resilience and the mitigation of climate change risks on social inclusion of Alexandria city. The research developed a theoretical and methodological background to guide the exploratory study of this city. Starting with the re-analysis of the results of the literature, analyzing the available secondary data and statistics, beside other tools. The research came out with several results about the multiple manifestations of climate change risks and varied losses in the city. The repercussions of climate risks on social inclusion, and building community resilience, the mechanisms of mitigating, the effects of climate change, and features of strength and weakness. Recommendations and issues that need more study in the future are discussed.

INTRODUCTION

The concept of building community resilience had attracted the interest of researcher in many disciplines. They made use of it in tackling many cities challenges. This concept had special status in the scope of adaptation to climate change risks, internationally regionally and locally. Egypt adopted special vision toward climate change issues, based on environmental policies and developmental goals. Sustainable development strategy, the Egyptian vision (2030), and National strategy for climate Change (2050) declared in May 2022, through UN cop 27. Building resilience and adaptation capabilities to climate change was its important goals. Because Alexandria city is more effects of climate change in Egypt,

DOI: 10.4018/978-1-6684-8963-5.ch001

this governorate Started from many years ago applied their programs and mechanisms for mitigating climate change effects. The current research defined its subject matter; Building community resilience and the mitigation of climate change risks on Social inclusion; an exploratory study on Alexandria city. The issues and goals of this research are developed in terms of Elliott perspective about sociology of Loss. Carmen and his colleges: the term community resilience, etc. The current research depended on the methodology of exploratory steps in carrying out its field study. The research had divided to different elements; Literature review, theoretical and methodological Background, the problem, finally, the three issues, conclusions and discussion.

LITERATURE REVIEW

Zehr (2013) published an article about The Sociology of global climate change. He argued that sociological research on global climate change (GCC) can be found in several subfields, but it has primarily emerged within the theoretical and substantive domain of sociology of environment. This review provides an overview of sociological literature on climate change and identifies key areas for further research and development. The review focuses on four broad areas: Social causes, Construction of the problem, relationship between GCC and inequality, and Social dimensions of mitigation and adaptation. He concluded that sociologies of mitigation and adaptation might fruitfully draw upon neighboring subfields and disciplined. Urban sociology and geography have much to offer in envisioning more sustainable future cities. Etc. In addition to disciplinary research, interdisciplinary collaboration and public and stakeholders engagement are necessary for sociologists to make a major impact on GCC Knowledge and solutions.

Elliott (2018) had published an article about the sociology of climate change as sociology of loss. He argued that climate change involves human societies in problems of loss, depletion, disappearance and collapse. Sociology endeavor to know about this particular from of social change. This article outlines the sociology of loss as a project for sociological engagement with climate change. He addresses four interrelated dimensions of loss that climate change presents: the materiality of loss, the politics of loss, Knowledge of loss, and practices of loss unlike sustainability. The sociology of loss examines what does, will or must disappear rather than what can or should be sustained.

Dapilah et al. (2020) published an article about the role of social networks in building adaptive capacity and resilience to climate change a case study from northern Ghana. They argued that increasing attention is being paid to the role of social networks in climate change research and new studies show that they farm an essential source of resilience. However, the role of social networks remains underexplored as these is only limited empirical evidence of their benefits, particularly for research on adaptation to climate change in developing countries. This paper provides a contribution to this field of research by examining how social networks faster livelihood diversification and resilience in a small rural community in northern Ghana. The findings show that, people in the studied community have experienced a range of climate change with negative impacts on agriculture in that last three decades. These climate change have forced community members to diversity their livelihood activities away from crop production and into off-farm and non-farm activities. This study shows how group activities and Social networks can also create adverse effects by enforcing exclusion and marginalization among certain groups in the community.

Samer Fawzy et al. (2020) had published article about Strategies for mitigation of climate change: a review. Climate change is defined as the shift in climate patterns mainly caused by greenhouse gas emissions from natural systems and human activities. Anthropogenic activities have caused about 1.5°

Building Resilience, Mitigating Impact of Climate Change on Inclusion

c of global warming above the pre- industrial level and this likely to reach 1.5° c between 2030-2052. In 2018 the world encountered 315 cases of natural disasters related to the climate. This article reviews the main strategies for climate abatement, namely conventional mitigation, negative missions and radiative forcing geoengineering change in climate indicators, namely temperature, precipitation, seal level rise, ocean acidification and extreme weather conditions have been highlighted in a recent report. There are three climate mitigation approaches; first conventional mitigation efforts employ decarbo. A second route constitutes of new set of technologies and methods recently proposed; such as potentially to capture and sequester. The article Concluded that there is no ultimate solution in tackle climate change and all technologies and techniques in this review if Techniqually and economically are viable should be deployed.

Cavelcante and Santos (2021) published an article about climate change mitigation policies Aggregate and distributional effects. They said that we evaluate the aggregate and distributional effects of climate change mitigation policies using a multi - sector equilibrium model with intersectional input - output linkages and worker heterogeneity calibrated to different countries. In the Us, workers with comparative advantage in dirty energy sectors who do not reallocate suffer a welfare loss 12 times higher than workers in non- dirty sectors, but constitute less than 1% of the labor force.

Yuchi et al. (2022) published an article about building social resilience in north Korea can mitigating the impacts of climate change on food securit" and referred to adaptation based on social resilience is proposed as an affective measure to mitigate hunger and avoid food shocks caused by climate change. But these have not been investigating comprehensively in climate sensitive regions. North Korea (NK) and its neighbours, South Korea and China, represent three economic levels that provide us with examples for examining climate risk and quantifying the contribution of social resilience to rice production. These findings highlight the importance of social resilience to mitigate the adverse effects of climate change on food security and human hunger and provide necessary quantitative information.

Tahmineh et al. (2022) published an article about application of machine learning and deep learning methods for climate change mitigation and adaptation. Considering Climate change as a global issue that must be addressed immediately, many articles have been published on climate change mitigation and adaptation But, New Methods are required to explore the complexities of climate change and provide more efficient and effective adaptation and mitigation policies. This paper aims to explore the most popular ML and DL methods that have been applied for climate change mitigation and adaptation. Another aim is to determine the most common mitigation and adaptation measures, actions in general, and in urban areas in particular, that have been studied using ML and DL methods. The results indicate that the most popular ML technique in both climate change mitigation and adaptation is the Artificial neural network. Moreover, among different research areas related to climate change mitigation and adaptation, geoengineering, and land surface temperature are the ones that have used ML and DL algorithms the most.

Carmen et al. (2022) published an article about building community resilience in a context of climate change the role of social capital. Argued that in spite of social capital is considered important for resilience across, social levels including communities, yet insights are scattered across disciplines. This meta-synthesis of 187 studies examines conceptual and empirical understandings of how social capital relates to resilience, identifying implications For community resilience and climate change practice Different conceptualization are highlighted, Empirical insights show that structural and socio-cultural aspects of social capital, multiple other factors and formal actors are all important for shaping the role of social capital for guiding resilience out comes.

Ripple et al. (2022) published an article about six steps to integrate climate mitigation with adaptation for social justice. They argued that climate change impacts are accelerating and there is an urgent need to address this global issue. Historically, mitigation and adaptation strategies have been treated separately, but there is now growing awareness of important synergies between them we highlight this synergies across sex key areas where humans need to make transformative changes in order to reduce the impacts of climate change, including energy, pollutants, nature, food, population and economy.

Hossain and Masum (2022) published research about corporate social responsibility helping mitigate firm-level climate change risk. They argued that the utility of corporate social responsibility (CSR) particularly during Crisis times has been a puzzle in the literature while climate change issues increasingly threaten corporate sustainability. We explore whether CSR provides corporate resilience against firm - level climate change risk (CCR) and find that CSR helps mitigate CCR sustainability. Moreover in the cross - sectional analysis we find that CSR is more helpful in reducing CCR for firms with higher environmental, social, governance (ESG) discloser and those located Republican Leaning States.

Marquand and Elsasser (2023) published an article about initializing climate change mitigation in the Global South: current trends and future research. They said that following the Paris Agreement, States and non- States actors have pledged countless commitments to mitigate climate change. Besides, efforts to take the climate crisis compete with other human development priorities. Withe the review, we explore the prospects and challenges of institutionalizing climate mitigation in the Global South. We (1) map the field in terms of concepts, methods, regions, sectors and topics (2) Suggest a differentiation between reforms - oriented, transformative, and failed attempts to institutionalize climate change mitigation and (3) propose future research agenda.

Aboudouh (2023) published an article about civil society and its role in climate change Issues. He argued that "climate change" represents one of the most important challenges facing human societies in the twenty- first century, and yet there is a great disconnect between our procedures for dealing with it and the seriousness of the threat it entails The important question is: How should we think about this issue?

How should we deal with the kinds of risks posed by human- caused climate change?

On this basis, this article attempts to emphasize the idea that the role of civil society in its various institutions should be strengthened to deal with climate change issues and its problems. And from this perspective, this article attempts to explore possible ways to do so, and the roles that civil society institutions can and should play in solving this problem.

A comprehensive review of the international literature on the issue of Climate Change during the decade from 2013 to 2023 showed that researchers in the Arab countries had paid little attention to this issue.

Despite the consensus among the researchers at the global level, that Changes in Climate are urgent issues, there was no important contribution has been made from researches until the year 2023. However, one research was added from Egypt, which suggested that there was a spatial and temporal gap in available literature.

In short, the researchers, in their studies, focused their attention on three main axes, some of which dealt with the risks of climate change, in terms of multiplicity and losses in the cities affected by these changes, the second focused on the repercussions of climate change risks on the structure of the city, as a community. The third dealt with establishing community resilience and supporting methods of mitigating the impacts of climate change on social integration. In fact, this is the main objective of the topic in the current research. The next part of this preface will deal with the problem, the study objectives, and will introduce its theoretical and methodological framework.

THE PROBLEM AND OBJECTIVES

In its report on the state of the environment, the Environmental Affairs Agency of the Ministry of the Environment in Egypt identified nine major risks of climate change to which Egypt is exposed in 2022, including the emergence of 5 most obvious ones in the city of Alexandria. Beginning with the increase or decrease in temperatures above their normal rates, and the rise in the sea level, which is expected to increase by 010 centimeters until the year 2100, as well as its effects on coastal areas, in addition to the increase in the rates of extreme weather events, such as dust storms and torrential rains, which affected and led to the deterioration of ecotourism. Moreover, the beach coasts are eroded, they suffered from a decline in the rate of public health and from the spread of insect supererogatory diseases, such as malaria, lymph nodes and various types fever (Al-Amir & Kamal, 2022).

In recent years, the city of Alexandria has faced a clear wave of risks resulting from climate changes. Also, the events cracking in the Alexandria Corniche, after the 2022 earthquake in Syria and Turkey, which the governor attributed to the erosion of the beach with high and bad weather waves (Al-Youm, 2023b). Besides, the public expressed their anger and anxiety, and they spread rumors among the population (Al-Youm, 2023a). Recalling what was stated by the former Prime Minister of England, at the United Nations Conference in the year (2021/26.Cop) that there are three global cities that are at risk of drowning due to climate changes, including the city of Alexandria (Al-Ain Al-Emirates, 2022). This prophecy found an echo and attempts to link it to some historical events, which supported by what was discovered underwater as a cultural heritage near the city's shores, and what proves that the drowning occurred in ancient decades of history.

Added to the above mentioned facts, what was reported on social networking sites in respect of information indicating the occurrence of a kind of forced social exclusion among some old families in Alexandria, who left the city of their own free will in anticipation of drowning, and established their commercial and tourism projects and settled in a number of Egyptian governorates, especially the city of Cairo (Rabie, 2022).

The Egyptian government has initiated huge procedures to confront the challenges and repercussions of climate change, foremost of which is the announcement in May 2022 and during the climate conference (Cop27) of the national strategy, and the adoption of policies, strategies and procedures for adaptation, resilience and mitigation of the effects of the risks of climate change.

However, there remains an urgent need at the level of scientific research, to shed light on what is happening in the city of Alexandria recently in terms of the risks associated with climate change, first of which is the risk of the disappearance of the city and the consequent social processes, such as social exclusion, inequality and lack of equity, those processes that result in many forms of inequality in human rights, a decent life, and the accompanying injustice and poverty, and the necessity to work on fairness for such.

An exploratory study will opens the way for other studies and issues that need the attention of researchers. Its title is *Building Community Resilience and Mitigating the Impacts of Climate Change Risks on Social Inclusion.*

The research problem and its subject can be developed into a set of goals and questions. And try to achieve it through this exploratory study for the city of Alexandria:

1. Detecting the risks of climate change in terms of multiplicity and losses in the city.
2. Identifying the repercussions of climate change risks on social inclusion.

3. Shedding light on building the community resilience and mechanisms to mitigate the impacts of climate change and achieve the desired adaptation.

THEORETICAL FRAMEWORK AND METHODOLOGY

The study will rely on the sociology of loss perspective of R. Elliott (2018) in clarifying the dimensions of climate change risks in terms of multiplicity and losses in community, which include disappearances, destruction, attrition and expulsion. This theoretical; orientation clarifies and expresses social processes such as human stability, political mobility, knowledge production, consumption practices, and others.

It may be guided by the conclusions of Dapilah et al. (2020) showed how the risks of climate change are reflected in social inclusion, and by clarifying the role of social networks and group activities in confronting all forms of social exclusion and marginalization resulting from these risks.

Based on the research of Carmen (2022), Hossain (2022) and Ripple (2020), it was possible for us to define the concept of community resilience as a set of mechanisms that contribute to achieving adaptation to climate change and help to mitigate its impacts, which combine soft tangible mechanisms such as various resources and policies, including, economic development, infrastructure, the ability to change performance and work, and soft intangible mechanisms, such as social capital, adaptive capacity building, communication and information network, sustainability, social awareness, and corporate social responsibility between the various sectors of society.

METHODOLOGICAL PROCEDURES

In this exploratory study of the city of Alexandria, we rely on a set of methodological procedures, beginning with the research literature review process, revealing the research movement and research gaps, through the method of reanalysis, following up previous studies, and re-dismantling and re-compiling their results. For consulting the different experts we will conduct in-depth interviews with cases of workers in the relevant ministries, administrative and technical departments in the Alexandria Governorate, and academic researchers interested in the issue of climate change in universities and scientific research centers. In addition, it is important to conduct case studies of clairvoyance from previous events related to climate change, and extreme cases, between loss and loss. And above all, using secondary sources and data such as statistics and others, with the help of visual ethnography and image and video data available on social media platforms will provide the research with valuable information.

The Risks of Climate Change in The City of Alexandria Between the Multiplicity of Manifestations of Change and The Variation in Losses

The available evidence in this regard indicates a variety of manifestations of climatic changes that the city has witnessed and is still in recent years, especially in the autumn season. With an increase of 3 degrees compared to the previous days. With the approach of the air depression that dominates the Mediterranean, (Figure 1) the temperature continued to rise and the formation of a mist on the surface of the Mediterranean Sea (Figure 2), this come with the meteorological authority warning of thunderstorms

Figure 1. Mediterranean depression is approaching Alexandria
Source: Mounir (2022)

Figure 2. Fog dominates Alexandria with high temperatures
Source: Mounir (2022)

Figure 3. Rising temperatures and forecasts of thunderstorms
Source: Mounir (2022)

in the evening reaching 80%, which is a high percentage and violated the Alexandria weather condition preceded by a drop in temperature.

The governor of Alexandria indicated that the amount of rain that fell in one day was more than 19 million cubic meters, and that the rains will increase every year and reach 10 times what we are facing now.

The governor also pointed out that the city was exposed to a wave of bad weather and the fall of medium-intensity rains, sometimes thundery.

The governor also referred to the integrated rainwater strategy project, as rain fell on Alexandria in 2015, estimated at 78 million cubic meters, while in 2021 it fell 96 million cubic meters.

The slums in the city and the surrounding villages have been living under this bad weather and continuous rains for decades, which increased after 2011 and helped by the continued encroachment on the infrastructure, and that they are located a meter below the surface of the earth, as the governor defined it.

Dr. Mohamed Abdel-Fattah Ragab, Secretary General of Pharos University, adds, after the economic impact of climate change and what is related to environmental scenarios, the need for the world to take precautions to protect future generations, because productivity is linked to the extent of human vulnerability to temperature, which continues to rise. Also, the relationship between fish mortality and the rise in temperature has an impact on fish stocks and the economic aspect. Add to that the losses that transport networks are exposed to in the event of floods and the threat these networks are exposed to.

There is also an effect of climate change on the spread of diseases and the infection of citizens with some diseases resulting.

Dr. Ayman El-Gamal, head of the Beaches Research Authority, had indicated that there is a temporary drowning of the city as a result of winter and the increase in rain, which the banks did not absorb. cause the earth to crumble.

In addition to the multiplicity of manifestations of previous climate changes, the Corniche being exposed to collapse as a result of high waves, which happens every winter along the Egyptian coasts from Rafah to Matrouh. The area decreases annually. Climate changes have led to a rise in sea level, and thus the decline of the beach helped by those storms, water currents, tides, and what happened slowly with the passage of time. Engineer Mohamed Ghanem, the official spokesman for the Ministry of Water Resources and Irrigation, explains that the subsidence of the Corniche was the result of the huge waves that occur with (Noh Al-Karam), and the people in Alexandria know about them because it is one of the storms that occur every winter. Five meters of waves hit the Corniche and affected the Corniche sidewalk and pulled sand from the bottom of the sidewalk and caused a 20-meter drop in this area.

The phenomenon of beach erosion was one of the natural coastal phenomena known to the world, which in turn contributed with the recent climatic changes in increasing the severity of the disappearance of parts of the beaches due to the melting of ice and thus the rise in sea level around the world, including the Egyptian beaches, albeit in varying degrees from one spot to another according to geographical features.

Dr. Hussein Abdel-Basir, Director of the Antiquities Museum at the Bibliotheca Alexandrina adds climate change has become a very important and very disturbing issue, this events threatening human existence as a whole in many parts of the world, where global warming and climate change lead to melting of ice and an increase in sea water level, which is reminiscent of the sinking of some places in the world, and effects of climate change have appeared on the city of Alexandria, Where the climate has changed rapidly, strangely and unfamiliar, and with the possibility of part of the Egyptian Delta sinking over time, what have an impact on the city and the Delta, and the disappearance of a large number of this heritage, before being submerged (Abdel-Basir, 2022).

In conclusion, the city of Alexandria witnessed multiple manifestations and different losses due to climatic changes, starting with high degrees and thunderstorms falling at a high rate, reaching 19 million cubic meters, and it will increase every year and double what we are currently facing, there are also waves of bad weather, and it has been estimated It fell from the rains of 2015 by about 78 million cubic meters, then 96 million cubic meters in 2021. The slums that lived in this bad weather were exposed to harsh weather for decades after 2011, with the continued encroachment on the infrastructure.

There are also economic impacts and losses resulting from climate change. Because productivity is related to the extent to which humans are affected by temperature, and then there is a relationship between fish deaths and high temperatures, and then on fish wealth. In addition to the losses that transport networks are exposed to, as well as there are effects of climate changes on the spread of diseases there is a temporary sinking of the city As a result of the increase in rain in a way that the banks did not absorb, and partial drowning due to the rise in sea level, the erosion of the beaches, the subsidence of the Corniche, and the slaughter of the beaches. And the increasing possibility of the disappearance of a large number of antiquities in the city.

The Repercussions of Climate Change Risks on Social Inclusion in Alexandria

We may have found some reflections of the dangers of climate change and its losses in the city of Alexandria, on social inclusion, due to forced migration, on the one hand, and the disappearance of the entire city from the Egyptian map on the other hand.

1. Forced Migration of City Residents and Their Exposure to Social Exclusion and Marginalization

Agence France-Presse (2022) indicated that until today hundreds of residents of the city (Alexandria) were forced to abandon their homes, whose walls were disturbed by the encroachment of water and torrential rains in 2015, as well as in 2022. Every year the city sinks by more than three millimeters, as does the dams built on the Nile River. Which prevents the arrival of silt that contributed in the past to the consolidation of its soil, with what resulted from gas extraction operations from offshore fields. The Mediterranean sea level is expected to rise by one meter within the next three decades, according to the worst forecasts of the United Nations Intergovernmental Panel on Climate Change, UN experts said that the level of the Mediterranean sea will rise faster than almost anywhere else in the world. According to the committee, this would drown a third of the highly productive agricultural lands in the Nile Delta, as well as cities of historical importance such as Alexandria, third of the city is threatened with drowning. And the head of the Egyptian General Authority for the Protection of Coasts, Ahmed Abdel Qader, told AFP, "Climate changes have become a reality that we live in, and not just warnings issued. The average citizen is now feeling the summer heat at higher degrees than he was used to, as well as the cold in winter, even for the best scenario; the Human Development Report for the year 2021 is expected." Issued by the Egyptian Ministry of Planning, in cooperation with the United Nations Development Program, that by the year 2050, the level of the Mediterranean sea may rise by one meter as a result of global warming, which results in some industrial cities and cities of historical importance such as Alexandria, Damietta, Rashid and Port Said being submerged. A sea level rise of half a meter could drown.

Thirty percent of the city of Alexandria, which will lead to the displacement of approximately 1.5 million people or more. It is expected that this will lead to the loss of 195 thousand people their jobs. Abdel Qader said that this catastrophe will have enormous repercussions on Egypt, which has 104 million people living in it. The city that was built by the Greek King Alexander the Great nearly 2400 years ago is the second most important city in Egypt because of its historical and archaeological dimension, in addition to that it includes the largest port in the country.

Since the sixties, the sea water has already advanced more than three kilometers, and in the eighties it swallowed the Rashid lighthouse, which dates back to the nineteenth century, as a result of the phenomenon of beach erosion. all over Egypt.

And in a study by Shaimaa Wael Rabie; Climate change and population distribution in Egypt, it seeks to shed light on the most important climatic and environmental changes in Egypt and the extent of their reflection on the economic and social determinants and motives that greatly affect the decision-making of internal migration in Egypt and change the trends of population movement and the formation of population distribution in Egypt.

With regard to the impact of climate change on population distribution and migration decisions. The International Organization for Migration has strongly supported defining migration as an adaptation strategy, or a possible way to determine adaptation to climate change. It views migration as a consequence of failure to adapt to the environment. Here we can consider migration as one of the coping and adaptation strategies used by families, which in turn respond to both push and pull factors, but how do climatic drivers (whether slow or fast such as floods and rainfall fluctuations) interact with non-climatic drivers (i.e. existing poverty, conflict and development deficit) To influence family immigration decisions. These decisions to move or stay have implications for the family's ability to adapt through remittances and changes in social capital.

And since the agricultural production in Egypt is concentrated mainly in the Nile Valley and the Delta, because the Nile is the main source of irrigation, this resulted in an increase in population density in the governorates of the Valley and the Delta, since there is a possibility of a link between desertification and all environmental problems of migration, cities in Egypt were Vulnerable to migration caused by climate change, and will suffer environmentally, socially, economically, and politically from climate change and rural-to-urban migration. As confirmation of this, the results of the last census in Egypt 2017 witnessed a significant change in the internal migration currents in Egypt. The total number of migrants from the coastal governorates and the delta reached about 52% immigrants, Alexandria governorate came first in the number of immigrants, followed by Eastern, then Damietta, then Port Said.

Another study conducted by the agency in 2021 confirmed that most of the population had migrated to Cairo and Giza, where temperatures are lower and job opportunities are greater. Migration concentrated in the productive age groups. This indicates that internal migration in Egypt is for the purpose of obtaining better job opportunities, and that the highest percentage of male migrants was also due to work, especially among intellectuals from rural to urban areas. The highest percentage of female migrants was due to marriage, followed by accompanying the family. The answer to the question was what are the predictions about the future of environmental migration in Egypt? Extreme weather phenomena, such as sea level rise and drought, will lead to forced migration and displace the Egyptian rural population to urban areas.. It is likely that the negative effects of migration will be on poor families who struggle to face various economic and material challenges in the wake of severe environmental and climatic risks, which reduces the capacity On adaptation. The eastern and western parts of the Nile Delta, including Alexandria, are likely to become hotbeds of out-migration due to reduced water availability and rising sea levels at the same time. Hence, places where water is better available will be hotbeds of inward migration due to climate change, including important urban centers such as Cairo.

The sea level rise will have a significant impact on the economic sector in Egypt, because a sea level rise of 0.25 meters will lead to a decrease of 60% of Alexandria's population of 4 million. In addition to 56% of the industrial sector in Alexandria. A rise of 0.5 may be more disastrous; Because 67% of the population, 65% of the industrial sector, and 70% of the service sector are below sea level. Moreover, 20% of the city's area will be destroyed...etc. Therefore, the economic and social consequences resulting from climate change require communities to adopt strategies A long-term integrated program that involves preserving the fertility of the land, increasing its productivity, rehabilitating and preserving it, rationalizing the use of water resources and managing them in a sustainable manner. This is to ensure the stability of the population and maintain their food security and thus their inclusion into society (Rabie, n.d.).

2. The Disappearance of the City of Alexandria Between the Threat of Drowning and Being Lost in the Waves of the Mediterranean

The controversy surrounding the sinking of the city of Alexandria can be traced, as the Emirati newspaper, Al Ain, asked on 11/3/2022 the question; Is Alexandria threatened with drowning, or is the sea swallowing up the bride of the Mediterranean? Over recent years, with the rise of voices warning of the issue of climate change, the spotlight has been shed on the Egyptian city of Alexandria, the bride of the Mediterranean, and the extent of the danger that besets it, and that during the Climate Summit COP 26 in Glasgow 2021, the former British Prime Minister Boris Johnson said We will say goodbye to entire cities like Miami, Alexandria and Shanghai, which will be in the midst of waves (Al-Ain Al-Emirates, 2022).

But the governor of Alexandria had confirmed in his intervention with the editor of Al-Masry Al-Youm on 10/2/2023, and talked about the fact that the Alexandria Corniche was cracked after the earthquakes in Syria and Turkey, that the beach was eroded due to the unusually high intensity of the waves in the past days. And because of the wave of bad weather, and after everyone saw what happened on the Alexandria Corniche in terms of cracks and fissures, and the spread of rumors and the state of anxiety, and how those began to exaggerate the matter and those who bid and say the earthquake came to Egypt. The governor said this talk has no basis in the reality (Al-Youm, 2023).

An interview with Dalia Al Hamshari under the title of an Egyptian expert; Climate changes will not drown the city, Dr. Abbas Shraki as the head of the Department of Natural Resources at the Institute of African Research and Studies, ruled out the scenario of the sinking of the city of Alexandria, stressing that the area threatened with drowning due to climate change phenomena is north of the Delta, and that the urban planning of the city of Alexandria guarantees its survival in the event of a rise in sea levels. This is because the city was built on a rocky barrier parallel to the shore, and the sand that the sea waves cast on the beach hardens and becomes elevated areas. In addition, the rock barrier in the Abu Qir area is 5 meters, while in the port it ranges between 20 and 30 meters. These heights protect the city of Alexandria from the danger of drowning.

In conclusion, the dangers of climate change in Alexandria have been reflected in the conditions of social inclusion, and pushed the city's residents to forced migration, which led them to a state of social exclusion and marginalization, due to the encroachment of water and torrential rains in 2015 and 2022.

During the years until 2050, 30% of the city may be flooded, 1.5 million people will be displaced, and 195 thousand people will lose their jobs. Considering forced migration as a result of failure to adapt to environmental changes, the results of the last census 2017 confirmed that the city of Alexandria ranks

Figure 4. Is Alexandria threatened with drowning?
Source: Al-Ain (2022)

first in immigrants from it. And that the negative effects of immigration on poor families are clear. The debate over whether Alexandria is threatened with drowning or the city being swallowed up by the sea ended with a temporary and partial drowning.

Building Community Resilience and the Mechanisms of Mitigating the Effects of Climate Change and Adapting to Its Risks: The Case of Alexandria

In May 2022, the Egyptian government announced the National Strategy of Climate Change, as an example of comprehensive planning for sustainable development, preparation for Egypt's hosting of the Climate Conference (Cop 27), and emphasis on policies, and the required procedures of adaptation, mitigation, and dealing with climate change and its effects, through Five primary goals, beginning with achieving sustainable economic growth, regulating energy efficiency, adopting sustainable consumption and production trends to reduce global warming emissions, building capacities and resilience to mitigate negatives effects, protecting citizens through improving health services, and preparing the health sector to confront diseases caused by change climate, preparing studies, workers training, educating citizens to preserve natural resources, environmental systems, state resources, and assets, infrastructure, and services.

Based on this, the Alexandria Governorate planned an action program, selected the mechanisms to mitigate the effects of climate change, and built the resilience of the community to adapt to the risks of these changes.

1. The Work Program of Alexandria City and the Effects Mitigating Mechanisms of Climate Change

It has already been mentioned that Alexandria Governorate in Egypt is one of the most cities directly affected by the phenomenon of climate change. Where the dates of the winter cores changed, and the amount of rain that fell in those cores varied, in addition to the erosion of the beaches or the phenomenon of erosion, which was evident last winter in Sidi Bishr beach.

The Egyptian state was aware of the danger of this phenomenon several years ago. And it started implementing marine protection projects in Alexandria Governorate, in coordination between Alexandria Governorate and the Ministry of Irrigation, where several projects were implemented to put concrete blocks and rubber barriers to protect beaches from erosion, and the governorate, in cooperation with the General Authority for Beach Protection, allocated 341 million pounds to establish marine protection works along the coast. In front of the Alexandria Governorate, with the aim of protecting the Corniche Road from storms and Cores, protecting the infrastructure and historical building such as Qaitbay Castle, and cultural as the Library of Alexandria, while expanding the Corniche Road, and increasing the places designated for tourists and vacationers, by increasing the beach area, as well as providing places for water sports, besides the implementation of economic, tourism and entertainment projects, to provide job opportunities, and stimulate tourism and investment projects, while taking care of the natural and social environment and its development.

Because the marine protection works lead to the development of some sandy beaches that have disappeared due to the expansions of the Corniche or the sea, while ensuring the quality of the beaches for vacationers, in terms of non-pollution and preventing the phenomena of eddies, clouds, and the intensity of waves that threaten their lives, as well as restoring the aesthetic shape of the beaches.

Building Resilience, Mitigating Impact of Climate Change on Inclusion

During the past years, the Egyptian Authority for the Protection of Beaches has implemented many major projects to protect beaches for 969 million pounds, including a project to protect the area in front of the Naval College for 67 million pounds, and the Bir Masoud and even Al-Mahrousa project with a length of 1600 meters to protect 20 kilometers of beaches for 200 million pounds, the project to protect the Corniche from El-Mahrousa for 335 million pounds, the project to strengthen and develop the corniche towards Mansheya and the Raml station with a length of 500 meters from the court complex, towards the Raml station for 100 million pounds, the project to protect Qaitbay Castle for 267 million pounds, and the strategic plan is being completed along the coasts of Alexandria.

The marine protection project for Qaitbay Castle, which was implemented by the General Egyptian Authority for Beach Protection, in coordination with the Ministry of Tourism and Antiquities, for 267 million pounds, was one of the most important of these projects. Mohamed Metwally, Director General of Alexandria Antiquities, indicated that it is an integrated marine engineering project to protect Qaitbay Castle from the influence of winter storms, high sea waves, and a 100% protective structure. The project includes four parts; The first: building a breakwater with a length of 520 meters, the second: a concrete walkway with a length of 110 meters, and the third: a sea shaft with a length of 30 meters. the fourth: a marine pier with a length of 10 meters.

Marine protection projects for beaches are being implemented in three stages. The first is from Bir Massoud to El-Mahrousa, with a length of 1,600 meters, and protects about 2 km of these beaches, for 189 million pounds. The second stage is from El-Mahrousa to San Stefano, and the length of the barrier is 570 meters, for 180 million pounds. The third stage is from Sidi Gaber to Al-Silsilah in the eastern port. The length of the barrier is 3,600 meters, and it protects about 4 km of beaches for 900 million pounds. This stage covers 8 beaches: (Abu Haif, Sidi Bishr, Al-Saraya, Sidi Gaber, Cleopatra, Sporting, Ibrahimia, and Camp Shizar).

In addition to a project to strengthen and develop the protection of the Corniche in the direction of Mansheya and the Raml station, to protect the historical Corniche wall. The Corniche road in the Mansheya area and the Raml station, for 103 million pounds. The castle, in addition to developing the area in front of the castle to protect it from high waves, to attract and refresh tourism, and support investment worth 235 million pounds (Mounir, 2022).

Al-Youm Al-Sabea added on 11/10/2022 that Alexandria is witnessing a package of procedures to reduce the risks of climate change. the governorate was keen to take several procedures within the framework of the World Climate Conference (COP 27), the foremost of which was participation in the national initiative for smart green projects and contributed 3 Projects, in addition to launching the "Alexandria without single-use plastic bags" initiative by intensifying awareness campaigns and providing support to reduce the risks and damages caused by plastic resources and their negative effects on health, the marine environment, and climate change in general. Beach cleaning initiatives and campaigns have also been intensified at the governorate level, in cooperation with civil society organizations, and the governorate has intensified its work for the presidential initiative "planting 100 million trees." So far, 24,000 trees have been planted at the level of neighborhoods, especially roads and traffic axes, in addition to starting to increase the surface areas. Greenery and planting fruit trees in all neighborhoods to be a natural outlet for citizens, and the governorate participated in the annual Green Cities Conference in the winter of 2022, which was held in Austria under the slogan "Make green a reality", to help cities become greener, as well On enhancing energy resilience in cities and financing their infrastructure.

And in implementation of the National Initiative for Smart Green Projects, as well as the Green City Initiative, Alexandria joined this project in 2021 with the Cairo Governorate, and the 6th of October

City, in cooperation with the European Bank, for reconstruction, the development of urban communities, and the improvement of the quality of life, by adopting a set of standard specifications, to benefit of alternative energy and reducing carbon dioxide emissions, a project that aims to include 100 countries around the world by 2024.

The governorate began working on many projects in the field of green infrastructure development, energy rationalization, and the use of insulating resources, through several axes, as the governorate imported electric buses, depending on environment friendly energy, and the Alexandria metro project is being implemented, which works with electricity as an alternative to the internal trains.

Alexandria also began expanding the use of solar energy and has already worked to open a wholesale market that operates with this energy in the Al-Ameriya area, west of Alexandria. Solar energy is also used in several official buildings such as Alexandria Port and Alexandria University. The governorate coordinates with specialized agencies to raise awareness of the risks that affect the environment as a result of climate change and to benefit from research projects to mitigate the impact of changes (Al-Youm, 2022).

The Governor of Alexandria held the first meeting with the Climate Change Committee during the month of February 2023, which he formed to serve as an early warning system to protect Alexandria from unbridled weather attacks and to identify and study the needs and demands for the development of this system, and its prediction of weather attacks and climate change. Monitoring the phenomenon of sea level rise through climatic measurements of atmospheric elements, providing devices, and installing them in places most vulnerable to weather changes. The committee is also specialized in preparing detailed studies of more damages that may arise from weather attacks that affect citizens, and the duties that must be taken towards the negative effects of climate change, as well as preparing a short-term and long-term plans to deal with the phenomenon. Also, capabilities building in combating the impact of climate change and sea level rise. Beside, benefit from the available international studies and experiences in this regard. The Egyptian state has spent and is still spending billions to protect Alexandria and address the mistakes and randomness that occurred in the city decades ago, and the committee relies on the cooperation of everyone, universities, civil society, institutions, and expert houses, to address this phenomenon that threatens society. A set of presentations were made in which most of the experts of the committee participated, such as the head of the National Institute for Astronomical Research (Mounir, 2023).

The Governor of Alexandria inaugurated the first session of the meeting of representatives of the World Bank at the Bibliotheca Alexandrina on May 7, 2023, to start work on the green cities project for the city of Alexandria. He stressed that the governorate has joined the Green Cities Program of the International Bank for Reconstruction and Development since 2019, which aims to support the transition of cities to green cities by reducing carbon. He mentioned that the governorate presents three projects, namely: the committee of the early warning system to predict tsunami waves, the use of robots in removing waste from the sea and oceans, and the work of 3D printing of electronic waste. He added that the state and the governorate support the existence of all forms of investment in the field of environmental protection from climate changes that occur in the world

The Alexandria Governorate has implemented huge rainwater management projects, as well as a project to protect beaches from the phenomenon of erosion, activating environmental initiatives and participating in the "Be Ready for The Green" and "Alexandria Without Plastic Bags", in order to reduce its risks to the environment in general, and the marine environment in particular. The governor called on the World Bank team to increase the World Bank's contributions to projects related to climate change in

Alexandria, as it is the most vulnerable and the oldest city in the world, and to speed up the establishment of projects. The Executive Director of the Southern and Eastern Mediterranean Region, the European Bank for Reconstruction and Development, the Director of the Green Cities Support Fund at the European Bank, a number of representatives of universities and institutes, the Environmental Affairs Agency, and other representatives of the concerned authorities in the governorate participated in this meeting.

2. The Building Mechanisms of Community Resilience (Alexandria) and Adapting to the Risks of Climate Change: Features of Strength and Weakness

We have previously identified mechanisms for building community resilience as a set of mechanisms that contribute to achieving adaptation to climate change and help mitigate its effects and risks. This includes hard tangible mechanisms such as resources, policies, economic development, infrastructure, and the ability to change performance and work, as well as soft intangible mechanisms such as social capital and building resilience, communication and information networks, sustainability, environmental awareness, and corporate social responsibility among the different sectors of society (government, private sector, civil society), and it may be useful for rely on this identification in reading the work programs and projects that the Egyptian government and the Alexandria governorate relied on as mechanisms to mitigate the effects of those climate changes, by using the method of (SWOT) analysis and highlighting the strengths and weaknesses as well as the positives and negatives that can be recorded on these mechanisms, perhaps it will help us to draw a set of results in the light of which another set of work programs and mechanisms can be proposed, and to emphasize another set of recommendations and proposals to support and enhance the efforts of the government and the governorate in this regard.

If we look at the set of programs and projects related to the resource mechanism, one billion and 341 million pounds have been allocated for the establishment and implementation of marine protection projects, 969 million pounds for beach protection projects, and 267 million pounds for the Qaitbay Castle protection project. Evidence of strengths and positive aspects in building the resilience of community in the city of Alexandria. The strength of this building was also evident through its keenness to develop the area in front of Qaitbay Castle in attracting and revitalizing tourism, supporting investment, as well as cooperating with the World Bank in developing urban communities and improving the quality of life.

In view of the intangible aspects of building the resilience of the city's community, it was evident that social capital was absent and weak, and communication and information networks were not invested sufficiently, while aspects of sustainability were confirmed through participation in national initiatives for smart green projects, Alexandria without plastic bags, and beach cleaning campaigns. Increasing green areas and planting fruit trees, expanding the use of solar energy in work areas such as the wholesale market, the port of Alexandria, and the University of Alexandria, and a capacity-building mechanism in the field of combating the impact of climate change and rising sea levels, and addressing errors and randomness that occurred in the city since Tens of years, and coordination with specialized agencies to raise awareness of the risks that affect the infrastructure, as a result of climate changes, and benefit from research projects, to manage and mitigate the impact of these climate changes, and activate the solidarity social responsibility between the various sectors of society, by participating in the meetings of the Early Warning Committee and the European Bank, and evidence of the strength of these aspects, benefit from the international partnership in the processes of sustainable development. In general, building community resilience (Alexandria) adhered to the basic goals announced by the Egyptian government in May 2022.

RESULTS AND DISCUSSION

1. Alexandria city had suffered in the last four years from multiple climate change manifestations and variations of losses such as light rates of rain, bad weather, and drop in temperature continued in slums the infrastructure, especially in summer. The economic Impacts of climate change, such as fish death and fish wealth, spread of diseases partial drowning of the beaches due to increase in the sea level, disappearance of large numbers of antiquities, and temporary sinking. All of these were the big losses of the city some of which referred to by Elliott 2018 the sociology of loss perspective especially disappearance part of the city, attrition the government resources and destruction of the infrastructure.

2. The climate change risks Alexandria city reflected on social inclusion status cue. The city came first in the number of internal migration. According to last census in Egypt 2014, so the negative effects of migration will be in poor families, then the social exclusion and the marginalization may happened. Add to that evidence refer to about the drawing of the city temporary partial which lead to many aspects of inequalities in human rights, quality of life, and suffering of poor as Ripple and hid collogues said in them studies.

3. The building of community resilience in Alexandria city had multiple effective mechanisms, such as financial resources, international partnership for development, cooperation with international Bank for development of when communities based improvement like quality, participation in local initiations for smart green projects, benefit of alternative energy and reducing carbon dix emissions and building capabilities in the field of climate change and its effects on the rise of sea surface social capital, network on information and communication, poor infrastructure and slums. So the building of community resilience in the city did not take in consideration all the mechanisms which Carmen and his collogues referred to.

RECOMMENDATIONS AND ISSUES NEED STUDYING IN THE FUTURE

About the non- effective mechanisms of community resilience in the city, it is expected depending on corporate social responsibilities, that special sector and civil society could participate and offer part of financial resources needed in this regard. The sustainable umbrella has to take in consideration social and economic aspects for enforcing social inclusion, and avoid internal migration. The youth generations in social clubs and universities could participate in the environmental awareness companies for social capital investments adaptation to risks.

There are some issues still need studying in the future; such as corporate social responsibility and mitigation climate change effects, the environmental awareness building capabilities for adoption climate change risks and the relation between climate change risks diseases diffusion.

REFERENCES

Carmen, E., Fazey, I., Ross, H., Bedinger, M., Smith, F. M., Prager, K., McClymont, K., & Morrison, D. (2022). Building Community resilience in a context of climate change. The Role of Social Capital. *Ambio, 51*(6), 1371–1387. doi:10.100713280-021-01678-9 PMID:35015248

Dapilah, F. (2020). The role of social networks in building adaptive capacity and resilience to Climate change: A case study from North Ghana. *Climate and Development, 12* (1), 42-56.

Elliott, R. (2018). The sociology of climate change as a sociology of Loss. European journal of sociology. *Archives Europeees de Socioligie, 59*(3) 301-337.

Hossain, A. T., & Masum, A. A. (2022). Does Corporate Social responsibility help mitigate firm Level Climate change risk? *Finance Research Letters*, *47*, 102791. doi:10.1016/j.frl.2022.102791

Marquand. J, F, A. & Elsasser, J, p . (2023) Institutionalizing climate change mitigation in the Global South: trends and future research Earth system Governance, 15, 100163.

Aboudouh, K, K.(2023). Civil society and its role Climate change issues. مقال منشور في مجلة آفاق مستقبلية، العدد الثالث يناير ٢٠٢٣، مركز دعم واتخاذ القرار . ج . م . ع.

Al-Amir, N. Kamal. (November, 2022). The Egyptian Vision towards Climate Change Political and Environmental Issues and Development Goals. *Journal of International Politics*.

Al-Masry Al-Youm, (July, 2023a). *The full truth behind the subsidence of the Alexandria Corniche, the collapse has nothing to do with any earthquake.* 2/7/2023.

Rabie, S. (2022). *Climate Change and Population Distribution in Egypt.* Central Agency for Public Mobilization and Statistics, Egypt.

Ripple, w, J, M, (etal). (2022) Six steps to integrate climate mitigation with adaptation for social Justice. *Environmental science & Policy, 128*, 41-44.

Tahamineh, L. (etal), (2022). *Application of Machine Learning deep Learning Methods for Climate change Mitigation and Adaptation.*

Yuchi, Z, Y. (et al), (2022). Building social resilience in North Korea can mitigate the impacts of climate change on food security. *Nature food, 3*(714 99- 511)

Zehr, S. (2013). The sociology of global climate change. *Wiley Interdisciplinary Reviews: Climate Change, 6*(2), 129–150. doi:10.1002/wcc.328

KEY TERMS AND DEFINITIONS

Building Community Resilience: Hard mechanisms such as: resources, economic development, infrastructure, and soft mechanisms such as: Social capital, capabilities, adaptation, social awareness, and corporate social responsibility.

Climate Changes Risks: depletion, disappearance, and collapse.

Social Inclusion: The opposite of inequalities, equity, and marginalization.

18

Chapter 2

Climate Change and Its Impact on Sustainable Development Goals in the Mediterranean Countries:
An Evaluation Study Through Geographic Information Systems (GIS)

Bdor Osama
Alexandria University, Egypt

ABSTRACT

The study was divided into three main sections. The first section is terminology related to climate change. The second is to introduce the sustainable development goals in relation to climate change. The author inquired about the positive or negative impacts of climate change in relation to achieving them in 2030. The last section contains the results of impacts through GIS software. The assessment processes include (1) the data preparation stage, (2) the data entry stage, and (3) the geographical processing of the data. This section also sheds light on sustainable development with regard to the social and economic aspect, as this aspect is reflected in the economy, politics and the environment. The researcher reveals (1) how economics and politics protect the environment and (2) how economics and politics help society protect the environment for future generations. She concluded by identifying the relationship between climate change and sustainable development as a cycle.

CONTEXT OF THE STUDY

Climate change is one of the major global challenges with diverse, significant, comprehensive, and even irreversible impacts on human and natural ecosystems. These impacts may affect and hinder the achievement of sustainable development. Investigating these impacts requires using GIS software in

DOI: 10.4018/978-1-6684-8963-5.ch002

Copyright © 2024, IGI Global. Copying or distributing in print or electronic forms without written permission of IGI Global is prohibited.

assessing vulnerability to risks associated with climate change, which is essential for successful planning and implementation of adaptation plans. GIS is a computer system for managing, processing and analysing data of a spatial nature for displaying and outputting geographic and descriptive information for specific purposes that help in planning and making various decisions. Due to GIS great capabilities to handle, process and analyse huge geospatial data, can help in simulating, understanding climate change as one of the dynamic real-world phenomena of diverse spatial dimensions, especially in the Mediterranean countries.

SCOPE OF THE STUDY

The conducted study investigates climate changes' impact in the 20 countries of the Mediterranean basin, namely (Egypt, Libya, Algeria, Tunisia, Morocco, Spain, Portugal, France, Italy, Slovakia, Bosnia and Herzegovina, Montenegro, Albania, Greece, Turkey, Cyprus, Syria, Lebanon and Palestine) these countries are located between longitudes 12°30W to 45°E and latitudes 20°N and 51°N, with a combined area of 9 million km^2 and the largest country in terms of area is Algeria and the least is Cyprus, and the population of the Mediterranean basin countries is more than About 500 million people, or 7% of the total world population (Rajdali, 2020).

Figure 1. The Boundaries of the study area
(CLIMATE AND ENVIRONMENTAL CHANGE IN THE MEDITERRANEAN BASIN, Current situation, and risks for the futurey, 2020) (Fawaz & Suleiman, 2015)

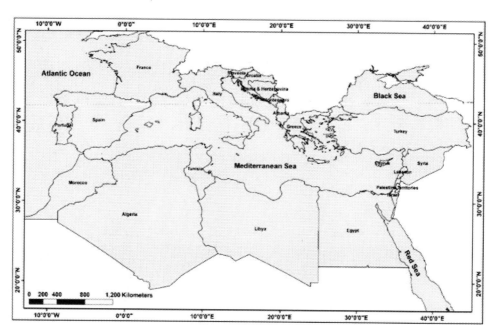

STATEMENT OF THE RESEARCH PROBLEM

The study aims to assess the nature and causes of climatic changes for the elements of heat and rain, for instance in Egypt. The study also determines the extent to which global and regional climate changes relate to climate changes for the elements of heat and rain in Egypt. During the twentieth century and studying its different climatic characteristics, the researcher relied on the regional approach to address climate changes in three spatial frameworks: the global, regional, and local framework. The coefficients for the decrease in the annual average temperatures during it ranged from -1.5 in the centre and west of the Western Desert, to -3.4 in the northern and southern outskirts of the country. Declining coefficients are -1.5 in the extreme southwestern corner of the country, while the trend coefficients are high -1.6 in the region extending around the central delta, Cairo, and the Bahiriya Oasis region.

RESEARCH QUESTIONS

The primary questions of this study are:

1. What are the factors that lead to climate change?
2. What effects that are a result of climate change?
3. What is the most important strategy for adapting to climate change?
4. How to achieve sustainable development goals regarding climate changes?

OBJECTIVES OF THE STUDY

The research mainly aims are to identify the expected effects of climate change regarding the best possible development ways in relation with adapting to climate changes and then the research seeks to:

1. Investigating the phenomenon of climate change and the factors leading to its occurrence.
2. Defining the effects of climate change.
3. Examining the most important strategies for adapting to climate change.
4. Achieving sustainable development goals concerning climate changes.

This study also illustrates the research sample's awareness of the effects of climate change on health security with its dimensions represented in: (public security, food security, and environmental security), and how to confront them. Pictures, causes of occurrence, dimensions of health security, identifying the bilateral relationships between the studied independent variables and the degrees of awareness of these effects, and knowing how to face these effects.

SIGNIFICANCE OF THE STUDY

The conducted study highlights the correlation between (1) factors, (2) the effects of climate change and (3) sustainable development goals. This research represents an attempt to expand the scope of evaluating

Climate Change and Impact on SDGs in the Mediterranean Countries

the effectiveness of financing programs to confront climate change considering the lack of such studies, especially Arab ones. The importance of research also increases considering the importance of the issue of climate change and the economic, social, and environmental challenges it imposes on all countries of the world in general, with special focus on developing countries.

THEORETICAL BACKGROUND AND REVIEW OF THE RELATED LITERATURE

This section illustrates the terminologies and highlights previous studies, concerning Climate Changes, factors, and effects of climate change in relation to sustainable development and its goals.

Climate Change

Climate changes are natural phenomena that occur every several thousand years, and they also represent a challenge facing humanity. Interest in them began at the end of the nineteenth century (Mahmoud, 2014). The United Nations Framework Convention on Climate Change (UNFCCC) defines climate change as: a change that is attributed directly or indirectly to human activity, that changes the chemical composition of the atmosphere at the global level, and that is in addition to the natural climate variability observed during similar time periods (Fahim M. A, 2013). Accordingly, the Convention distinguishes between climate change that is due to human activities, and climate variability that is due to natural causes (Fawaz & Suleiman, 2015).

The Intergovernmental Panel on Climate Change (IPCC) defines the term climate change as referring to a change - of statistical significance - in the average state of the climate or a change in its characteristics that extends over a long period, usually decades or more, and climate change may be attributed to climate processes. natural internal or external influences or continuous human changes in the composition of the atmosphere or in the use of land (Fourth Assessment Report, Climate Change, 2007).

The phenomenon of climate change is also known as an imbalance in the usual climatic conditions such as temperature, wind patterns, and rain that characterize every region on Earth. and the amount and types of rain, in addition to the possibility of possible extreme climatic developments, which lead to environmental, social, and economic consequences that have a wide impact that cannot be predicted (Al-Saee & Al-Qahtan, 2016). What is observed in the composition of the Earth's atmosphere, in addition to the natural variability of the climate, over similar periods of time (Hassan K. E., 2021). The researcher illustrates relevant terms concerning the term climate changes, for instance: Climate Variability, Climate Fluctuation, and Climate Oscillations.

Causes of Climate Change

At the root of climate change is the phenomenon known as the greenhouse effect, the term scientists use to describe the way that certain atmospheric gases "trap" heat that would otherwise radiate upward, from the planet's surface, into outer space. On the one hand, we have the greenhouse effect to thank for the presence of life on earth; without it, our planet would be cold and unlivable.

But beginning in the mid- to late-19th century, human activity began pushing the greenhouse effect to new levels. The result? A planet that's warmer right now than at any other point in human history, and getting ever warmer. This global warming has, in turn, dramatically altered natural cycles and weather patterns, with impacts that include extreme heat, protracted drought, increased flooding, more intense storms, and rising sea levels. Taken together, these miserable and sometimes deadly effects are what have come to be known as climate change.

NATURAL CAUSES OF CLIMATE CHANGE

Some amount of climate change can be attributed to natural phenomena. Over the course of Earth's existence, volcanic eruptions, fluctuations in solar radiation, tectonic shifts, and even small changes in our orbit have all had observable effects on planetary warming and cooling patterns.

But climate records are able to show that today's global warming—particularly what has occured since the start of the industrial revolution—is happening much, much faster than ever before. According to NASA, "[t]hese natural causes are still in play today, but their influence is too small or they occur too slowly to explain the rapid warming seen in recent decades." And the records refute the misinformation that natural causes are the main culprits behind climate change, as some in the fossil fuel industry and conservative think tanks would like us to believe.(https://www.nrdc.org/stories/what-are-causes-climate-change#natural)

ASTRONOMICAL CAUSES OF CLIMATE CHANGE

The astronomical theory of paleoclimates aims to explain the climatic variations occurring with quasi-periodicities lying between tens and hundreds of thousands of years. Such variations are recorded in deep-sea sediments, in ice sheets and in continental archives. The origin of these quasi-cycles lies in the astronomically driven changes in the latitudinal and seasonal distributions of the energy that the Earth receives from the Sun. These changes are then amplified by the feedback mechanisms which characterize the natural behaviour of the climate system like those involving the albedo-, the water vapor-, and the vegetation- temperature relationships. Climate models of different complexities are used to explain the chain of processes which finally link the long-term variations of three astronomical parameters to the longterm climatic variations at time scale of tens to hundreds of thousands of years. In particular, sensitivity analysis to the astronomically driven insolation changes and to the CO_2 atmospheric concentrations have been performed with the 2-dimension climate model of Louvain-la-Neuve. It could be shown that this model simulates more or less correctly the entrance into glaciation around 2.75 Myr BP, the late Pliocene-early Pleistocene 41-kyr cycle, the emergence of the 100-kyr cycle around 850 kyr BP and the glacial-interglacial cycles of the last 600 kyr. During the Late Pliocene (in an ice-free – warm world) ice sheets can only develop during times of sufficiently low summer insolation. This occurs during large eccentricity times when climatic precession and obliquity combine to obtain such low values, leading to the 41-kyr period between 3 and 1 Myr BP. On the contrary in a glacial world, ice sheets persist most of the time except when insolation is very high in polar latitudes, requiring large eccentricity again, but leading this time to interglacial and finally to the 100-kyr period of the last 1 Myr. Using CO_2 scenarios, it has been shown that stage 11 and stage 1 request a high CO_2 to reach the interglacial level. Moreover, the insolation pattern at both stages and modeling results lead to conclude that stage 11 is a better analogue for our future climate than the Eem. Although the insolation changes alone act as a pacemaker for the glacial-interglacial cycles, CO_2 changes help to better reproduce past climatic changes and, in particular, the air temperature and the southern extend of the Northern Hemisphere ice sheets. Insolation and CO_2 scenarios for the next 130 kyr lead to an interglacial which will most probably last particularly long (50 kyr). This conclusion is reinforced by the possible intensification of the greenhouse effect which might result from man's activities over the next centuries.(A. Berger,2004)

EVIDENCE OF CLIMATE CHANGE

The evidence of climate change is becoming obvious every day near impossible for someone to doubt the adverse effects of climate change that we are experiencing on daily basis in many places, including rainfall variability, desertification, melting of the glacier, global warming, forest fire and rising sea level. (Yakubu Gambo Hamza,2020)

1-   Seasonal Rainfall Variability:

The conceptual basis for variation in the amount of precipitation (rainfall) w Other elements linked with duration, amount, distribution, intensity and distribution of precipitation, particularly when the variation is extremes as climatic condition differs from one place to another also observed.

Warming increases the rate of evaporation of soil moisture which results to dryness in soil, this may trigger the potential incidence and harshness of droughts that widely detected in several places around the world. However, the atmospheric moisture was carried out by the wind in the atmosphere to place where storms favored. Mostly storms covered a distance of about four times the radius of other in dimension, and collect in the water vapor, to produce precipitation . This may lead to a significant increase in one place and an intense decrease in other places across the globe.

2-   Desertification:

Desertification is one of greatest environmental devastation facing humanity today. It results in water scarcity, loss of biodiversity and ecosystem instability. The United Nations Convention to Combat Desertification(UNCCD) 2018-2030 Strategic Framework was set with a vision to abate, reduces, and converses desertification/land degradation and alleviates the effects of drought in areas affected and make every effort to. Climate change is one of a number of factors that are considered to contribute towards desertification. It is crucial to note that desertification is a man-made phenomenon that is intensified by climate change as a consequence of deforestation and bush burning. The fact that reduces of annual precipitation leads to severe drought while heavy rainfall as a result of climate change will lead to land degradation and loss of vegetation cover which serves as a shield in curbing and addressing desert encroachment.

3-   Melting of glacier:

Modifications in glaciers are clear pointers of the projected climate change . A study on the glacier status of over three decades by studying the glacial retreat of 82 glaciers, area shrinkage of 7,090 glaciers

4- <u>Global warming:</u>

The maximum and minimum temperature for the period of 1901-2003, rainfall for the period of 1987- 2002 and sea surface period of 1901-2003 were studied. The results of the study reveal that winter-time increase in the mean temperature over India is about 1.0°C, meanwhile, Pre-monsoon period, Monsoon and Postmonsoon period rise were observed to be 0.3°C, 0.4°C, and 1.1°C respectively (Dash, 2015). In the same vein, sea surface temperature based on the research shows a dramatic increase in trends of sea surface temperature from the period of 1901-2003.

However, an increase in temperature in most parts of the world is the greatest impact of climate change which can result in either negative or positive ecological impacts. The increasing temperature has led to increased land-based ice instability and its melting . The thawing of the Arctic, cool and cold temperate ice, the increased rainfall in some parts of the world and expansion of the oceans as the water warms has the greatest impact on sea-level rise, coastal flood and erosion.

5- <u>Forest Fire:</u>

Forest fire is one of the environmental issues which continue attracting concern globally, it is widely believed that climate change plays a significant role in influencing the intensity of this environmental catastrophe. Forest resilience is the capability of the forest to recover to a pre-disturbance state and is strictly rely on abundant tree redevelopment . Both temperature and water scarcity have a negative impact on trees growing in their early life stages (seedlings and saplings), forest firmness to disturbances under warm climate remains unclear, from the year 2000 weather and climate are major factors influencing the rate of fire activity. The changes in climate mostly occur as a result of anthropogenic activities, warmer weather is expected in the future which could accelerate the severity and intensity of forest fire. Even though there are will be great spatial and temporal dissimilarities in the fire activity response to climate change.

6- <u>Rising Sea Level:</u>

The sea level is rising due to a significant increase in warmer climatic conditions as a result of anthropogenic activities. Therefore, the majority population living in coastal areas are facing threat from sea-level rise which affects their socio-economic development, water quality and deteriorating health condition, when care is not taking resulting in a disease outbreak. The increasing rates of sea-level rise caused by global warming within the 21st century are predicted to accelerate flooding in a low-lying coastal environment, though the impact is more disastrous in developing

Climate Variability

Climate Variability refers to variations in the average and other statistics of the climate at all time and spatial scales, and climate variation is attributed to natural internal processes within the climate system

Climate Change and Impact on SDGs in the Mediterranean Countries

internal variability or to variations in external natural or anthropogenic radiative influences external fluctuations (Fourth Assessment Report, Climate Change, 2007).

As defined by the Encyclopaedia of Global Warming, the term climate variance is often used to denote deviations of climate statistics over a specific period of time, such as a specific month, season, or year, from long-term climate statistics related to the same period of the year, where the term variance expresses Climate for the inherent feature of the climate, which is the change through time, and the degree of climate variation can be described by the differences between the long-term statistics of meteorological elements calculated for different periods (Dutch, 2010).

Climate Fluctuation

Climate fluctuation means the rapid change in the weather condition during a short period of time, such as the weather condition that accompanies the storms of the Khamaseen in Egypt (Zahran, 2007).

Climate Oscillations

The Encyclopaedia of Global Warming defines *Climate Oscillation* as a fluctuation in a climate element during which the element tends to transition gradually and smoothly between successive maximum and minimum values (Dutch, 2010).

Types of Climate Change

Climate Changes are divided according to the nature of their occurrence and their source into two types. The first type is *Regular Climatic Changes* that occur in the atmosphere periodically so that the amount and time of occurrence can be determined (Eissa, 2007). The second one is *Irregular Climatic Changes*: They occur in the atmosphere, but it is difficult to determine their amount or times and places of their occurrence, such as a rise or fall in the temperature in the summer or winter than their rates during the same time of the year for a period and then return to normal.

Climate Changes are a fact, even if the levels of this change differ in terms of daily changes, even change at the century level, and even more than that, passing through months, seasons and years, and these years also include a series of changes (Youssef, 1982), and this type is divided into two parts, *Natural Irregular Changes* and *Abnormal Irregular Changes*.

Factors leading to Climate Changes

Scientific reports issued by the Intergovernmental Panel on *Climate Change* suggest that what is happening now in terms of global warming is the result of human activities; This is due to several reasons: burning fossil fuels, cutting down forests, industrial activities, using of means of transportation, providing buildings with energy, and food production.

Burning Fossil Fuels

Burning fossil fuels such as coal, gasoline, diesel fuel, gas, and others to produce energy. Burning fuel is currently the main cause of emissions, whether using this fuel to produce electricity or to rotate the engines of factories or various means of transportation, in addition to industrial processes (Climate Change, n.d.).

Cutting Down Forests

cutting down forests that absorb carbon to produce wood or use land in agricultural or industrial activities or for construction and expansion of cities and roads, as nearly 12 million hectares of forests are destroyed every year, causing the accumulation of carbon emissions in large quantities in the atmosphere, in addition to greenhouse gases - which are called anthropogenic gases - due to the inability of natural systems (trees and oceans) to absorb and store them in excess of natural rates, and deforestation, along with agriculture and other changes in land use, is responsible for nearly a quarter of greenhouse gas emissions (Climate Change, n.d.).

Industrial Activities

Industrial activities that have led to an increase in the proportions of some gases in the atmosphere above the natural rate, such as: carbon dioxide, methane, and nitrous oxide., and a group of perfluorocarbon gases, which are used in many industrial processes that need cooling such as air conditioners, and these synthesized gases act similar to natural carbon gases and trap the heat rising from the surface of the globe, and thus trapping heat / energy within the atmosphere in greater proportions than normal rates leads to An imbalance in the Earth's climate.

Using of Means of Transportation

Transportation is one of the human activities that consumes fossil fuels in large quantities, and as a result it is responsible for nearly a quarter of global energy-related carbon dioxide emissions, as most cars, trucks, ships, and planes run on fossil fuels, which makes transportation a major contributor to greenhouse gas emissions, especially carbon dioxide emissions. Carbon monoxide. Road vehicles account for the bulk of combustion of petroleum-based products, such as gasoline and diesel in internal combustion engines, but emissions from ships and aircraft also continue to increase.

Providing Buildings With Energy

Residential and commercial buildings consume more than half of the electricity globally change, and as they continue to rely on coal, oil and natural gas for heating and cooling, they emit Including large quantities of greenhouse gases, and the increased demand for energy for heating and cooling, with the increase in the possession of air-conditioning devices, and the increase in electricity consumption for lighting and electrical appliances, have contributed to an increase in energy-related carbon dioxide emissions from buildings, which negatively affected climatic conditions and contributed to causing change it (Climate Change, n.d.).

Food Production

Food production emits carbon dioxide, methane and other greenhouse gases in different ways, including deforestation and clearing of land for agriculture and grazing, digestion processes of cattle and sheep, production and use of fertilizers and manure to grow crops, energy use to run farm equipment or fishing boats, Usually using fossil fuels, as well as food packaging and distribution processes, all of which make food production a major contributor to climate change, directly or indirectly.

Sustainable Development

In 1987, the Brundtland Commission of the United Nations defined sustainability as "meeting the needs of the present without compromising the ability of future generations to meet their own needs." Today, there are 140 developing countries looking for ways to meet their development needs, but this is accompanied by an increasing threat because of the negative effects of climate change, so great efforts must be made to ensure that development today does not negatively affect future generations.

Sustainable Development Goals Concerning Climate Changes

The efforts made by the United Nations take place on September 25, 2015, through the development of 17 goals that represent the sustainable development agenda for the period between 2015 and 2030. The Sustainable Development Goals are a global call and a global framework for action and coordination of efforts. To eradicate poverty and complete hunger, and to ensure education for all, gender equality, empowerment of women, good health, and well-being, ensuring prosperity for all and other desired goals to be achieved during the period 2015-2030.

The first objective is concerned with achieving sustainable economic growth: This objective means achieving low-emissions development in various sectors, by increasing the share of renewable and alternative energy sources in the energy mix, and expanding them by establishing wind farms, solar power plants, producing energy from waste and expanding the use of energy. vitality, in addition to developing new technologies to accommodate the use of renewable energy sources such as smart control systems and exploring new alternative energy sources such as green hydrogen and nuclear energy, in addition to increasing the use of renewable energy to generate electricity within industrial facilities, applications of solar thermal energy in industrial processes, and disposal gradual transition from coal to low-carbon fuels.

The second objective of sustainable development is to build resilience and the ability to adapt to climate change, by mitigating the negative effects associated with climate change: This depends on protecting citizens from the negative health effects of climate change, through improving health services, increasing the health sector's preparedness to confront diseases caused by climate change, and preparing Studies, training health sector workers, and educating citizens. Preserving natural resources and ecosystems from the effects of climate change by improving their ability to adapt and promoting the adoption of an approach based on linking efforts to address biodiversity loss, climate change, land degradation and desertification, and conservation of reserves.

Improving the governance and management of work in the field of climate change is the third objective. This requires defining the roles and responsibilities of the various stakeholders in order to achieve the strategic goals, and improving Egypt's position in the international ranking of climate change measures to attract more investments, climate financing opportunities, and reforming the sectoral policies necessary to accommodate Interventions required to mitigate the effects of climate change, adapt to it, and strengthen institutional, procedural and legal arrangements such as the Monitoring, Reporting and Verification (MRV) system.

Fourth Objective focuses on improving the infrastructure for financing climate activities: To achieve this goal, efforts will be made to promote local green banking, green credit lines, and innovative financing mechanisms that give priority to adaptation measures such as green bonds, private sector participation in financing climate activities, promotion of green jobs, and compatibility With the Guidelines of Multilateral Development Banks (MDBs) for Climate Financing, building on the success of existing climate financing programmes, as well as promoting scientific research and technology transfer, managing knowledge and awareness to combat climate change, and raising awareness about climate change among various stakeholders (policy makers). decisions, citizens, and students).

The final Objective regarding sustainable development is Promoting scientific research, technology transfer, knowledge management, and raising awareness to combat climate change: Scientific research represents an important component of preparing to address the consequences of climate change, as well as benefiting from global expertise.

The Relationship Between Climate Change and Sustainable Development

The relationship between climate change and sustainable development is multiple and different, and in general, both interact in a circle of mutual effect and influence. Mankind, with its continuous contribution to the exacerbation of the problem of climate change, poses a clear threat that impedes the achievement of sustainable development goals that seek to achieve integrated economic growth and social justice within a sustainable environment (Adaptation to Climate Change in the Context of Sustainable Development: A Workshop to Strengthen Research and Understanding, 2006). The interrelationship between development and climate change has become more evident in the current era, as economic growth is no longer acceptable (Saadet B., 2015) unless it leads to preserving the environment and not depleting its natural resources. To guarantee the rights of future generations, studies indicate the impact of climate change on agricultural wealth and energy. seas and oceans, health, and others. However, their influence extends to the entire world, which requires close international cooperation to reduce the effects of climate change from a human perspective, preserving the environment on the one hand, and achieving balanced development rates that guarantee the rights of all countries, present and future, on the other hand, in order to Achieving that development that meets the needs of the present without harming the needs of the future (Saadet B., 2015).

Hence, it can be said that the economy reflects the path of humanity throughout history and the size of the relationships that bind it. The industrial revolution that the world witnessed following the scientific renaissance in the 18th century AD. Since that time, no one would have appreciated its negative effects at that time, as much as its positive effects were tangible in various social and economic aspects. So, the same human mind and thanks to the technological progress it has achieved, will now return to sound the alarm, and declare that the world is now suffers from the remnants of contracts, Progress has risks that outweigh its benefits, since it first degrades what it has built. Toxic gases emitted from factories, waste and excessive and irrational consumption of natural resources are among the most important direct causes of environmental pollution and the exacerbation of global warming.

The Impact of Climate Change on the Sustainable Development Goals

There are many manifestations of climate change that scientists have observed, represented in the acceleration of the average Earth's temperature at an unprecedented rate compared to before, and the rise in the temperature of the polar and glacial regions from their normal rates, which led to the acceleration of the melting of ice sheets and glaciers, and thus the rise in global warming. The sea level, the increase in the severity, intensity and frequency of unbridled weather events such as: hurricanes, torrential rains, drought, hot and cold waves, dust, sand and ice storms, forest fires, etc., and the increase in the acidity of the oceans from their natural levels, and the following is a presentation of some manifestations of climate changes in the basin countries The Mediterranean and its impact on achieving the goals of sustainable development.

Geographic Information Systems and its Applications

Geographic information systems (GIS) are computer tools used to store, analyse, and display geographic information such as: maps, aerial photos, weather observation data, terrain, and geological data. wide range of geographical challenges (Using Geographic Information Systems to Analyze Climate Change Data, Economic and Social Commission for Western Asia, 2019). Tracing the climate changes can be reached through various procedures, including aerial photographs, earth temperature analysis, and observing the movement of ice.

GIS can be used to analyze temperature and precipitation data over specific periods of time and monitor any changes that occur. This can be used to determine the expected climate model in different regions and predict future changes. Aerial photographs and digital technologies can be used to determine patterns of change in different regions and to predict the impact of global warming. Geological and geophysical data can be used to determine ground temperatures and their impact on climate changes. Aerial photographs and digital technologies can be used to determine patterns of change in ice and the impact of climate changes on them.

In general, the use of geographic information systems allows better analysis of climatic and thermal data and the identification of the main factors that affect them, and thus the development of appropriate strategies to confront climate changes and mitigate their impact.

Several techniques are used to analyse and monitor temperatures and climate changes using geographic information systems, where the techniques used in analysing and monitoring temperatures and climate changes using geographic information systems (Using Geographic Information Systems to Analyze Climate Change Data, Economic and Social Commission for Western Asia, 2019), including:

- Remote sensing: Satellites and other remote sensing devices are used to collect data related to temperature and climate changes in different regions.
- Image analysis: Aerial and satellite images are used to analyze the pattern of vegetation cover and its temporal changes, which enables the detection of changes in temperature and climatic variations.
- Numerical modelling: Computer modelling is used to analyze expected climate changes and predict their impacts.

- Geographical classification: Geographical classification is used to analyze different regions based on specific criteria, to identify areas that are vulnerable to extinction, and to identify coastal areas that are threatened with drowning.
- GeoWeb Applications: GeoWeb application technologies are used to provide access to data on temperature and climate changes on the Internet.

DATA AND METHODOLOGY

This section illustrates the research design by clarifying the methodology employed to analyse the data. The corpus of the study reveals the selected data utilised to investigate the factors and effects of climate changes in relation to sustainable development through a descriptive analysis. The researcher retrieves the selected data from institutional official public domains available on internet that afford authentic, accurate, and documented data. The collected data represents mainly a comparison between to stages: (1) a stage that represents climate changes in a period between 70s and 80s till 2020, and (2) a stage that illustrates effects of climate change between 2020 till 2080. The primary reason for selecting institutional public domains and making such comparison is to avoid prejudice concerning the findings of this study.

ANALYSIS AND DISCUSSION

As mentioned, this study investigates the impact of climate change on the sustainable development goals. There are many manifestations of climate change that scientists have observed, represented in the acceleration of the average Earth's temperature at an unprecedented rate compared to before. The rise in the temperature of the polar and glacial regions from their normal rates, which led to the acceleration of the melting of ice sheets and glaciers, and thus the rise in global warming. The sea level, the increase in the severity, intensity and frequency of unbridled weather events such as: hurricanes, torrential rains, drought, hot and cold waves, dust, sand and ice storms, forest fires, etc., and the increase in the acidity of the oceans from their natural levels, and the following is a presentation of some manifestations of climate changes in the basin countries The Mediterranean and its impact on achieving the goals of sustainable development.

Change in Temperature

The change in temperature is one of the most important manifestations of climatic changes that affect humans and the environment at the present time, especially since the change in it leads to a change in other climate elements such as evaporation, relative humidity, rain, and the consequent change in wind movement and directions to identify On the temperatures in the Mediterranean countries, as a comparison between the current period between the years (1970-2021) and the future period (2021-2080).

During the current period (1970-2021), the average temperature in North African countries ranged between 15 to 25 degrees in most of the northern and central parts. As for the southern parts, especially in Egypt and Algeria, the average temperature rises above 25 degrees. As for the countries located in in southern Europe, the average temperature ranged between 10 and 15 degrees, except for some parts

Figure 2. The change of the temperature in the Mediterranean countries between (1970-2021) and (2021-2080) (annual average)
Source: https://www.worldclim.org/data/index.html

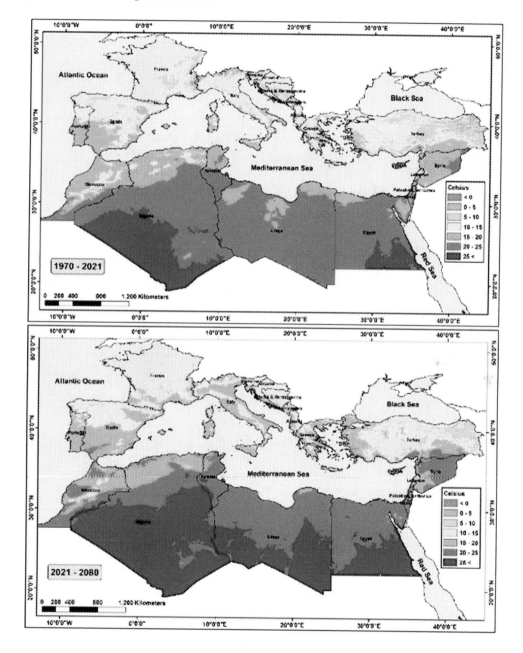

where the average temperature is less than 5 degrees, as in the northern parts of France and Italy, as well as in eastern Turkey.

It is expected that during the period (2021-2080) the continuation of the rise in temperature, as the Mediterranean countries suffer from a rise in temperature that exceeds the acceptable rate by 1.6 degrees Celsius, while the global average did not exceed one degree Celsius, and this rise in temperature is expected to increase. to reach from 2 to 3 degrees Celsius in 2030 and from 3 to 5 degrees Celsius by 2080, studies in the Mediterranean basin showed an annual increase in the general trend of temperatures estimated at about 0.03 degrees Celsius, which also exceeds Global averages, and urban areas are usually warmer than the surrounding rural areas due to human activities - especially at night - in what is known as the urban heat island factor (Island Heat Island), which enhances the increase in the frequency and intensity of heat waves in the Mediterranean countries under the influence of this factor (Risks Associated with Cimate and Environmental Changes in the Mediterranean Region, 2019)

As a result, the temperature will rise on average to exceed 25 degrees Celsius as a general average in large parts of most countries in the Mediterranean region, especially in Algeria and Libya. As for European countries such as: Spain, Portugal and Italy, temperatures will rise and the areas where temperatures drop will be less than 5 degrees. Celsius as it will happen in Turkey and France.

Change in the Amount of Precipitation

The change in temperature affects the amount of annual precipitation in any region, especially in areas that are located around large water bodies such as the Mediterranean countries, as an increase of one degree in the average global temperature would lead to a decrease in precipitation by approximately 4%. In most parts of the Mediterranean region, especially in the southern regions, and an increase in the global temperature by 1.5 degrees Celsius leads to an increase in dry periods by 7%, and it is expected that the intensity of extreme precipitation will increase by between 10% and 20% in All seasons of the year except summer (Risks Associated with Cimate and Environmental Changes in the Mediterranean Region, 2019).

In the following comparison between the distribution of the amount of rainfall over the Mediterranean countries during the period (1970-2021) and the amount of precipitation expected during the period (2021-2080), it becomes clear that:

- The annual average amount of precipitation in most Mediterranean countries ranges between 38 mm and 150 mm, during the period (1970-2021), except for some parts in which the average amount of precipitation exceeds 180 mm, such as: northwest Spain and north-western Italy.
- It is expected that the amount of precipitation will decrease in some North African countries, especially Morocco, Tunisia, and Algeria, by 30% by 2080, but it is expected that there will be an increase in the amount of precipitation in European countries, notably in France, northern Italy, eastern and central Turkey, by a rate ranging between 10-10%. 20% of the current amount of precipitation, and therefore it can be said that the change in the amount of precipitation varies greatly from one country to another because of many natural factors that affect it, such as temperature, surface features, wind directions and speed, and the proximity to water bodies.

Change in Sea Surface Temperature

Climate changes are reflected in the Mediterranean in terms of the nature of the marine and ecological environment. The average sea surface temperature rise is estimated at about 4.0 °C per decade during the period between 1985 and 2006, with an increase of +0.3 °C per decade in the Western Basin and +0.5 °C per decade for the eastern Mediterranean basin.

Figure 3. A comparison between the distribution of the amount of rainfall over the Mediterranean countries during the period (1970-2021) and the amount of precipitation expected during the period (2021-2080) (annual average)

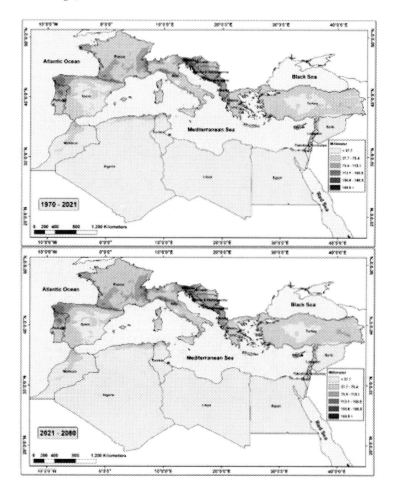

The increases in the surface temperature of the Mediterranean fluctuate throughout the year, but they mainly occur from May to July, and the maximum annual increase of 0.16 degrees Celsius was observed in the Tyrrhenian Sea, the Ligurian Sea and the Adriatic Sea, and near the African coast, as witnessed The Aegean Sea records the maximum changes in its surface temperatures during the month of August.

- It is expected that the average temperature of the Mediterranean surface will rise by 1.8 to 3.5 degrees Celsius during the period (2080-2021) compared to the period (1982-2021).

The Balearic Islands, the Greek northwest, the Aegean Sea, and the Levant Sea (the Levantine Sea) have been identified as the areas that witness the most severe rises in the surface temperature of the Mediterranean waters.

- This change in sea temperature will affect the integrity of marine ecosystems through the disturbances in plankton ecology, an increase in jellyfish outbreaks and a decrease in fish stocks, and in

Figure 4. A comparison view for the change in sea surface temperature between (1982-2022) and (2022-2080) (annual average)

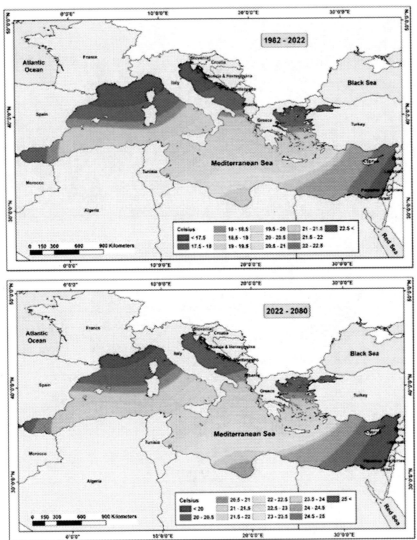

general the changes and modifications it causes in the physiology, growth, reproduction, crowding and behaviour of Marine organisms.
- Among the expected future effects of climatic changes in general and temperature changes in the Mediterranean are cases of major reorganization of the distribution of living organisms, loss of species of organisms, decrease in marine productivity, increase of non-native varieties of fish and marine organisms, and possible extinction of other species, all of which negatively affect the economies of many. It is one of the Mediterranean countries, where a large part of its population depends on fishing and related industries (CLIMATE AND ENVIRONMENTAL CHANGE IN THE MEDITERRANEAN BASIN, Current situation, and risks for the futurey, 2020)

Climate Change and Impact on SDGs in the Mediterranean Countries

- Also, the change in temperature leads to marine organisms approaching the tourist beach areas, which results in many damages to the summer areas, as happens in cases of sharks entering the northern Egyptian coasts at some times of the summer season as a result of the high water temperature; Which causes panic among tourists and vacationers at times, which negatively affects tourism activity and weakens its contribution to the economy, and weakens the economy's ability to face environmental changes resulting from climate changes.

Figure 5. The land's area that might be exposed to flooding in the Nile Delta by the year 2100
(CLIMATE AND ENVIRONMENTAL CHANGE IN THE MEDITERRANEAN BASIN, Current situation, and risks for the futurey, 2020)

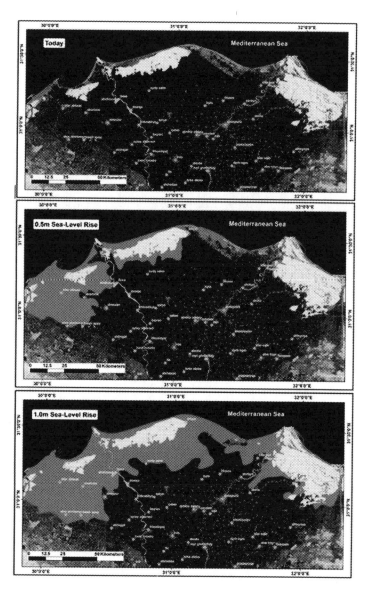

Change in Sea Level

Sea level rise is considered one of the most subsequential phenomena resulting from climatic changes. The sea level in the Mediterranean basin recorded an annual rise of 1.1 mm between 1975 and 2006. An annual increase was observed during the past two decades, amounting to about 3 mm annually. Future projections about the average sea level rise from 52 to 190 mm in the year 2100, depending on the methodology used, and the acceleration of ice melt in Greenland and Antarctica represents a great risk of additional sea level rise with the possibility of a rise of several meters, even with the hypothesis of not exceeding warming. The global limit of increase is estimated at 1.5 degrees Celsius (Weston, 2019).

In Egypt, for example, just one meter rise in sea level could cause an estimated 970 km2 to sink in the Nile Delta, affecting 9% of the country's population and about 13% of its arable land. For instance, the area Lands that will be exposed to drowning from the Nile River Delta in the event of a 0.5 meter sea level rise or 1 meter sea level rise by the year 2100, which will cause significant environmental, economic and urban damage in that region (Magdy, 2022).

Extreme climatic events and the negative effects of climate change also threaten the infrastructure of houses, roads, bridges, and buildings for various economic and residential establishments. For example, sea level changes affect the sewage networks of countries bordering the Mediterranean because most of these networks do not rise much above sea level. Most of the vital roads in coastal cities are located near the beaches, in addition to the pressures of internal migration resulting from climate change, which increase pressures on basic infrastructure services in urban areas (Muhammad, 2023).

Desertification and Widespread Drought

The term was coined by the French ecologist Auberville, who used it in the early 1950s to describe the process of degradation of tropical forests in Africa, and it wasn't until more than two decades after Auberville's warning that the topic reached a political agenda. An extended drought in the West African Sahel region in the early 1970s, leading to the convening of the United Nations Conference on Desertification in Nairobi in 1977 (World Day to Combat Desertification and Drought, 17 June/Home, n.d.).

After the historic Rio Earth Summit in 1992, the United Nations Convention to Combat Desertification (UNCCD) entered into force in 1996 defining desertification as the process of land degradation in arid, semi-arid and dry sub-humid regions of the world, which is the result of natural phenomena (such as climate change). and anthropogenic factors (World Day to Combat Desertification and Drought, 17 June/Background, n.d.).

UNCCD uses the Aridity Index Appendix (calculated as the mean annual precipitation percentage (P) stands for potential annual evaporation-transpiration (PET)) to define drylands, which includes those areas with an Aridity Index between 0.05 and 0.65 (excluding the polar and subarctic regions) (the UNCCD files, 2018).

The Mediterranean region is one of the regions expected to be greatly affected by desertification resulting from climate changes and human factors, especially since the Mediterranean region is home to more than 500 million people, but it contains only about 7.9% of the world's agricultural land, which means that there are Pressure on the agricultural resources available in that region, especially with the increasing population (Regdali, 2020).

The environment of the southern Mediterranean countries, especially Algeria, Morocco and Libya, suffers from severe sensitivity to desertification, due to the fragility of the environment in those countries and the presence of a large number of people in it, which puts pressure on the available natural resources, especially water resources and agricultural lands, which are characterized by a low area in those countries, which makes them More exposed and affected by desertification resulting from climate change.

The suffering of some countries in the northern Mediterranean basin, especially Turkey and some parts of Italy and Spain, from desertification, but to a lesser degree than in the countries in the southern Mediterranean due to the nature of wetlands in those countries and the expansion of the area of agricultural lands in some of them, and the precipitation factor has an important role in reducing the impact of desertification.

There is evidence that climate change will affect the Mediterranean Sea in different ways. However, all climate models expect that the region will become drier and hotter because of the severity of climate changes that will lead to increased drought. The link between climate change and desertification is represented in the increase in arid regions over calculating the wetter areas, which will be accompanied by lower crop yields. In addition, agriculture will suffer from a short growing season, and plants will be exposed to heat stress during flowering and rain during seeding operations. Moreover, other consequences of climate change causing increased erosion, flash floods, and instability of slopes which negatively affects the agricultural areas in it, and there are other economic sectors that will suffer from the high temperature and the drought that it causes. For instance, the tourism sector as an industry based on: marketing for rest, entertainment and enjoyment might be unavailable in the event of high temperatures and changing climatic conditions, especially in coastal areas.

Figure 6. Mediterranean countries' sensitivity to desertification
Source: (Risks associated with climate and environmental changes in the Mediterranean region, preliminary assessment by the Network of Experts on Climate and Environmental Changes in the Mediterranean Region -   2019)

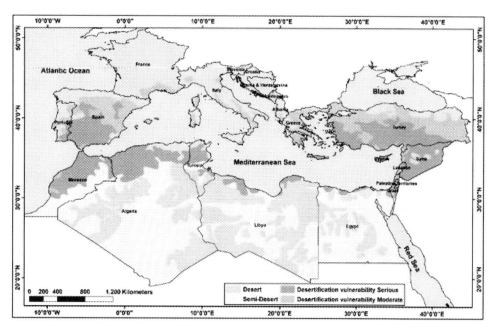

As mentioned, direct and indirect causes would affect the ability of countries in the Mediterranean basin region - especially the countries in the south and east of the Mediterranean - to commit to achieving sustainable development goals due to their lack of resources and their environmental sensitivity to climate changes, which increases the burdens on the economies of those countries and reduces from its response to development and development at various levels.

Due to the lack of agricultural productivity affects the food security of the state and increases pressure on other food resources, and the deterioration of tourism resources or their natural environments will lead to a decrease in economic returns due to the decrease in tourism attractions, especially if they depend on the environmental and climatic nature and their biological diversity.

The Impact of Water Resources

The amount of available water decreases in the Mediterranean basin due to a decrease in precipitation and an increase in temperature. In addition, it is expected that it will witness a significant decrease in the Mediterranean region, at a rate ranging between 2 and 15%, in exchange for an increase in temperature by two degrees Celsius, which in turn is one of the largest rates of decrease worldwide. It is expected that the length of the drought period will increase, and its intensity will increase dramatically.

Countries located in the south and east of the Mediterranean, which fall within the range of arid and semi-arid climates, are more vulnerable to water scarcity and fluctuating quantities, and the inhabitants of river valleys will be more vulnerable to chronic water shortages, even if global warming is limited to an increase below two degrees Celsius, and it is expected that the amount of water will decrease. Water resources in Greece and Turkey are less than 1,000 cubic meters per person per year by 2030. It is also expected that the available water resources per capita will decrease in the southeastern part of Spain and the southern beaches, which are currently suffering from an acute shortage to less than 500 cubic meters annually (UNESCO World Water Assessment Programme, 2009).

The classification of Mediterranean countries according to the annual per capita share of water from natural resources(see Figure 7), it is clear that:

- Most North African countries suffer from a low per capita share of natural water resources, ranging from 500 liters annually to less than 1000 liters of water annually, which puts pressure on the available natural resources considering climate changes.
- The small number of countries that have an abundance of water in the Mediterranean basin, and all of them are located in the north of the Mediterranean basin, and they are countries in which the average per capita share is more than 5000 cubic meters annually, and therefore the impact of climate change on them will be weak unless the population size increases or the level of demand for irrigation water increases. Or increase the demand for water in other human activities. (the Network of Experts on Climate and Environmental Changes in the Mediterranean Region, 2019).
- It is worth noting that the efficiency of dealing with water resources and their good utilization has a great impact on maximizing their utilization and increasing their sustainability. Which helps to reduce the negative effects of climate change, by following up on Figure (8), which shows the efficiency of the Mediterranean countries in dealing with the water resources available to them in 2018, as this indicator is calculated through the dollar return on each cubic meter of water that is used in three activities Essential, which is agriculture, industry and services, which is an indicator

Climate Change and Impact on SDGs in the Mediterranean Countries

Figure 7. Classification of Mediterranean countries according to the annual per capita share of water from natural resources
Source: (Risks associated with climate and environmental changes in the Mediterranean region, preliminary assessment by the Network of Experts on Climate and Environmental Changes in the Mediterranean Region - 2019)

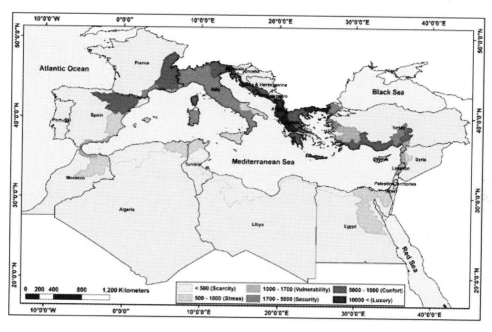

Figure 8. The efficiency of Mediterranean countries in dealing with available water resources in 2018
Source: (CLIMATE AND ENVIRONMENTAL CHANGE IN THE MEDITERRANEAN BASIN, Current situation and risks for the futurey, by MedECC,2020)

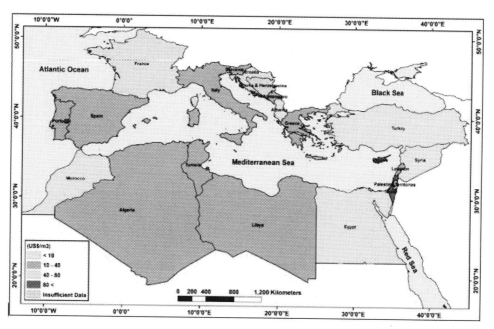

proposed by the Food and Agriculture Organization in 2017, and through the figure it is clear that the Mediterranean countries are divided into four groups:

a) The first group: Morocco, Syria, Egypt, Turkey, and Albania, with a yield of less than $10 per cubic meter of water.
b) The second group: Libya, Algeria, Tunisia, Portugal, Spain, Italy, Slow Viña, Montenegro, Greece, Lebanon, with returns ranging from 10 to 40 dollars per cubic meter of water.
c) The third group: includes France, Bosnia, and Herzegovina, with a return of 40-80 dollars / cubic meter of water.
d) The fourth group: it includes Palestine and Cyprus, with a return of more than 80 dollars / cubic meter of water.

This reveals that most Mediterranean countries do not exploit water in a way that achieves the greatest benefit for it, as a result of the high percentage of wasted water during consumption, and the adoption of traditional methods in manufacturing processes in some countries, and agricultural methods, especially irrigated agriculture, lead to the loss of a large amount of water. without taking full advantage of it.

Food Resources

Climate and environmental changes, in addition to economic and social changes, constitute a major threat to food security throughout the Mediterranean region and in the various production sectors. Soil degradation and erosion are among the most important factors affecting the agricultural and livestock sectors in the Mediterranean basin, where climatic extremes such as drought can cause Heat waves and heavy rains lead to unexpected losses in the production of agricultural crops, as well as a decrease in the agricultural area and fluctuation in the productivity of the area unit of agricultural crops, especially in the river deltas, which are among the most important areas of agricultural production, such as the Nile River Delta, for example, where it is expected to affect Climate changes on the acre productivity of agricultural crops, as a result of high temperatures, which cause a severe decrease in the productivity of

Table 1. Climate change and its impact on food production

data	Percentage of decrease in acre productivity %			% increase in water consumption
the crop	1.5m	2m	3.5m	3.5m
Wheat		-9	-18	2.5
barley			-18	(-2)
maize			-18	8
sorghum			-19	8
the rice			-11	16
soybean			-28	15
sunflower			-27	8
tomatoes	-14		-50	14
sugar cane			-25	2.5
cotton		17	-29	10

Figure 9. Distribution of cities that are expected to be affected by climate change in the Mediterranean region

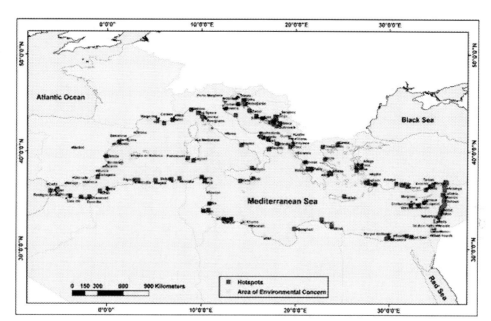

most of the main food crops in Egypt, in addition to an increase in water consumption for them, which is shown in Table (1), and from it the following becomes clear:

Also, the high temperature affects the safety of food in storage places due to the accelerated growth of parasites that secrete toxic substances on them, which negatively affects the health of the population consuming those foods, and the change in the amount of rainfall will lead to water stress for some crops such as vines in many countries. Regions in the Mediterranean countries such as Spain and Italy, where the expected increase in droughts associated with water stress will lead to a decrease in grape productivity in terms of quantity and quality.

Therefore, climate changes threaten the food security of the Mediterranean population, which stimulates the continued efforts of countries to confront climate changes to achieve sustainable development goals and at the same time increases economic and social pressures on countries with fragile economies that cannot bear the impact of climate changes.

Cities Affected by Climate Change

Climate and environmental changes, as well as economic, social and political instability in some countries threaten human security in the Mediterranean region in various ways. More than 40% of the coastline has been built in the Mediterranean basin, so that a third of the population lives - that is, about 150 One million people - near the sea, and the infrastructure is usually close to the average sea level, due to the lack of storms and the limited range of tides. Coastal cities, ports and infrastructure, in addition to the low-level wetlands and beaches of the Mediterranean region.

All these challenges impede the countries they face from achieving the goals of sustainable development in light of the economic crises and inflation that most Mediterranean countries suffer from at

Figure 10. The expected death rate resulting from heat waves in the European countries bordering the Mediterranean in 2085 (The first biennial updated report of the Arab Republic of Egypt submitted to the United Nations Framework Convention on Climate Change 2018, 2018)

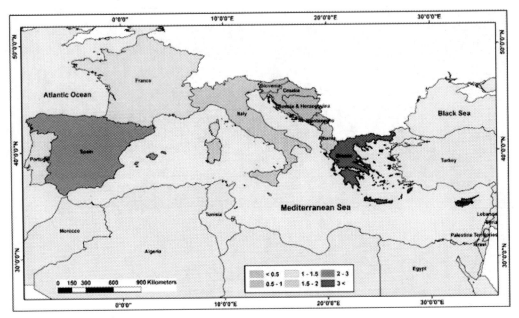

the present time, which prompts those countries to rearrange their economic and social priorities while trying to reduce the negative effects of current climate changes in light of their available capabilities. materially, cognitively and technically.

Human Health

Man is affected by climatic and environmental changes in the society in which he lives, whether urban or rural, and the negative effects of those changes are reflected in the health, psychological and social aspects of man. The direct effects include those associated with high temperatures, increased ultraviolet radiation, drought, and other extreme weather purifiers such as storms and floods. Heat can cause illness or death when high ambient temperatures, associated with high relative humidity, do not allow the body to dissipate its heat normally (Causes and effects of climate change, n.d.).

It is expected that the European population that will face the risk of heat stress in the Mediterranean basin will increase in the coming years at a rate of 4% annually, and the increase may reach from 20-48% in 2050, according to the considered social and economic scenarios. The effect of heat on the death rate will be related to the factors Social and economic factors are stronger than its association with exposure to high temperatures, due to the impact of these factors on the population's susceptibility to infection.

Accordingly, this will lead to a negative impact on the human forces in society, and make the work environment uncomfortable, which directly and indirectly affects worker productivity and production in general in various sectors, which extends its impact to the various aspects of the economy in that society and increases pressures. The consequences of climate changes, and attempts to achieve sustainable development goals are difficult unless local and international efforts come together to work on facing the social and economic challenges of climate change.

The Most Important Strategies for Adapting to Climate Changes

The strategies that adopt the approach of adapting to climate change, limiting its causes and controlling the damages resulting from it have varied, especially since these damages negatively affect the economic, social and health aspects of man and cause many damages to the natural environment with all its elements in general, and the following are some examples of those strategies at the Global, regional and local level.

At the Global Level

The United Nations sponsors international action to confront climate changes through encouragement and financial, technical and informational support to countries seeking to reduce the effects of climate changes on their economy and environment through a set of alliances and strategies:

1- Energy Efficiency Alliance (Three Percent Club).
2- Coalition to Get Rid of Coal.
3- The Urban Climate Investments Initiative.
4- Decarbonize the shipping sector (Zero Coalition).
5- The vision of the Insu Resilience Group Partnership in 2025.
6- The Global Climate Change Program of the International Federation of Red Cross and Red Crescent Societies.

At the Regional Level

All Mediterranean countries adopted the Mediterranean Strategy for Sustainable Development at the 19th Meeting of the Contracting Parties to the Barcelona Convention (Athens, Greece, 9-12 February 2016). This agreement aims to:

1- Ensure the achievement of sustainable development in the marine and coastal areas.
2- Enhancing resource management, food production and food security through sustainable forms of rural development.
3- Planning and managing sustainable Mediterranean cities.
4- Addressing climate change as a matter of priority.
5- Transition to green and blue economy
6- Improving governance within the framework of supporting sustainable development, and to achieve these goals, this strategy has adopted several initiatives to overcome climate change.
7- Promotion of the "Environmentally Friendly City" Award: Sponsored by the Government of Turkey, the Istanbul Environmentally Friendly City Award was awarded twice in a row at the Twentieth Meeting of the Contracting Parties (2017) and the Twenty-first Meeting of the Contracting Parties(2019).

8- Creation and Promotion of the Mediterranean Business Award for Eco-Innovation": in 2020: the RACSP launched the Mediterranean Green Entrepreneurship Award in the context of the EU-funded SwitchMed programme.

At the Local Level

The National Climate Change Strategy puts the quality of life of the Egyptian citizen as a priority, in line with the first strategic objective within Egypt's strategy for sustainable development, as the strategy's vision is formed in a way that guarantees the protection of citizens from the effects of climate change, in parallel with preserving the state's development in a sustainable manner and the preservation of natural resources. Where the vision of the National Climate Change Strategy is to effectively address the effects and repercussions of climate change, which contributes to improving the quality of life for the Egyptian citizen, achieving sustainable development and sustainable economic growth, as well as preserving natural resources and ecosystems, and strengthening Egypt's leadership at the international level in the field of Climate change, where five main goals were identified, from which twenty-two sub-goals were branched, each containing a number of directions that would contribute to achieving the sub-goals; The objectives have been prepared so that the main objectives 1 and 2 are the two that most require interventions from different sectors, and they are the most influential on the axes of reducing greenhouse gas emissions and the ability to adapt to climate change. These two axes are of the utmost importance, as they are mutually reinforcing processes that must be achieved together to ensure mitigation. The effects of climate change on the country, especially the poorest and most affected areas. The first two goals are followed by three other goals, which are equally important for implementation, as they serve as basic and important ingredients towards achieving the first and second goals. These goals revolve around the following:

1- Achieving sustainable economic growth with low emissions in various sectors: The energy field is one of the largest sectors contributing to greenhouse gas emissions, accounting for about 64.5% of the total greenhouse gas emissions according to According to the updated report every two years, which was prepared in 2018, these emissions result from Burning natural gas and petroleum products to produce energy.

2- Building resilience and the ability to adapt to climate change and mitigate the negative effects associated with climate change: Egypt's updated Vision 2030 states that the human being is the center of development as one of the governing principles of the strategy and based on this principle D The importance of protecting citizens from the negative health effects of climate change is evident, especially on the cusp of - Covid-19 pandemic, which demonstrated the importance of good preparation for emergency health changes and concerted efforts in various relevant sectors to try to manage the crisis well and limit the negative effects on citizens.

3- Improving the governance and management of work in the field of climate change: where the concerted efforts are considered a major factor in the success of achieving the strategic objectives, and the importance of institutional integration between the various sectors and ministries is increasing in a complex issue such as the issue of climate change. The third sub-objective (a) Determining the roles and responsibilities of the various stakeholders in order to achieve the strategic objectives, all the importance of defining and distributing roles in a manner commensurate with the current state of business conduct and the intent of integration with it.

Figure 11. Contribution of greenhouse gases from each sector to total emissions in 2015
(The first biennial updated report of the Arab Republic of Egypt submitted to the United Nations Framework Convention on Climate Change 2018, 2018)

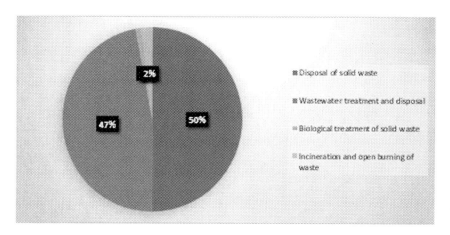

4- Improving the infrastructure for financing climate activities: The banking sector is one of the most important sources of financing projects in the private and public sectors. The environmental and social aspect can be added to it to transform it To green business, following the example of the international banking community such as the World Bank and other international sources of financing, while giving sufficient attention to micro, small and medium enterprises to benefit from financing opportunities. In 2020, Egypt issued the first offering of green bonds, at a value of $750 million. Thus, Egypt was a pioneer in issuing these bonds at the level of the Middle East and Africa. The green bonds aim to attract investors working on sustainable projects, as they are intended for projects related to the environment and climate. during which projects are financed such as those related to renewable energy, energy efficiency, waste management and transportation cleaning, climate change adaptation and other projects of environmental, social and governance importance.

Greenhouse gases affecting climate change (Egypt -   case study)
A greenhouse gas inventory was prepared in accordance with the 2006 IPCC guidelines for the time series between 2005 and 2015.
The inventory of greenhouse gases includes four sectors:
First: energy:
The energy sector represents 64.5%, which is the highest percentage of total emissions For the year 2015, estimated at about 210,171 gigagrams of carbon dioxide equivalent.
The energy sector contributed 87% of carbon dioxide emissions of the total emissions in the country, 3% of the total methane emissions, and 2% of the total nitrous oxide emissions. Energy sector emissions are generated mainly from...

1)   Fuel combustion activities (97%)
2)   Emissions from fossil fuels, mainly oil and natural gas (3%)

Figure 12. Emissions per energy sector category in 2015
(The first biennial updated report of the Arab Republic of Egypt submitted to the United Nations Framework Convention on Climate Change 2018, 2018)

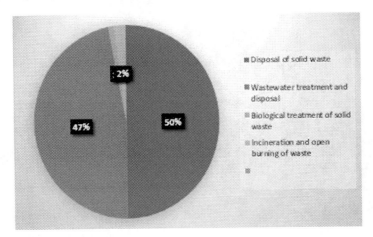

The uncertainty analysis conducted using the 2006 IPCC program resulted in a total of 3% for the total energy sector inventory and 4% for the prevailing uncertainty over the period Extending from 2005 to 2015.

Second: Industrial processes and use of products:

The industrial processes and product use sector is responsible for 12.5% of total greenhouse gas emissions estimated at 40,664 gigagrams of carbon dioxide equivalent in 2015.

This sector is responsible for 12% of carbon dioxide emissions and 12% of total nitrous oxide emissions. It produces emissions The sector is mainly from

1- Mining industries 54%
2- Chemical industries 18%
3- Metallurgical industries 17%
4- Using products as alternatives to substances that deplete the ozone layer 11%

The analysis resulted in a null Ascertainment conducted using the 2006 Intergovernmental Panel on Climate Change (IPCC) programme. A total of 14% for the industrial process sector and the use of products, and 27% for uncertainty. During the period 2005 to 2015

Third: Agriculture, forestry, and other land uses

The agricultural sector, forestry, and other land uses contribute 14.9% of total gas emissions The greenhouse capacity in 2015 was estimated at 48,390 gigagrams of carbon dioxide equivalent. Emissions are released The main sectors of this sector are:

1- Intestinal fermentation
2- Animal manure management
3- Flooding rice cultivation
4- Agricultural soil management
5- Open burning of agricultural waste.

Figure 13. Emissions per category for the industrial process and product use sector in 2015
(The first biennial updated report of the Arab Republic of Egypt submitted to the United Nations Framework Convention on Climate Change 2018, 2018)

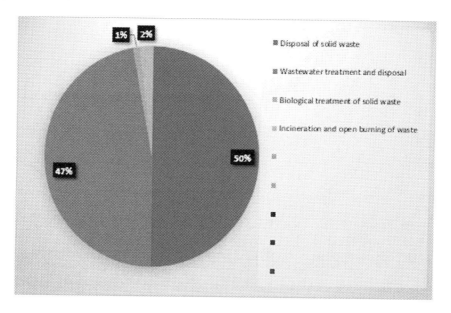

It is the largest contributor to total greenhouse gas emissions. The sources of CO_2 emissions are on land (66%), followed by livestock (34%). An uncertainty analysis was conducted for the activity data based on expert judgment. It ranges between ± 15%, while the uncertainty of emission factors ranges between ± 50%.

Fourth: Waste.

Figure 14. Emissions per sector category of agriculture, forestry, and other land use in 2017
(The first biennial updated report of the Arab Republic of Egypt submitted to the United Nations Framework Convention on Climate Change 2018, 2018)

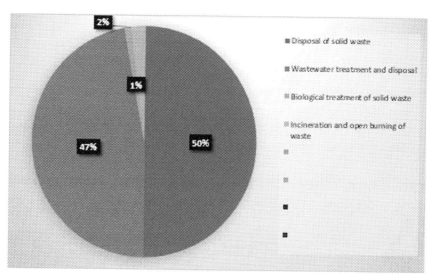

Figure 15. Emissions for each waste sector category in 2015
(The first biennial updated report of the Arab Republic of Egypt submitted to the United Nations Framework Convention on Climate Change 2018, 2018)

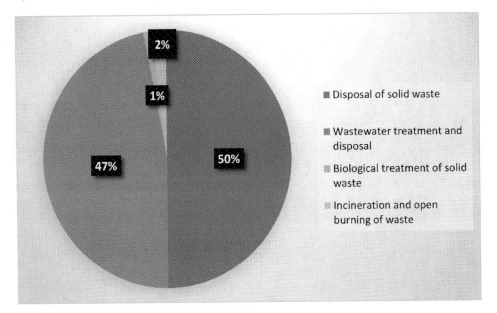

The waste sector contributes about 8.1% of greenhouse gas emissions estimated at 26,389 Giga in 2015.Gram of carbon dioxide equivalent. The sector's emissions result mainly from

1- Disposal of solid waste
2- Treatment of domestic and industrial wastewater

With minor contributions from biological waste treatment Solid waste, incineration and open burning of solid waste. An analysis has resulted Uncertainty measured using the 2006 Intergovernmental Panel on Climate Change (IPCC) programme. For a total of 83% of the state of uncertainty prevailing during the period extending from 2005 to 2015. It is likely that This is due to the large uncertainty in activity data for industrial wastewater and emission factors. The height that was used.

FINDING AND CONCLUSION

This study tackles the impact of climate change and sustainable development in the Mediterranean Basin. The researcher analyses the impact change concerning the following:

- The diversity of the manifestations of climatic changes in the Mediterranean region, most notably the rise in temperatures, the change in precipitation rates, and the change in sea surface temperature.

Climate Change and Impact on SDGs in the Mediterranean Countries

- Among the most important causes that led to the occurrence of climate changes are: burning fossil fuels, cutting forests, industrial activities, the use of means of transportation, supplying buildings with energy, and food production.
- Despite the clarity of the goals of sustainable development and their great importance, however, achieving them requires concerted local, societal and international efforts to confront the causes of climate change, and deal with its impact on the economic, social, health and environmental aspects.
- The countries most affected by the impact of climate change, according to expectations, are the countries of the south and east of the Mediterranean, because they are the most vulnerable countries to the change in high temperatures and the spread of desertification and the presence of many low areas on their coasts, which could be exposed to the risk of drowning as a result of the rise in sea level, which means They are countries with fragile environments, and most North African countries suffer from a low per capita share of natural water resources, in addition to that they do not make good use of water, as a result of the high rates of wasted water during consumption, which puts pressure on the available natural resources in light of climate changes.
- It is expected that the number of Europeans who will face the risk of heat stress in the Mediterranean basin will increase in the coming years at a rate of 4% annually, and the increase may reach from 20-48% in 2050, according to the social and economic scenarios used in the forecast studies.

REFERENCES

United Nations. (2006). *Adaptation to Climate Change in the Context of Sustainable Development: A Workshop to Strengthen Research and Understanding.* United Nations Department of Economic and Social Affairs. New Delhi,: The United Nations.

Al-Droubi, A., Janad, I., & Al-Seb, M. (2008). Climate Change and its Impact on Water Resources in the Arab Region. *Arab Center for the Studies of Arid Zones and Dry Lands, (ACSAD), Arab Ministerial Conference on Water.* Research Gate.

al-Kashef, T. M.-F. (2014). *Modeling Climate Change in Egypt, A Study in Applied Climate Geography, Using Geographic Information Systems and Remote Sensing.* College of Arts -   South Valley University.

Al-Saee, S.-D. F., & Al-Qahtan, M. S. (2016). Studying some of the Environmental, Economic and Social Effects of Climate Change on the Fisheries Sector from the Perspective of Specialists. *Journal of Agricultural Economics and Social Science, 7*(2).

Al-Taher, F. A.-H. (2008). Climate Changes and their Impact on Food, Water and Energy Shortages and the Role of Standards in Mitigating this impact. *National Conference on the role of standards in facing climate changes and food, water and energy shortages, 25.* Cairo.

Baraka, A. I. (2019). *The impact of climate change on the natural, economic and social environment (the Republic of Chad as a model).*

Berger A., & Loutre, M. F. (2004). Astronomical theory of climate change. *Journal de Physique IV.*

Causes and effects of climate change. (n.d.). The United Nations. https://www.un.org/ar/climatechange/science/causes-effects-climate-change

Climate Change. (n.d.). The United Nations. https://www.un.org/ar/climatechange/science/causes-effects-climate-change

Consult, I. (Ed.). (2022, May). The National Strategy for Climate Change in Egypt (2050).

Dutch, S. I. (2010). *Encyclopedia of global warming* (Vol. 245). Salem Press.

Eissa, M. M. (2007, December). New statistical study for Global temperature. *Meteorological Research Bulletin, 22*, 17.

Fahim, M. A. H. M. (2013). Climate Change Adaptation Needs for Food Security in Egypt. *Science Pub, 11*(12). https://www.sciencepub.net/nature

Fawaz, M. M., & Suleiman, S. A. (2015, September). An Economic Study of Climate Change and Its Effects on Sustainable Development in Egypt. *The Egyptian Journal of Agricultural Economics, 25*(3), 3.

Hamza, Y. G., Ameta, S. K., Tukur, A., & Usman, A. (2020). Overview on Evidence and Reality of Climate Change. *IOSR Journal of Environmental Science, Toxicology and Food Technology (IOSR-JESTFT), 14*(7).

Hashim, S. M., Issa, M. M., & Maghribi, N. (2021, June). Climate and its Impact on Human Comfort in the Nile River Delta in Egypt for the Period (1986-2005), a Study in Applied Climate. *Journal Research,* (6).

Hassan, K. A.-W., Yassin, B. R., & Kazem, A. A.-Z. (2019). The Impact of Climate Changes on the Sustainable Development of Water Resources: an applied study in Basra Governorate. *Peer Journal, 44*(4).

Hassan, K. E. (2021). *Climate Change and the Global Goals for Sustainable Development.* Al Jazeera Library.

Jadallah, A. M., & Abdel-Meguid, E. M. (2021). Rural women's awareness of the effects of climate change on health security and how to confront them: A study in the village of Sanhour Al-Madina, Desouk Center, Kafr El-Sheikh Governorate. *Agriculture Economics and Rural Development, Al-Jam'iya Scientific. Agriculture Sciences, 7*(1).

Madani, M., Abdel-Gayed, S., & Murad, M. (2011, January). *The Future Effects of Climate Change on the Agricultural Sector in Egypt: Cost Estimation.* Research Gate.

Magdy, S. (2022, November 4). Climate Cahnge and Sea Level Rise Threaten Food Security in Egypt. *Independent Arabia.* https://www.independentarabia.com: https://www.independentarabia.com/node/388946/

Mahmoud, H. A.-M. (2014). An Analytical Economic Study of the Current Situation and the Future of Wheat Self-Sufficiency in Egypt. *The Egyptian Journal of Agricultural Research, 92*(2), 781–801. doi:10.21608/ejar.2014.156315

Muhammad, M. S. (2023). *The Impact of Climate Change on Sustainable Development and the Labor Market in the Arab World.*

Rajdali, M. (2020). The Mediterranean Basin, Advantages and Challenges of Sustainable Development. 5.

Refaat, A. (2023, January). Evaluation of the Effectiveness of Financing Programs to Confront Climate Change. *College of Politics and Economics,* (17).

Regdali, M. (2020). *The Mediterranean Basin.* Advantages and Challenges of Sustainable Development.

Risks Associated with Cimate and Environmental Changes in the Mediterranean Region. (2019). *Network of Experts on Climate and Environmental Changes in the Mediterranean Region.*

Saadet, B. (2015). *The Effects of Climate Change on Sustainable Development in Algeria (a Prospective Study)* [PhD Thesis, Boumerdes, Algeria: Faculty of Economic, Commercial and Facilitation Sciences, University of M'hamed Bougherra].

Speer, I., Pazsegaran, M., & Heidegger, M. M. (n.d.). *Adapting to Climate Change - The New Challenge for Development in the Developing World* (K. Essin & R. Asi, Eds.).

The Arab Strategic Report. (2010). Center for Political and Strategic Studies. Cairo: Al-Ahram.

The UNCCD files. (2018, June). UNCCD. https://www.unccd.int/: https://www.unccd.int/sites/default/files/2018-06/GLO%20Arabic_Full_Report_rev1.pdf

UNESCO. (2009). *World Water Assessment Programme.* UNESDOC. WWW.UNESCO.ORG: https://unesdoc.unesco.org/ark:/48223/pf0000374903_spa?posInSet=1&queryId=N-EXPLORE-6f43efa9-f9da-451e-8948-31b90a8bd1e8

Using Geographic Information Systems to Analyze Climate Change Data, Economic and Social Commission for Western Asia. Beirut: ESCWA - United Nations. Retrieved from www.escwa.un.org

Weston, P. (2019, May 23). *Sea Levels Could Rise More Than Two Meters by 2100.* Independent Arabia. https://www.independentarabia.com/node/27226

Weston, P. (2019, May 23). *Sea Levels Could Rise More Than Two Meters by 2100.* Independent Arabia. https://www.independentarabia.com: https://www.independentarabia.com/node/27226/مستويات_ سطح_البحر_يمكن_أن_ترتفع_أكثر_من_مترين_بحلول_2100

Weston, P. (n.d.). *Sea Levels Coud Rise more than Two meters by 2100.*

World Day to Combat Desertification and Drought, 17 June/Background. (n.d.). The United Nations. https://www.un.org/ar/observances/desertification-day/background

World Day to Combat Desertification and Drought. (n.d.). United Nations. https://www.un.org/en/observances/desertification-day

Youssef, A. A. (1982). *Climatic Characteristics of the Heat Element in Egypt during the Twentieth Century: a Study in Climatic Geography.* Faculty of Arts, Ain Shams University.

Zahran, Z. B. (2007). *Rains on the North African Coast, a Study in Climatic Geography.* Department of Geography, Faculty of Human Studies, Al-Azhar University.

Chapter 3
Climate Change in World and Egyptian Cinema:
A Sociocultural Study of Highlighted Values

Arij Elbadrawy Zahran

Independent Researcher, Australia

ABSTRACT

The chapter focuses on the climate changes and environmental transformations that swept the world through two aspects: a realistic aspect and an artistic one. The first section of the chapter identifies the problem of climate change and its causes, most of which are attributed to human action, in addition to natural disasters. The second section presents the international and global action towards these environmental problems, the global conferences that were held and the agreements that were concluded in this regard, and the importance of regional institutions and constitutions was also highlighted. The artistic aspect was dealt through the third and fourth sections of the chapter. The third section presents cinema art, film sociology, and the term environmental cinema. The fourth and final section deals with climate changes as presented by international cinema films - especially American cinema - and Egyptian cinema, and tried to explain the reason for the difference from a sociological point of view.

INTRODUCTION

This chapter pursues to study and address the issue of climate change and environmental risks as discussed by world cinema; whereas issues of climate change, threats to the climate and environmental disasters i.e.: global warming, forest fires, sea-level rise, and floods attracted part of the attention of international cinema (i.e.: The Day After Tomorrow, Interstellar, Dark Waters)

The United Nations Framework Convention on Climate Change (UNFCCC) defines climate change in its first paragraph as: "a change of climate which is attributed directly or indirectly to human activity that alters the composition of the global atmosphere and which is in addition to natural climate variability

DOI: 10.4018/978-1-6684-8963-5.ch003

Copyright © 2024, IGI Global. Copying or distributing in print or electronic forms without written permission of IGI Global is prohibited.

observed over comparable time periods." This definition refers to the causes of climate change, which the humankind considered the main actor, in addition to the natural factors. The Intergovernmental Panel on Climate Change (IPCC) defined climate change as "a change in the state of the climate that can be identified (e.g., using statistical tests) by changes in the mean and/or the variability of its properties, and that persists for an extended period, typically decades or longer. Climate change may be due to natural internal processes or external forcings such as modulations of the solar cycles, volcanic eruptions and persistent anthropogenic changes in the composition of the atmosphere or in land use." This definition adds the characteristic of the continuity of the phenomenon of climate change, which, although its causes are temporary, its negative effects will extend for generations to come. After the series of reports issued by IPCC, scientific opinions have clearly agreed that climate change is clear and unambiguous, and that most of the climate changes observed over the past 50 years, are very likely; 90% confidence; human-caused. Hence, it is a matter of a huge balance between the Earth and its components of the atmosphere and its layers and elements, and the cosmic space in which the Earth swims. *(Al-Hawari, 2019, p. 12)*

In general, this chapter may fall - in addition to being related to cultural sociology - under the branch of social morphology, which is the study of the environment and climate and its impact on the nature of society and its social entity, including the main patterns of the groups of society and the relationship of the environment to social organization, and population and their numbers and densities. *(Al-Dahry, 2010, p. 171)* It focuses on geographical and environmental factors and their impact on the nature of society, such as the study of climate and topography on the quality of social life, especially as it played an important role in human social life since its very existence, where the individual was an integral part of the environment in which he lived. He lives in it and follows its laws and provisions. *(Mansour, 2016, p. 22)*

The term Social Morphology is used in three different ways. In the widest meaning it is the study of the structure of social groups and in this sense all social sciences have a morphological aspect. The sociologist Maurice Halbwachs use it in a very restricted sense- only demography falls within his definition- but this definition has not been widely accepted. There is a third definition, narrower than the first and wider than the second, which has gained general acceptance: 'The study of social facts in their material substrata' (A. Cuvillier). This definition is derived from Durkheim: 'social life rests on a substrate which is fixed in size and form, composed of the individuals which make up society, the utilization of the soil and the nature and configuration of all sorts of elements which affect collective relation.' Thus, in practice, social morphology includes human geography and demography. *(Duverger, 2020)*

We see that the geographical nature of USA with the occurrence of floods, hurricanes and other climate changes and natural disasters have played a role in formulating the themes of the films that dealt with these phenomena, while the moderate climate nature of Egypt has made such issues somewhat far from the Egyptian culture and thus they were excluded from the Egyptian cinema.

Manifestations of Climate and Environmental Changes

The climate change problem is attributed to those huge human activities such as: fuel burning, biomass burning, greenhouse gases and aerosols production, which affect thermal radiation. Also changing the methods of land use from cultivation, irrigation, deforestation (assarting), and industrial forestation, physically and biologically, affect the characteristics of the earth surface; worth to mention the influence of the increasing and expanding cities that lead to the formation of urban heat islands with strong local effects. *(Al-Attiyah, 2020, p. 50)*

The changes to our environment are so profound and so clearly forced by human activity (particularly in western countries) that some scientists are even proposing that humans have provoked the development of a new geological epoch: The Anthropocene. The concept of the Anthropocene was popularized by Paul Crutzen and Eugene Stoermer in 2000 when they argued that human activity was working as "a major geological force" that was (and is) altering ecosystems profoundly. The Anthropocene constitutes a radical change, as transitions between geological periods do not happen frequently: the previous, post- glacial period- the Holocene- lasted between 10,000 and 12,000 years. The concept of the Anthropocene does not only refer to the radical environmental disruption brought about by greenhouse gas emissions- it also alludes to whole array of human activities and their impact on the natural functioning of the planet's ecosystems. Accordingly, Crutzen and Stoermer mention events and activities such as population growth, resource consumption (specially water), the burning of fossil fuels, urbanization, land usage, the use of fertilizers in agriculture, species extinction, and the release of "toxic substances in the environment". Apart from these, later scientific studies also refer to dam construction, mining, landfills, sediment movement, and the terraforming that cities require. The idea of the Anthropocene, therefore, suggests that, apart from generating the emissions that produce climate change, humans also perform other activities that unsettle natural forces severely and present environmental challenges for human and non- human life. As Steffen, Crutzen, and McNeill note, the Anthropocene begins in the late eighteenth century with the Industrial Revolution, although they indicate that the impact of human activities has been particularly forceful from the 1950s to the present, a period that they refer to as "the Great Acceleration". *(Gomez- Munoz.,2023)*

The consequential risks of climate change form many challenges, especially for the least developed countries (LDCs) and societies with fragile and conflict-affected situations (FCS). *(Megahed, not dated, p. 134)* Although the African countries are the least responsible for climate change for their low contribution in the global greenhouse gas emissions, yet Africa bears the largest amount of the repercussions of climate change *(Al-Sabahi, 2022, p. 175)*

The UN Food and Agriculture Organization (FAO) confirmed that global climate change and its further impacts on global water security will have serious repercussions on the global economy and global food security in 2030, especially in impoverished areas, resulting in wars and conflicts over water resources, and more than 155-600 million people will suffer from water shortage as temperatures continue to rise according to the current global rates. *(Al-Ta'i, 2021, p. 221)* And due to the diversity of environment elements, including the marine, air, and wild environment, the agreements are characterized by specialization in addressing the environmental system for each of these elements, which are as follows:

Marine Environment-Related Agreements: International Convention for the prevention of pollution of the sea by oil, London 1954, as amended in 1962, 1969, 1969 and 1971. It is worth noting that this agreement despite its importance, was not joined by any of the countries bordering the Red Sea except Egypt and KSA. Additionally, The Brussels Convention of 1969 relating to intervention on the high seas and civil liability in oil pollution casualties, the Oslo Convention of 1972 designed to control the dumping of harmful substances from ships and aircraft into the sea, the London Convention of 1972 relating to the prevention of marine pollution by dumping of wastes and other matter, and the Paris Convention of 1974 relating to the prevention of marine pollution from land-based sources. *(Muhammad, 2014, p. 171)*

Wild Environment-Related Agreements: The International Plant Protection Convention (IPPC) Rome 1951, the African Convention on the Conservation of Nature and Natural Resources signed in Algeria 1968, the Ramsar Convention on wetlands of international importance and waterfowl habitats 1971, the Paris Convention concerning the Protection of the World Cultural and Natural Heritage signed in

Climate Change in World and Egyptian Cinema

1972, the Convention on the Conservation of Migratory Species of Wild Animals signed in Bonn 1979. *(Muhammad, 2014, p. 171-172)*

Air Environment-Related Agreements: Among the international agreements for the protection of the air environment are the Geneva Convention of 1979 related to long-range transboundary air pollution, and the Vienna Convention of 1985 for the protection of the Ozone Layer, in addition to many international agreements related to the use of atomic energy. *(Muhammad, 2014, p. 172)*

Ecosystem studies under the supervision of the International Biological Program (IBP) recently led to a new assessment of the characteristics of different ecosystems and their functional relationship to human groups. The mathematical models and patterns based on the data of the global biological program have enabled environmental scientists to study natural and induced changes in environmental systems and to reach more accurate predictions regarding the environmental consequences arising from human activities. In other words, it is possible to predict environmental changes that may occur in the future. *(Al-Saadi, 2020, p. 254)*

Global Warming is one of the global environmental challenges to international security and its immediate effects will first hit the world's poorest and most vulnerable people, who are already living the consequences of global warming. In our divided world, global warming deepens the gap between the rich and the poor, depriving people of the opportunity to improve their lives. Looking to the future, we find that climate change poses a risk of environmental disasters that have security repercussions for the world. *(Al-Ta'i, 2021, p. 210)* Global Warming is one of the most controversial phenomena among scientists, researchers, those interested in the environment, and the public, and this interest went beyond to other circles reaching to governmental and non-governmental entities, as well as international entities. The reason for this is attributed to the dangerous and destructive effects of this phenomenon on both humans and the elements of the environment. It has been noted recently that there is an increasing interest by the audio-visual and print media in the manifestations of climate change, as it is an issue that occupies a large part of the interests of public opinion. *(Nabhan, 2012, p. 15)*

People usually use the terms climate change and global warming interchangeably, assuming that they denote the same thing, but there is a difference: global warming refers to an increase in the average temperature near the surface of the earth, while climate change refers to changes that occur in the layers of the atmosphere such as temperature, rainfall, and other changes that are measured over decades or longer periods. It is preferable to use the term climate change when referring to the influence of factors other than a rise in temperature. Climate change may result from the following: natural factors such as changes in the intensity of the sun, or slow changes in the Earth's rotation around the sun, natural processes within the climate system, such as changes in the water cycle in the ocean, human activities that change the composition of the atmosphere; such as: burning fossil fuels and land surface such as deforestation, urbanization and desertification. *(Megahed, not dated, p. 134)*

In conclusion, environmental pollution may appear to be a local problem, but it is considered a global one. Pollutants under the influence of many factors do not know political borders. They are characterized by their ability to move from one location to another in the short or long term. *(Megahed, not dated, p. 135-136)* Thus, it is necessary to protect nature and preserve the environment through: Environmental Safety and Environmental Education: many international entities have developed special legislation related to the concern for environmental protection through teaching environmental sciences at all educational levels *(Mazahrh and Shawabka, 2010, p. 59)*

International and Government Efforts to Confront Climate Change

Climate Change has become one of the global environmental challenges that mostly attracts increasing attention of international organizations, governments, environmental organizations, the media, research centers and citizens worldwide, due to its serious and devastating effects on many societies. Developing and poor countries are among the countries most affected by this phenomenon, which is attributed to the fact that the governments of these countries do not have enough economic capabilities to protect their citizens from the effects of this globally growing-risk problem. *(Al-Kufi and Al-Ta'i, 2015, p. 133)*

The environment cause was of great importance in the UN Economic and Social Council (ECOSOC). The Committee of Sustainable Development was established following the UN General Assembly Resolution No. 47/191 in 1992. This committee consists of 53 countries whose members are elected by ECOSOC. Within the framework of ECOSOC's interest in forests, the United Nations Forum on Forests (UNFF) was established by virtue of ECOSOC Resolution No. 35/2000 dated December 18th, 2000, and consequently a cooperation partnership related to forests was established with international and regional organizations and authorities concerned with the forests. This partnership supports the UNFF and achieves sustainable forest development and enhancing cooperation among members. *(Kaseb, 2020, p. 92)*

The manifestations of international concern can be clarified through holding various international environmental conferences, and the agreements concluded in this regard as follows:

United Nations Conference on the Human Environment, Stockholm 1972: Among the concerns of this environmental conference was biodiversity, as the declaration issued by this conference stated that it is necessary to preserve for the benefit of the present and future generations the natural resources of the earth, including air, water, soil, animals and plants, especially typical samples of natural ecosystems.*(Kaseb, 2020)* This conference is also known as the "First Earth Summit". *(Abu Dayyah, 2010, p. 53)* Among the most important outputs of this conference: it is the beginning of the development of international environmental law, developing the United Nations Environment Program to coordinate and assess the measures of global issues, a preliminary definition of the concept of sustainability by linking the environment and development, and an emphasis on the indirect responsibility of developed countries. *(Fikri, 2021, p. 448)*

The World Commission on Environment and Development (1987): It laid the foundations for the concept of sustainable development *(Abu Dayyah, 2010, p. 55)*.

The Basel Convention on the Transboundary Movements of Hazardous Waste: it was signed in Basel - Switzerland to control transboundary movements of hazardous waste and entered into force in 1992 while USA did not sign it. *(Abu Dayyah, 2010, p. 55)*.

The United Nations Conference on Environment and Development (Second Earth Summit 1992): Representatives of 172 countries and 108 presidents of countries attended the 2nd Earth Summit held in Rio de Janeiro - Brazil, - that is why it is called the Rio de Janeiro Conference or Earth Summit *(Abu Dayyah, 2010, p. 56)*

The Convention on Biological Diversity: which became effective in 1993, under which countries have the right of sovereignty over living species. This Convention resulted in the Cartagena Protocol on Biosafety, endorsed and signed in January 2000 in Montreal by 103 countries, which authorizes a state to prevent the import of genetically modified organisms. *(Abdul Hamid, 2012, p. 157-158)*

The Convention to Combat Desertification: The convention was adopted in 1994. *(Abdul Hamid, 2012, p. 157-158)*

Kyoto Protocol of 1997: This agreement forms the basis of global efforts to combat global warming, as it was adopted at the "Rio de Janeiro" conference in June 1992, and entered into force on March 21, 1997, and in November 1998, it was endorsed by 176 countries. *(Samir, 2013, p. 111-112)* The agreement also called for effective cooperation in the fields of education development, training programs and public awareness in the field of climate change; with the aim of reducing greenhouse gas emissions. *(Abu Dayyah, 2010, p. 54)*

The UNFCCC Copenhagen Summit (2009):

Paris Agreement 2015: It includes ambitious action before and after 2020

COP25 in Madrid 2019: Participants stressed the urgent need for action to combat global warming, but without reaching agreement on the main points of response to the climate emergency and the calls of environmental activists. *(Fikri, 2021, p. 450)*

COP27: held in Egypt from 6-18 November 2022. Egypt has worked on a succession of meetings with African commissioners concerned with climate change, as Egypt hosted the African Group of Negotiators on Climate Change in Sharm El-Sheikh 2018-2019, as well as chairing the Meeting of the Committee of African Heads of State and Government on Climate Change (CAHOSCC) and the African Ministerial Conference on the Environment in 2015 and 2016, then meeting of the Minister of Foreign Affairs with the African Group of Negotiators on Climate Change on June 8th 2022, during his presence in Bonn, Germany, in conjunction with the 56th session of the meetings of the 2 subsidiary entities of the UNFCCC. *(Muhammad. 2023, p. 102)*

Regional institutions play an important role in solving transboundary problems related to climate change. East African countries have developed numerous protocols related to transboundary resources (water, energy, land and biodiversity), and there are policy documents presenting strategies on climate change. However, most of these policies have not been translated into regulatory principles and protocols or endorsed by all countries in the region. *(Al-Sabahi, 2022, p. 172)*

Nevertheless, the constitutional duties in the field of the environment are numerous and topped by: the duty to protect and preserve the environment, and this duty involves several other duties: the duty to preserve and improve the environment, the duty of prevention and precaution, and the duty to compensate for environmental damage:*(Abu Al-Majd, 2015, pp. 98-101, adapted)*

Sociology of Film

The cinema was the first among the new communications media developed in this century to mature into an art form. An art form might be characterized as a medium an artist would consciously choose for purposes of putting forward his vision. *(Jarvier. 2013, p.15)*

Cinema plays a very important role in conveying the data of thought and life in a language based on a common understanding, with more effective and efficient tools in shaping the thoughts and sentiments of the masses. Cinema is considered one of the most important means of media, advertising, general guidance and propaganda, in addition to its important role in the entertainment, educational and cultural aspects and its goals that cannot be limited to the social, religious, political and other fields. *(Kamel, 2018, p. 38)*

Cinema is an art form that borrows from other forms of art, from the theatre, music, photography, etc. It is 'impure'. A similar impurity manifests itself in the relationship between cinema and 'social reality'- between the cinematic network and the social field. The cinema is a place of intrinsic indiscernibility between art and non-art. No film, strictly speaking, is controlled by artistic thinking from beginning to end. Artistic activity can only be discerned in a film as a process of purification of its own immanent non- artistic character. Cinema is necessitated but at the same time is strives for autonomy and 'purification'. *(Diken.and Laustsen. 2007, p.6)*

The cinema, in other words, enters the social arena not by a mimesis of class-conflicts or movements of a collective unconscious, but as that form of social relation which the consumption of narratives and images changes. And it does so in order to block or displace the 'real contradictions' of history and society not into 'imaginary resolution' (levi-Strauss), but into effects of disavowal and substitution. Films are not versions of (bourgeois) historiography; rather, they act upon another history: that of commodity-relations and their modes of production and consumption. It is this physical reality, which the cinema, in Kracauer's phrase, attempts to redeem, by inserting itself as a quasi- magical power, a fetish – object, into the reified and abstracted relationships which characterize our 'society of the spectacle'. *(Grainge., Jancovich. & Monteith.,2007, p.145)*

Cinema is "a human fact" as Morin stated, thereby becoming a part of social reality. Whether the project outlined here falls under the traditional categories of film sociology is hard to say. Nevertheless, it points to a relevant dialogue between aesthetic theory and sociology, which, compared to the vast area of film studies in general, has been somewhat neglected in recent years. *(Jerslev. 2002, p.118)* Sociology has always, both before and after its encounter with cinema, been haunted by the question of representation, of how to relate representation and reality. It has often dealt with this question in terms of correspondence: if representation can correspond to the represented reality more or less, then the task must be to construct as precise representations as possible. In this perspective, cinema represents the social world only in an indirect, artistic way. Indeed: we sociologists might consider it our duty, perhaps, to 'rewrite' the misguided vision of the filmmakers by imposing on the film a more sociological overlay. The films would be richer and more realistic if the character would behave in a manner that indicated their implication within social worlds that are wider, fuller, and older than their own individual lives. To do this requires that we stray outside the boundaries of the fictive world created by the filmmaker. *(Diken.and Laustsen. 2007, p.4)*

Pluralistic approach incorporates elements from memory studies and trauma theory to examine the uses of personal testimony and reenactments as means of bearing witness to history. In this approach, psychoanalytical concepts borrowed from trauma theory add a further dimension to more familiar social and political analyses, the latter including elements of gender theory employed to examine the representation on screen of women's role in historic political resistance. This is then conjoined with the methods of film studies to closely analyze specific films and to explore what some have understood as a tendency in trauma texts to favour a modernist aesthetic over 'realism'. In contrast to that pluralist strategy, a 'strong programme' for the sociology of film would aim to prioritize sociological theories and methodologies in comprehending the workings of the system of cinema, including those aspects of the cinematic institution which are part of its own self-under-standing. The latter, of course, would include film criticism and, indeed, products of the discipline of film studies. A recent example of this stronger use of sociology, though not one that reflexively examines film studies itself, is Hughey (2014) which marshals an array of carefully elucidated methods in examining white saviour films and their contribution to a 'post-racial racial ideology' in American society…what is to be done, then, to further the historically neglected sociology of cinema? There is no simple answer, but in seeking a framework in modern sociology within which to develop a strong programme, Bourdieu's work is of immediate relevance. Of course, his ideas have already had some isolated influence in film studies and, more often, cultural studies. At one point

Climate Change in World and Egyptian Cinema

his expression 'cultural capital' gained a good deal of general currency, particularly in the later 1980s when La Distinction was first translated (Bourdieu, 1986). *(Inglis & Almila,2016, p.489-490)*

Three cases can suggest how the concept of the social field productively reframes existing areas of inquiry and debates within film studies. First, the instance of middlebrow in classical Hollywood cinema would seem to corroborate a sociology-of-taste approach but at closer examination is characterized by a more complex dynamic. Second, documentary ethics is a subfield often removed from reception study, yet Bourdieu's work gives a valuable perspective on the normative claims of ethics. Third, the social field helps reconcile some of the methodological positions in film festival studies. These case studies are not exhaustive, but they do show possibilities of Bourdieuian study beyond a purely reception approach. Richard Jenkins (1992) remarks that Bourdieu is a good thinker for thinking, by which he means that the thinker's impact may be less in his particular insights than in modelling a kind of reflexive socio-logical practice. This claim is certainly debatable, but some version of it is appropriate for film studies. The predominance of sociology of taste spring from a clear methodology for analyzing movies, but the concept of the social field means something less instrumental: the social field is not a methodology for study of cinema per se but rather an imperative to think relationally between social agents, cultural creators and the cultural object (cinema) itself. The value of the model is that it straddles complexity and simplification. *(Austin, 2016, p.36,37)*

This convergence of Hollywood and sociological ideas bears examination. Sociology had been established in the United States since the late nineteenth century, but the field developed the attributes of an established academic field only during the first half of the twentieth century, culminating in a postwar boom. Springing from sociology's growth as a discipline, the subfield of social problem sociology emerged, too, gaining its own journal, social problems, in 1950. The notion of a "sociological" and eventually (by the 1940s) "social problem" film came to define what would become, at least among film scholars and some critics, a genre. The connection between the "social" of the social problem film and sociology's concept of the social might at first view seem tenuous. Indeed, to date, while film historians have made valuable discursive readings of 1940s cinema generally and of earlier (1910s and 1930s) social problem films, few have paid sustained attention to the ideas of 1940s and postwar sociology. However, this criti-cal period of growth and intellectual firmament in U.S. based sociology fostered Hollywood's desire, at least in certain quarters, to inscribe sociology on film. Rather than simply depict social issues, social problem films in the 1940s acted as a type of popular social science. *(Cagle, 2017, p 14-15)*

According to paper by Bob Mellin, (Wall-E) assumes that the apocalyptic warning found in documen-taries such as An Inconvenient Truth are valid, and as such we can be comforted by the movie's claim that the environmentally degraded planet in (Wall- E) can be restored to the garden that it once was. Near the end of the closing credits of the film, we see Wall- E and EVE, who have seemingly made an escape from the degraded city where they first met, holding hands in a green, pastoral landscape, reminiscent of a new Eden- like the original closing of Blade Runner, or as explored in Silent Running. In his 1967 touchtone work The Machine in the Garden, Leo Marx advanced the now commonplace argument that pastoralism was foundational to the American experience with the Anglo- colonizers, who originally perceived North America as literally a new Eden. However, one wonders if contemporary audiences are satisfied in the same way by these old forms of pastoralism, as argued in a recent book on Hollywood and ecological cinema by Robin Murray and Joseph Heumann (2009). Meanwhile, Hollywood Utopia (Brereton, 2005) presumes They still are. However, to help resolve this question, audience/ reception studies are needed to test such a hypothesis either way. *(Brereton, 2012, p.149)*

What affect did (The Day After Tomorrow) actually have on cinema goers? Was it an effective way of engaging with a section of the public that might be less easily reached through more conventional forms of science communication? Did it alter the way people viewed the science of climate change? Survey work conducted with cinema audiences in a number of countries- the USA, the UK, Germany and Japan- revealed mixed reactions. Ambiguous and ambivalent indications of attitudinal and behavioral change were revealed among respondents who had viewed the film. Analysis focused on four key social and behavioral issues: people's perception of the Likelihood of extreme impacts; their concern over climate change versus other global problems; their motivation to take action; and the locus of responsibility for the problem of climate change. Some changes in concern, attitude and motivation were found in a number of viewers. Seeing the film changed some people's attitudes, at least in the short term. These viewers were significantly more concerned not only about climate change, but also about other environmental risks such as biodiversity loss and radioactive waste disposal. While the film increased general anxiety about environmental risks, viewer experienced difficulty in distinguishing science fact from dramatized science fiction, in particular, the dramatic portrayal of climate change in the film reduced viewers' belief in the likelihood of extreme weather events occurring as a result of climate change. Although the film may have sensitized viewers, and perhaps even motivated some of them to act on climate change, the indications were that the public did not feel they had access to information about what action they could take to mitigate climate change. In addition, the research suggested that any increase in concern about climate change induced by the film appeared short lived, with most viewers treating the film purely as entertainment. Overall, the film sent mixed messages about climate change to viewing audiences and cannot be said to have induced the sea-change in public attitudes or behaviour that some advocates had been hoping for. The Day After Tomorrow is just one of a number of high- profile popular devices for communicating climate change in recent years which have been claimed to have had a powerful effect, either intentionally or unintentionally, on public opinion. In December 2004, just a few months after The Day After Tomorrow was released, the best- selling novelist Michael Crichton published his fictional thriller State of Fear. *(Hulme, 2009)*

Such films fall within the so called Environmental Cinema. Environmental cinema has been broadly defined as cinema about the environment or environmental issues. The films discussed in this chapter, however, offer an understanding of "environment" that takes into account the relationships that are constitutive parts of an environment. These films express an ecology of interdependence that best defines them as instances of eco-cinema. From this perspective, humans live symbiotically within and "compliantly as members of Earth's communities". *(Willoquet- Maricondi. 2010, p. 65)* It may also be called: Ecological Cinema or Green Cinema.

Environmental Cinema and Climate Issues

Although sf films have dealt with environmental issues at least since the 1950s and recent films share some of their concerns about the environment with twentieth- century sf movies, climate change constitutes an unprecedented challenge in terms of scope and scale. This spate of recent films tends to focus on these time- specific concerns. Following studies on climate change (Frame and Allen, Beck 2009, Giddens, Vanderheiden, Golub and Marechal, Klein). (Gomez- Munoz.,2023)

Barring certain exceptions, International Relations as a discipline has relatively under- explored the interface of climate change and aesthetics. However, there does exist a wealth of scholarship of the various aspects of climate change and its interaction with pop- culture and movies or eco-cinema. As a corollary

Climate Change in World and Egyptian Cinema

to the emergence of climate change fiction, there has emerged a number of monikers and neologisms associated with movies specializing in climate change like petro-fiction, solarpunk, Anthropocene fiction, solar fiction, climate trauma cinema, eco-trauma cinema, crisis cinema, Anthropo-cinema, Envirothon, etc. (Singh and Marwah. 2023, p. 51)

Over the last couple of years, awareness of environmental problems like climate change has greatly increased. This is mostly due to the growing attention of the media and to the emergence of the genre of eco-cinema. Although there is no exact definition of which films can be regarded as belonging to this genre and although there is some dispute among eco-film critics, Rust and Monani state that a few essential points have been agreed on. The first is that "all cinema is unequivocally culturally and materially embedded' that the views of our culture will always influence the way we make films and therefore also the way in which we depict for example different relationships between people or of course also environmental problems. The second point is that "cinema provides a window into how we imagine (a certain) state of affairs, and how we act with or against it". Thirdly, Rust and Monani mention that though some films might be regarded as being less ecocritical than others, all films can be seen as ecocritical in certain ways and have the potential of showing us new perspectives of the connection between cinema and the physical world around us. (Westhues,2014, p.2)

Sociologists and environmentalists agree that the threats gathering on the horizon are too significant for nationalism to overcome because of its narrow-mindedness. The world faces many threats, including environmental pollution, misuse of agricultural lands, and waste that threatens the world with a starvation which the world has never known before. In addition, we will soon suffer from a shortage of potable water, and no one will be safe from the epidemics that will necessarily result from the lack of sanitary means. Speculative fiction could warn and multiply our chances with the help of cinema. The science fiction cinema has become one of the fabrics of international cinema, which falls within the framework of important films that have remarkable technical competence and compete for prizes in film festivals as well as in the Oscars, in addition to the existence of specialized film festivals only to display this type of film, and to compete with each other. In film libraries, there are many cinema directories through which one can identify the most important Sci-Fi films within the global cinema production in general. For example, the book "1000 Movies You Must See Before You Die" included 32 films of Sci-Fi among other genres all over the world. Here are these films in chronological order with the name of the Director, Author if the text is taken from literature and the name of the producing country:

- 1902: A Trip to The Moon (Georges Méliès – France)
- 1927: Metropolis (Fritz Lang – Germany)
- 1929: Man With A Movie Camera (Dziga Vertov - Soviet Union)
- 1931: Frankenstein (James Whale – based on Mary Shelley's masterpiece novel "The Modern Prometheus" - USA)
- 1933: King Kong (Merian C. Cooper & Ernest B. Schoedsack – USA)
- 1935: Bride of Frankenstein (James Whale – USA)
- 1936: Modern Times (Charlie Chaplin – USA)
- 1936: Things to Come (William Cameron Menzies – adaptation of H. G. Wells novel - UK)
- 1951: The Day The Earth Stood Still (Robert Wise – adaptation of Harry Bates' short story – USA)
- 1956: Forbidden Plant (Fred M. Wilcox – USA)
- 1956: Invasion of The Body Snatchers (Don Siegel – USA)
- 1963: The Nutty Professor (Jerry Lewis – USA)

- 1964: Dr. Strangelove (Stanley Kubrick – adaptation of Peter George's novel "Red Alert" - UK)
- 1968: Planet of The Apes (Franklin J. Schaffner – based on Pierre Boulle novel - USA)
- 1968: 2001 A Space Odyssey (Stanley Kubrick – based on Novel by Arthur C. Clarke – UK)
- 1968: Night of The Living Dead (George Romero – USA)
- 1971: A Clockwork Orange (Stanley Kubrick – based on Anthony Burgess novel – UK)
- 1972: Solaris (Andrei Tarkovsky – based on novel by Stanisław Lem - Soviet Union)
- 1977: Star Wars (George Lucas – USA)
- 1977: Close Encounters of The Third Kind (Steven Spielberg – USA)
- 1979: Alien (Ridley Scott – UK)
- 1979: Mad Max (George Miller - UK)
- 1980: The Empire Strikes Back (Star Wars) (Irvin Kershner – USA)
- 1982: E.T. The Extra-Terrestrial (Steven Spielberg – USA)
- 1982: Blade Runner (Ridley Scott – USA)
- 1984: The Terminator (James Cameron – USA)
- 1985: Brazil (Terry Gilliam – UK)
- 1986: The Fly (based on short story by (David Cronenberg - George Langelaan – USA)
- 1989: Batman (Tim Burton – USA)
- 1990: Edward Scissorhands (Tim Burton – USA)
- 1993: Jurassic Park (Steven Spielberg – based on novel by Michael Crichton - USA)
- 1999: The Matrix (Lana Wachowski, Lilly Wachowski – USA)
- 2006: El Laberinto Del Fauno (Pan's Labyrinth) (Guillermo Del Toro – Spain & Mexico)

From this list, we see that American cinema has taken the lead, followed by UK. Sci-Fi has been linked primarily to countries in which science flourishes. *(Kasim, 2018, adapted)*

Following the release of "The Day After Tomorrow" (Emmerich) in 2004, several films have used the extrapolative power of science fiction to deal with one of the most pressing global issues: climate change. Since climate change is a group of threats that do not respect national boundaries, affect all countries (to varying degrees), and are sometimes produced by actors hundreds or thousands of miles away from the places that suffer the worst consequences, cosmopolitanism offers a particularly suitable perspective to approach these phenomena. SF films about climate change go from the desert landscapes of "Young Ones" (Paltrow 2014), the deadly cold of "Snowpiercer" (Bong 2013), and the waste in "Wall-E" (Stanton 2008) to the more spectacular catastrophic events of 2012 (Emmerich 2009) and the galactic searches for resources and habitats in Avatar (Cameron 2009) and "Interstellar" (Nolan 2014). Apart from presenting environmental and geographical changes, many of these films suggest that one of the most significant transformations that climate change brings about is the need for large groups of human beings to migrate, find homes far from home, reorganize social structures, survive lethal weather conditions, and even attempt to live in outer space. That is, they point to the biopolitical implications of climate change. These Sci-Fi films tend to deal with radical environmental transformation through disaster packed spectacular rides or post-apocalyptic scenarios *(Gomez- Muñoz, 2023)*

Film solutions varied between realistic solutions to environmental issues and the search for solutions in space. For example: the movie "Animal" directed by the French Cyril Dion sought to reach tangible solutions that lead to changing conditions for the better and preserving the planet, by searching for an understanding of what is happening on the planet in terms of changes that led to the destruction of environmental diversity and the extinction of thousands of species of animal from the face of the earth.

Climate Change in World and Egyptian Cinema

"Interstellar" 2014, focused on discovering space as a solution to the environmental crisis. The film takes place in a miserable future where mankind struggles to survive due to climate conditions that were embodied in the form of dust, and then the film follows a team of astronauts who search for a new home for mankind, as they travel through a dodder hole in search of a new place to live. "Geostorm" film, presented the use of satellites in addition to international cooperation as solutions to the crisis. It dealt with the state of the planet after global warming that made the earth completely uninhabitable, motivating the international community to cooperate to build a system of satellites surrounding the planet and controlling the weather. *(Al-Arabawi, June 12, 2022 and Muhammad, October 17, 2022)*

The Oscar-winning Australian filmmaker Eva Orner shed light on the unprecedented, catastrophic and deadly Australian bushfires of 2019-2020 known as the "Black Summer" through "Fire". The documentary tells the story from the perspectives of those involved and directly affected by the fires, as well as activists and scientists. It exposes the Australian government's inaction and the irreversible damage these catastrophic events have created. The tsunami flood also occupied the attention of some filmmakers who dealt with it artistically through a number of films, including the movie: "The Perfect Storm", starring George Clooney and Mark Wahlberg, as the story is about Captain Billy Tyne, who leads his crew on a fishing mission, but his ship is caught in an intense storm and puts them in a mortal danger. As for the movie "The Impossible", starring Naomi Watts, it tells the story of "Henry" family, which intends to spend the Christmas holidays in "Thailand", but the tsunami disaster was not announced to the media, which caused the separation of family members and a struggle for survival at all costs. From the Indian capital, New Delhi, the film "Invisible Demons" by director Rahul Jain presents a cinema treatment with his camera that reflected the problem of air pollution in that big city and its impact on the population. "In Search of Snow Leopard", which was shot entirely in the Tibetan plateau, it is about the journey of a wildlife photographer who is fond of nature and animals, accompanied by a writer and a poet. They set out together in nature in search of an animal that was believed to have been extinct for a long time (the snow leopard). The film is full of human philosophy and raises questions about nature, humans, and animals. *(Al-Arabawi, June 12, 2022)*

Some films also shed light on the responsibility of companies in climate change, including "Dark Waters", which was distinguished not only by its focus on the climate, but by showing the tactics used by companies to hide researches that prove the damage they cause to people and the environment, and how they prevent government decisions that would negatively affect their business, impeding the continuation of their environment polluting activities. *(Al-Arabawi, June 12, 2022)* Also, the movie "Erin Brockovich" is based on a true story, and tells the story of the legal dispute that arose between Erin and a gas and electricity company in California. It tells the story of Erin's life and her investigation into the pollution issue caused by that company, which in turn endangered the lives of the city's residents and made them vulnerable to poisoning and fatal diseases *(Al-Arabawi, June 12, 2022)*.

Some of these films had a future view of the world after the exacerbation of disasters, including: "Alita: Battle Angel" and "Ready Player One", which presented future visions of mankind after major cities became waste landfills, and the movie "Aquaman", which presented the future war between mankind and marine creatures that revolted due to pollution of the seas. Some films presented the environment as an obstacle for the heroes, so they have to face it and resist its harsh conditions to survive natural disasters, including "The Revenant", which won 3 Oscars for Best Actor Leonardo DiCaprio, Best Direction by Alejandro Gonzalez Iñárritu and Best Cinematography Emmanuel Lubezki. Produced in 2016, the film tells the struggle for survival amidst harsh winter weather and snowy winds. "San Andreas", in which its heroes face the largest earthquake in history. The movie "2040" inspires optimism and focuses on creative solutions rather than falling into captivity of the urgency and gloom of problems by presenting innovative alternatives to meet the challenges of climate change. Cases of optimism in this regard include renewable energy as solar roof tiles, shifts towards renewable agricultural practices, and the diversified use of seaweed as a guarantee and resource for food security *(Al-Arabawi, June 12, 2022)*.

Possibly one of the bleakest takes on the climate crisis, "First Reformed" follows a heavy-drinking priest (Ethan Hawke) as he spirals into despair following the suicide of a local radical environmentalist. Hawke's character becomes consumed by the prospect of devastating climate change, eventually plotting violent action. In the end, there is the possibility of salvation – but First Reformed, written and directed by Taxi Driver's Paul Schrader, does not offer any easy answers when it comes to the environment. *(Chilton, April 19, 2021)*

"Soylent Green" is one of the oldest and most outstanding movies that addressed the nightmarish truths of climate change. Its events occur in the crowded New York City in 2022. The classic movie depicts the world as an uninhabitable place due to global warming leading to drying oceans and depleted human resources. It is based on the 1966 novel by Harry Harrison "Make Room". *(Muhammed, October 17, 2022)*. In many ways, 1995's "Waterworld" was ahead of its time, a bombastic, high-budget thriller that touched on plenty of contemporary anxieties about the future of our climate. It was famously a box office flop, one which its star, Kevin Costner, has been living down ever since. The money is mostly there on screen, however; Waterworld remains one of the most memorably realized visages of the potential world-altering effects of global warming. *(Chilton, April 19, 2021)*

"Don't look up" comes in the list of the best movies that dealt with climate change, starring Leonardo DiCaprio. The movie tells the story of a science professor and his youngest student who make an astounding discovery of a comet orbiting within the solar system with a direct collision course with Earth. Together they need to convince the government as well as the public of the existential threat and the catastrophic consequences that it may occur for mankind. *(Al-Arabawi, June 12, 2022)*

Michael Shannon puts in a layered, tortured performance in "Take Shelter", Jeff Nichols's 2011 film about a young family man who experiences visions of the apocalypse. We never quite know if he's losing his mind, or dead on the money. As the character descends into paranoia and fear, we know there is always the possibility that his terrors are grounded in fact – that the catastrophe he so dreads is eminently plausible, no matter how hard we try to look away. *(Chilton, April 19, 2021)*

As for the Egyptian cinema and the interest in climate issues, it came as a secondary issue within the events of the film. It was also an expression of natural disasters that are commensurate with the nature of the Egyptian climate and environment, such as: the earthquake and the Nile flood, or it was dealt within corruption issues such as chemical waste, crop pollution, crop burning, Nile pollution, and epidemics. The issue of overpopulation came as one of the causes of pollution. Among these films, we mention, for example:

Youssef Chahine's "Ibn El-Nile" Son of the Nile 1951, embodied the people's construction of bridges - mud and wooden dams that are built to hold water from reaching out to houses - and the establishment of camps for the flood victims. Chahine benefited from the 1946 flood, as it was one of the most powerful floods in Egypt and caused a bang in the pages of newspapers and magazines due to the large number of victims, which were very similar in number to the victims of the 1887 flood that washed away corpses from Sudan and Upper Egypt to Cairo. *(Al-Deken, 6/11/2022, adapted)* Here, cinema was affected by the phenomenon of floods, but it did not leave a huge impact due to the low occurrence of floods in Egypt and the association of water with goodness and development in the Egyptian culture. It

Climate Change in World and Egyptian Cinema

is not classified as a natural disaster, as the Egyptians, since ancient times, build bridges and cut canals to benefit from the Nile waters.

Likewise, the phenomenon of earthquakes did not find an echo in the Egyptian cinema, as it is an incident that Egypt went through once in the 1992 earthquake, which had dire effects on Egypt and the following aftershocks that were not severe and did not cause losses. Therefore, the 1992 earthquake was shown in some Egyptian TV series as a secondary issue. The TV series "Arabesque", whose events end with the occurrence of the earthquake, and "Man Allazi La Yoheb Fatma" Who Doesn't Love Fatima it was the cause of the death of one of the characters in the series. as it was casually referred to in the TV series "Bent Esmaha Zat" A Girl Named Zat and "Abu Omar Al-Masry", while recently, specifically in Ramadan 2019, it was dealt with as a case within the issues of the TV series "Earthquake". As for cinema - which is the subject of our research - the movie "Korsi Fil Kolob" Tumbledown (2001) showed one scene at the beginning of the events. There are no other films that discuss or present this disaster. *(Mai Ezzat, February 23, 2023, adapted)*

There are some films that dealt with fires, but not on the grounds that they are one of the causes of climate change, but were dealt with as an issue in itself or as a means of revenge. Movies that contain fire scenes: "Lan Abki Abadan" I Will Never Cry (1957) there is a scene of the wheat crop burning. There are several scenes within the events of the movie "Shey Min Al-Khouf" A Taste of Fear (1969), as well as the scene of the factory on fire that shelters the workers within the events of the movie "Bayieat Alkhubz" The Bread Peddler (1953), while during the scene at the end of the movie "Mal Wa Nisaa" Money and Women (1960), a large fire appears that ignites in the warehouses of a private hospital. There are films dealing with the Cairo Fire: "La Waqt Lil Hob" No Time for Love, "Rod Kalby" Back Again, "Ghoroob Wa Shrooq" Sunset and Sunrise" and "El Bab El Maftuh" The Open Door. *(Elsayyed, May 13, 2016, adapted)* Fires continued as a means of revenge even in comedy films such as the movie "Matab Sena'y" Speed Bump.

The Egyptian cinema also dealt with radioactive materials, but in the context of corruption and crime issues. For example: In "Eish El Ghorab" The Mushroom (1997), We see businessman who work in shady business (weapons trade and drug trafficking) buys a radioactive plutonium cargo and tries to sell it to countries that encourage terrorism. He hides plutonium in one of the caves of Sinai and under tight guarding. He puts some drugs in the cave as a camouflage for guards to let them think that they are guarding the drugs. The Egyptian authorities are seeking to arrest him and seize the plutonium for fear of its explosion and threatening the region as a tourist destination for decades. (Wikipedia) It was preceded by "Tasrih Bialqatl" License to Kill (1991) and "Anbar El Moot" The Death Ward, released in cinemas on January 23, 1989, which tells the story of Tawfik Al Sharbatly who imports a spoiled shipment of radioactively contaminated baby food, exposing thousands of children to death. It is worth noting that the Basel Convention on the Control of Transboundary Movements of Hazardous Wastes was signed in 1989 and came into force in 1992. Hence, the impact of society in cinema appears.

"Kharaga Wa Lam Ya'ud" Mission Person discusses the overpopulation as one of the causes of environmental pollution. Overpopulation is the cause of climate changes according to Paul J. Crutzen and Eugene F. Stoermer in their book "Anthropocene"

The issue of climate change as a major issue in cinema appeared in a comical way through the movie "Hamlet Freezer" The Fraser Expedition, which attributed these climate changes to human causes due to the action of an international gang that controls the climate. In addition to the comic handling of the issue, it was raised as part of crimes and corruption issues, and not as a climate problem whose dangers are raised or solutions are proposed.

There is a scarcity of those films that address climate and environment issues as a major issue, but; there were some films that appeared accompanying Egypt's hosting the COP27, as well as the result of the COVID-19 that swept the world. However, it takes the nature of documentaries or short movies.

We find the documentary short movie titled "Climate Changes All of Us" with the participation of Youssra and Ahmed Amin, and directed by Marwan Hamed. It dealt with climate change at the COP27 Climate Conference, but it came in English and was not released to the Egyptian public.

We find the Egyptian short movie "Elbattaria Da'eifa" Battery Low, which warns of a catastrophe threatening mankind after the dangerous climate changes the world has witnessed. This film was screened at the National Film Council, as it monitors the natural disasters that the world has been exposed to in recent years, such as: the drought of rivers, the outbreak of fires, and the increasing phenomenon of desertification that threatens of food poverty. The film tells the story of Dr. Adam, the Egyptian scientist who lives in USA, and after being infected with the Corona virus, he decides to return back to Egypt with his wife, Dr. Lubna, who devotes herself to accompanying him on his treatment journey, and they decide to return to life in a place far from pollution and viruses. They chose the city of "St. Catherine" located in the south of Sinai, west of Egypt, which has remained in its nature for thousands of years, in an attempt to search for a lifeline for mankind in a message to the world from the country of peace and tolerance. Among the people and Bedouins of Sinai and from the top of St. Catherine's Mountains, the filmmakers are trying to make their shout to pay attention to the planet that is approaching an end if human behavior continues to pollute the environment and natural life with various pollutants that negatively affected the entire planet. The movie story is written and starred by Atef Abdel Latif, co-starring Dina Salah, screenplay, dialogue and treatment by Joseph Fawzy, directed by Sameh Maher, co-starring Bedouins of Sinai. *(Dwair, November 14, 2022)*

CONCLUSION

The research concluded that international cinema has paid attention to environmental issues and climate change. We find British, French, German, Spanish, Australian and Soviet films. However, American cinema was more productive of such genre of films. While climate issues were shown in Egyptian cinema as a secondary issue, or they were shown in one or several separate scenes, not with the aim of presenting the environmental issue itself, but to serve the main issue of the film. The climate issue was included in the Egyptian cinema within the political issues, such as those films that dealt with Cairo Fire, or within the issues of corruption and crime, such as those issues that dealt with food contamination or uranium smuggling.

International cinema has paid attention to climate issues through Sci-Fi films to confront environmental issues and the end of the world. Some American films have tended to make the American hero the savior who will keep the planet from collapsing at the end of the world.

Natural disasters were embodied in world cinema as tsunami disaster, and here international cinema and Egyptian cinema share the reflection of climate on cinema. Egyptian cinema ignored the issues of climate change and environmental issues because of the nature of its moderate climate, which was largely devoid of natural disasters such as floods, hurricanes, and devastating earthquakes. Unlike world cinema, which dealt with, for example, bushfires.

International and Egyptian cinema participated in shedding light on the role of corruption in environmental pollution, such as Dark Waters, Erin Brockovich and the Egyptian film "The Death Ward".

Cinema seeks to change the social reality. Although the realistic approach to the climate issue by the industrialized countries is characterized by an attempt to evade bearing the burdens of environmental reform, even though it is the main cause of this issue and does not abide by the agreements. Though, the cinema treatment of through "The Day After Tomorrow" (2004), for example, which released during the George W Bush presidency, was a pointedly political work underneath its popcorn movie trappings, a cry of frustration at the US government's environmental inaction. Scientifically, it's all over the place, but the core idea in its premise – that human sustainability is pushing the world to the precipice of disaster – remains chillingly true. *(Chilton, April 19, 2021)*

REFERENCES

Abdullah, A. (2016). *Ozone*. Arab Press Agency.

Abdullah, A (2012). *Eltaghayorat Elmonakheyah*. Arab Press Agency.

Abu Dayyah, A. (2010). *Elbey'aa Fi 200 Soal*. Dar Alfarabi Publishing and Distributing.

Al-Arabawi, R. (2022) *Cinema Al-Monakh Towage Kawareth Eltabeya'a Behazehe Eltareeqa*. Akhbar Al-Youm Portal.

Al-Attiyah, A. (2010) *Asbab Eltaghayor Elmonakhy,* [Thesis, Aleppo, Faculty of Arts and Humanities, Geography Department].

Al-Dahry, S. (2010). *Asaseyat Elm Elegtema' Elnafsy Eltarbawy Wa Nazareyatoh*. Amman: Dar Alhamed Publishing and Distributing.

Al-Deken, R. (2022). *Elcinema Elmasreya. Hal Ghabat Kadaya Elbeya'a*. Elseyasa Eldawleya https://www.siyassa.org.eg/News/18395.aspx

Al-Hawari, A (2019). Monakh Alard 2150. Bibliomania Publishing.

Al-Kufi, H. (2015). Zaherat Al-Ehterar Al-Kawni w Elaqhatha Benashatat Alensan wal Kawareth Altabey'ea. Academic Book Center

Al-Saadi, H. (2020). Elm Elbey'aa, Jordan: Dar AL-YAZORI for Publishing and Distribution

Al-Sabahi, N. (2022). Eltaghayor Elmonakhy wa Atharoh Ala Elsera'at Fe Sharq Africa, Egypt: Al Arabi Publishing and Distributing

Al-Ta'I, T. (2021). Tahadeyatoh gheir Altakleedeyah w Afaqoh Almostaqbaleyah. Egypt: Dar Academics for Publishing & Distributing Co.

Austin, G. (2016). *New uses of Bourdieu in film and media studies*. New York: berghahn

Brereton, P. (2012). *Smart cinema, DVD Add- Ons and New Audience pleasures*. UK: Palgrave Macmillan.

Cagle, C. (2017). *Sociology on film; postwar Hollywood's prestige commodity, New Jersey*. Rutgers university press.

Diken, B. (2007). *Sociology through the projector*. London and New York: Routledge Taylor & Francis group.

Duverger, M. (2020). *Introduction to the social sciences*. Routledge.

Dwair, M. (2022). "Elbattaria Da'eifa", Egyptian short movie addresses pollution and climate change. *Sky News Arabia*.

Gomez- Muňoz, P. (2023). Science fiction cinema in the twenty- first century; Transnational futures, cosmopolitan concerns. New York: Routledge

Grainge, P. (2007). *Film Histories an introduction and reader*. Edinburgh: Edinburgh university press.

Hamid, A. (2012). *Salah*. Algafaf w Altasahor, Almakhater w Aleyat Almokafha, Arabian Heba Nile for Publishing & Distribution.

Hulme, M. (2009). *Why We Disagree About Climate Change; Understanding controversy, inaction and opportunity*. Cambridge, New York: Cambridge university press.

Inglis, D & Almila, A. (2016). *The SAGE handbook of cultural sociology*. Los Angeles/London/ New Delhi: SAGE reference

Jarvier, I. (2013). *Towards a sociology of the cinema (ILS 92)*. New York: Routledge.

Jerslev, A. (2002). *Realism and 'Reality' in film and media*. Museum Tusculanum press, University of Copenhagen

Kamel, S. (2018). *Sorat Elsahafy Fil Cinema; Mashahed Sahafeya Fil Aflam Elarabeya Khelal Elfatra Men 1952 Hata 2009*. Egypt: Al Arabi Publishing and Distributing.

Kaseb, M. (2020). *Almasoleya Aldawleya Lehemayet Altanawoa Alehyaey w Beat Alfada' Alkharegy men Altalawoth Fe Etar Almoaahadat Aldawleya, Cairo*. Egyptian publishing house and distribution.

Chilton, L. (2021). *10 best movies about climate change, from Avatar to The Day After Tomorrow*. Independent website Arabic Edition

Magd, A. (2015). Altanzim Aldostoury Lelhoqoq w Alhorreyat Alektesadeya: Derasa Tatbeykeya Ala Alnezam Aldostoury (Altaadelat Alakhera w Afaq Altanmeya). Cairo, The National Center for Legal Publications.

Mansour, E. (2016). *Elmadkhal Ela Elm Elegtema*. Amman: Dar Arabian Gulf Publishing House

Mazahrh, A. (2016). *Elbey'aa Wal Mogtama'*. Amman: Dar Alshorooq Publishing and Distributing.

Megahed, N. (n.d.). *Eltarbeyah Ala Qeyam Elmowatanah Alalameya Lemowagahat Mogtama' Elmakhater*. Alexandria: University Education Publishing and Distributing.

Muhammed, L. (2022). *8 Aflam Alameya Tosalet Eldoa Ala Tahdeedat Eltaghayorat Elmonakheya Legawaneb Elhayah*. Youm 7 Portal.

Muhammed, M. (2014) Eldabt Eledary Wa Dawroh Fi Hemayet Elbey'aa, Comparative Study, Riyadh: Law and Economics Library

Muhammed, N. (2023). *Khetat Eltanmeya Elmostadama: Derasa Fil Elaqat Elmasreya Elafrikeya.* Al Arabi Publishing and Distributing.

Nabhan, Y. (2012). *Elehtebas Elharary Wa Ta'theeroh Ala Elbey'aa.* Amman: Dar Konooz for Publishing and Distribution

Samir, F. (2013). *Hemayet Albeaa w Mokafhet Altalawoth w Nashr Althaqafah Albe'eya.* Dar Al-Hamed Publishing.

Singh, S. & Marwah, R. (2023). *Politics of climate change: Crises, conventions and cooperation.* USA: World Scientific publishing

Westhues, M. (2014). *Climate change and environmental documentary film; An analysis of "An Inconvenient Truth" and "The 11 Th Hour."* GRIN publishing.

Willoquet- Maricondi, P. (2010). *Framing the world: Explorations in Ecocriticism and film.* University of Virginia press Charlottesville and London.

Chapter 4
Climate–Resilient Crops and Tribal Women Empowerment:
A Model of Odisha Millets Mission in India

Prageetha G. Raju

 https://orcid.org/0000-0001-7074-5196

Rainbow Management Research and Consultants, India

ABSTRACT

The state of Odisha in India has a grim malnutrition problem. The growing risk of climate change is rendering rice-wheat cropping unsustainable given the resource intensiveness. The present chapter narrates the potential of nutrient-rich, resource-efficient, and climate-resilient millets to address the malnutrition among tribal populations. The dual challenge of malnutrition coupled with growing food insecurity, environmentally unsustainable agricultural practices on account of climate change vis-à-vis empowerment of tribal women was solved by the government initiative called the Odisha Millets Mission (OMM). OMM aimed to promote the cultivation of millets in the state to boost the livelihood concerns, and nutritional deficiencies. Women empowerment in tribal areas was taken up because vulnerability to climate change is exacerbated by inequalities and marginalization related to gender, low income, and other socio-economic factors. Solutions to combat climate change will be more effective if they can address this reality.

INTRODUCTION

Climate change poses frightening challenges to agricultural production and food security. In India, ending food security remains high on the list of development priorities, as the country's rapid economic growth has not always translated into reducing hunger and correcting malnutrition. Sadly, India has the second highest number of undernourished people in the world (FAO 2015). According to the latest Global Hunger Index (2019)[1], (GHI) India is 102 out of 117 countries. In fact, India's neighbouring countries like Sri Lanka (66th), Nepal (73rd), Bangladesh (88th) and Pakistan (9th) have better GHI[2] values and scores.

DOI: 10.4018/978-1-6684-8963-5.ch004

Copyright © 2024, IGI Global. Copying or distributing in print or electronic forms without written permission of IGI Global is prohibited.

It is difficult to identify the causality of India's growing food insecurity, since the country is the second largest producer of fruits, vegetables and food grains.

To substantiate, in India, even after seven decades of independence, tribal people suffer from extreme inequity in health outcomes arising from food insecurity[3]. Government interventions following the passage of the National Food Security Act, failed to eradicate food and nutritional insecurities[4] completely. Green Revolution has prioritized wheat and rice (fine grains) cultivation and undermined the millet cultivation as they were coarse unlike the fine grains (Bhatt et al, 2016). Though, rice-wheat cropping system is extremely resource intensive and unsustainable in the long run, millets were perceived to be poor man's food and it became an orphaned crop as it lacked institutional support for its cultivation. The poor forcibly shifted to paddy cultivation in spite of high agricultural costs (decline in groundwater levels and water-logging) and also began consuming the fine grains deviating from their staple millets diet leading to micronutrients deficiency. National Family Health Survey found prevalence of stunting and wasting (underweight and under-height) among children of Scheduled Tribes followed by Scheduled Castes due to malnutrition in various states including Odisha state. The challenge of food insecurity is expected to grow in future due to climate change. It is observed that climate change affects women more than men given their vulnerabilities due to cultural norms and socio-economic factors, such as systematic violence, domestic work overload, and minimal access to basic human rights and so on.

The current instance focuses on the 42 million-person state of Odisha in India, where 70% of the population depend on agriculture as the source of income. *The state is most vulnerable to climate change as it is located in the country's east coast with a 480 kilometres of sensitive coastline making it prone to frequent cyclones, and coastal erosion.* With the monsoon dictating the timing of agriculture, the land productivity of the non-irrigated, rain-fed agriculture that the tribal women engage in is directly impacted by irregular, late, or insufficient monsoons. Majority of the tribal women in Odisha rely on agriculture, forestry, and fishing for their lives, livelihoods, and living conditions. Prolonged droughts/ rains devastate food supply and dried-up/flooded water sources thus destroying livelihoods.

The challenge before Government of Odisha state was three-fold:

- How to face the dual challenge of malnutrition coupled with growing food insecurity, and environmentally unsustainable agricultural practices on account of climate change?
- What types of crops should be promoted that can be climate-smart as well as reduce nutritional insecurities and outcomes?
- How to come up with regenerative strategies for climate justice?

To answer the above, a strategic approach begins with an assessment of the current scenario: our strengths, weaknesses, opportunities, and threats to achieve sustainable nutrition security for all.

In the above backdrop, government of Odisha in India resolved to revive millets on a large scale as a response as well as a solution to the existing/growing nutritional and climatic concerns under its flagship initiative. In order to save the farmers from climate stresses, there is an imperative need for promotion of climate-smart agricultural practices among the farmers. Cultivation of millets is considered to be as one of the climate smart agricultural practices. Being climate resilient and needing less water and being super nutritious, millets are becoming the crop of choice for Odisha's tribes. The revival of millets was for three reasons – a) to deal with climate stresses and b) to deal with extreme iniquity in health outcomes of tribals arising from food insecurity c) *to improve livelihoods of vulnerable rain-fed farmers.* Thus, Odisha Millets Mission (OMM) was launched in 2017.

Women empowerment was also taken up by OMM because vulnerability to climate change is exacerbated by inequalities and marginalization related to gender, ethnicity, low income and other social and economic factors. Tribal women were found to be anaemic due to malnutrition and pregnant women often suffered miscarriages; adolescent girls had no knowledge about menstrual hygiene. Solutions to combat climate change will be more effective if they can address this reality *and thus empowerment of women amongst the tribal population of the state was also addressed by OMM.*

OBJECTIVE

Given the above, the present paper elaborates on the following:

- Does OMM respond to the present nutritional and agricultural insecurities and inconsistencies?
- How does revival of millets empower tribal women?
- Is OMM a social innovation?

METHODOLOGY

The present study uses a persuasive case study design to narrate the origin of Odisha Millets Mission in Odisha state of India. The challenges of climate threat are identified and the solution, i.e., the OMM is presented as a single powerful example that can be emulated in developing and less developing economies.

The data is gathered from secondary sources supported by a strong review of literature. The present study narrates the first phase of OMM implementation only (2016-21) and its impact.

LIMITATION

The study concludes with the implementation of first phase of OMM where Ragi (finger millet) cultivation is the focus alongside women empowerment. The second and third phases happened after 2021 and this doesn't fall under the scope of this case.

REVIEW OF LITERATURE

The literature is presented under the following headings:

Case Study Method

Case studies have become widespread among researchers and other scientists (Tomas, 2011). Merriam (1998) stated that the case study does not claim any specific data collection methods, but focuses on holistic description and explanation. Within this focus, the case study can be further described as particularistic, heuristic, or descriptive. Soafer (1999) opined that a case study presents a comprehensive picture of a phenomenon that is intricate and gets closer to a solution. Pilot and Beck (2010) reported that the primary emphasis for a qualitative research is not to be able to generalize the conclusion, but to provide detailed information so that readers understand the aspects in the research of the case study. Siggelkow (2007) asserted that since a single case study cannot hold readers' attention; it can be made persuasive, only when it can provoke new thoughts and ideas rather than criticise the existing theories. The present study is presented as a descriptive case with a persuasive tone that sees climate smart agriculture as a means to revive livelihoods, build nutritional security and achieve women empowerment and has proven that climate change requires a gendered response.

Climate Change and Associated Injustice: How Men and Women Are Affected by It

Climate change is a global phenomenon, but its effects are being shaped by deep-rooted gender inequality. The United Nations defines climate change as long-term changes in temperature and weather patterns. Since the 19th century, human activity has been a major contributor to climate change, primarily through the burning of fossil fuels such as coal, oil and gas.

Soil erosions, heat waves, famines, rising sea levels causing intrusion of salt water, and storms and cyclones disproportionately upsets women because most men migrate out and the entire workload of the house falls on women (Mc Carthy, 2020). Women and girls continue to experience the greatest impacts of climate change as widening existing gender inequalities pose unique threats to their livelihoods, health and safety. (UN Women, 2022). To substantiate, women live in more poverty given less access to human rights, land, and ability to move; systematic violence escalates during unstable times adding to their misery. Household responsibilities like cooking, cleaning, gathering resources, and childcare get harder with climate change. (UNFCC, 2023). Globally, women are becoming more dependent on natural resources, but their access to them is declining. In many areas, women bear a disproportionate responsibility for food, water and fuel security

Since agriculture is the primary source of employment for women, during times of drought and irregular rains, women work hard as farmers and gatherers to ensure income and resources for their families. This increases the pressure on girls to leave school to help their mothers cope with the growing burden.

Climate change is a "**threat multiplier**", meaning, increasing social, political and economic tensions in fragile and conflict-affected areas. As climate change fuels conflicts world over, women and girls get increasingly vulnerable to gender-based violence, including conflict-related sexual violence, trafficking, child marriage and other forms of violence. Long-standing gender inequalities have resulted in inequalities in information, mobility, decision-making, access to resources and training. As a result, women and girls are deprived of emergency relief and assistance, during disasters, further jeopardizing their livelihoods, well-being and recovery, thus creating a vicious cycle of vulnerability to future disasters. Studies suggest that extreme heat is increasing stillbirth rates and climate change is increasing the prevalence of vector-borne diseases such as malaria, dengue and Zika, which are associated with worse maternal and neonatal outcomes.

Climate change poses enormous risks to farmers and undermines progress in poverty alleviation, food security and sustainable development. High temperatures and heat waves, changes in precipitation patterns, and severe weather events are seen in some parts of India too. By the end of the 21st century, temperatures in India are expected to increase by 2.5-5°C (Chandra, Karkun, Matthew, 2021). Smallholder farmers, who are heavily reliant on rain-fed agriculture, become vulnerable to the adverse impact of climate change due to lack of technical or financial support throwing them into abject poverty.

Resilience to climate is thus called for to help people anticipate risk, adapt, absorb, and recover quickly. Climate resilience is the ability of farmers, agricultural communities and agricultural value chains to maintain or improve their position in the face of climate change (Kabeer, Peterson & Waldron, 2021).

Climatic Conditions in Odisha

Odisha is the eleventh most populous state in India with 42 million people; agriculture is the backbone of over 70% of the population. World Bank (2016) reported that poverty is still very high as it has the largest tribal population living in the hills and highest malnutrition levels in the country. Panighrahi (2016) substantiated that although Odisha is endowed with rich natural resources, in the recent years the land, water and air of the region has been subjected to environmental stress through recurring natural hazards, agro-industrial activities and climate change. Sensitive coastline, cyclones, and coastal erosions make the state the most vulnerable to climate change (Patnaik, 2016), with the monsoon controlling the course of farming. Erratic (late or insufficient) monsoon directly impact the land productivity of the non-irrigated rainfed agriculture cultivated by the tribal women.

An interaction with the local farmer from Sundargarh, Pabitra Bhuin, a 50 years old farmer poured his woes saying, *"in addition to the rice fields, I also grew vegetables and corn. There is no irrigation system here in the tribal area. Also, rice fields are often destroyed by abnormal rainfall caused by climate change. It was not possible to grow high-yielding vegetables or corn."*(Lahangir, 2021)

Tribal Women of Odisha

Odisha state has 62 distinct groups of tribes of which Kondhs form the largest group and are largely concentrated in the areas of Phulbani, Rayagada and Kalahandi districts. The other popular tribal groups are *Sauras, Bonda, Santhals, Gonds, Bhumias, Orams, Koyas,* and so on are largely found in Gajapati, Ganjam, Mayurbhanj, Keonjhar, Jajpur, Balasore, Bhadrak, Sambalpur, Jharsuguda, Sonepur, Deogarh, Dhenkanal, Anugul, Jharsuguda, Sundergarh, and Kandhamal districts.

For the majority of the women in Odisha, their lives and livelihoods depend on agriculture, forestry, and fishing. The tribal women's organisation Orissa Nari Samaj (ONS) reports that the overworked tribal women and girls of the region suffer from unequal distribution of workload as they perform much of the agricultural labour such as foraging and timber cutting, shifting cultivation and settled agriculture, conservation of seeds, growing of food for domestic process, preparing, storing and processing of food, gathering of the forest products, collecting fodder and fuel and providing labour, besides fishing, farming, and hunting as well as entire domestic duties. Furthermore, women also suffer from physical and mental abuse and alienation but fear and stigma prevent them from reporting and seeking assistance.

Government Initiatives

The Government of India laid down three landmark policy expressions for safeguarding the interests of the tribal population – i) the Constitution of India, ii) Panchsheel Principles and iii) PESA Act as per the Ministry of Health and family welfare website.

Due to the rampant use of chemical inputs in agriculture and mechanisation of agriculture, the unemployment problem has become acute in Odisha, particularly among the tribal women (Patel and Jha, 2007). Despite policy protections, tribal people still experience substantial health inequities due to

Climate-Resilient Crops and Tribal Women Empowerment

escalating food insecurity even after seven decades of independence. During a focus group discussion conducted in Tada village in Bissamcuttack block of Rayagada district, mothers of the village showed concern about the changing diet.s of their school-going children (ashram schools, mission schools or government schools and hostels) from traditional millets food to rice and wheat, and associated nutritional defects. Rice and Wheat are resource intensive crops and depend a lot on climate and is unsustainable with tribals who reside on hills and slopes.

Coming to subsistence of the tribal populace, rains wash away the fertile top soil from the slopes destroying their livelihood. In spite of numerous central schemes running in the Central Indian Tribal Belt for poverty alleviation, (drought prone areas programme, flood control and mitigation, MGNREGA, etc), a need to revisit the interventions considering the ground realities regarding food insecurity and modify the programme design to adapt to the challenges of climate change in the region is essential. National Food Health Survey 4 shows that Odisha ranks in the top 10 of the most affected states of under-five child malnutrition on all the three indicators of wasting, stunting, and underweight. In terms of deprivations and marginalisation, the state is home to the highest population of tribals in the country. By establishing the "Special Programme for Promotion of Millets in Tribal Areas," also known as the Odisha Millest Mission, the state government of Odisha was the first to revive the production and consumption of millets (OMM). But, know-how about the food insecurities and nutritional outcomes is essential at this stage to design a sustainable solution.

Government of India has taken several initiatives under different policies formulated from time to time to promote millet cultivation such as, Nutritional Security through Intensive Millets Promotion (INSIMP) and Rainfed Area Development Programme (RADP) which are part of Rashtriya Krishi Vikas Yojana (RKVY), and Integrated Cereals Development Programmes in Coarse Cereals based Cropping Systems Areas (ICDP-CC) under Macro Management of Agriculture (MMA). Besides, the National Mission for Sustainable Agriculture (NMSA) adopted by Department of Agriculture and Cooperation, Ministry of Agriculture Government of India in 2014, has the objective of enhancing agricultural productivity especially in rainfed areas focusing on integrated farming, water use efficiency, soil health management and synergizing resource conservation. The programme has a mandate of improving millet production in the country. NMSA derives its mandate from Sustainable Agriculture Mission which is one of the eight Missions outlined under National Action Plan on Climate Change (NAPCC).. (Ministry of Agriculture and Farmers Welfare, 2018). Each scheme was mulled over and there emerged a comprehensive program for tribals called the Odisha Millets Mission (OMM).

The potential of millets is used as a key policy response to the present nutritional, climatic, and agricultural concerns through OMM.

Food (In)security and Nutritional Outcomes

According to the report on 'The State of Food Security and Nutrition in the World 2019' by FAO, 19.6 million people are undernourished in India, which is 1.5% of the population. This fact should be observed beyond hunger and aim at ensuring access to nutritious and sufficient food for all.

Indian Constitution, did not explicitly recognize the right to food, post-independence, but, starvation deaths in Odisha (Kalahand, Bolangiri and Koraput districts) in the early 2000s prompted the National Human Rights Commission to consider the "right to food", a fundamental right (Article 21). Xaxa (2014) argued that the right to food provisions in the Constitution requires a reading of Article 21 plus, two other articles (Article 39 and Article 47) to understand the obligation of the Indian state regarding

75

the right to food security. Article 39 of the constitution, (enshrined as one of the Directive Principles), is fundamental in the governance of the country, requiring the State to direct its policy towards securing that the citizens, men and women equally have the right to an adequate means to livelihood. On the other hand, Article 47 of the constitution, considered it to be the duty of the state to raise the level of nutrition and the standard of living and to improve public health. The People's Union for Civil Liberties (PUCL) filed a writ petition in the Supreme Court, asserting that the "Right to Food" is essential to the "Right to Life" (Article 21). During the on-going litigation, the Supreme Court issued several interim orders to implement central schemes as legal entitlements (Interim Order, 2001). These include the Public Distribution System (PDS), *Antyodaya Anna Yojana* (a public scheme to create food security and end hunger in India), the Midday Meal Scheme, and the Integrated Child Development Services (ICDS).

Government of India's department of food and public distribution (DFPD) reports that, in 1997, the central government launched a targeted public distribution system focusing on the poor which benefitted an estimated 60 million poor households, with 720,000 tonnes of food grains (such as wheat, rice and sugar) annually. In 2008, the court ruled that Below Poverty Line (BPL) households are entitled to 35 kg of food grains per month at a subsidized price. Further, to address the threats of food insecurity as well as to meet the Millennium Development Goals, the Government of India enacted the National Food Security Act (NFSA) in July 2013 which gave 67% of the population (75% in rural areas and 50% in urban areas) the legal right to receive food at heavily subsidised prices (DFPD website). The notable point here is that most government interventions followed by the passage of the NFSA, have failed to end food and nutritional insecurities. Corruption contributed to beneficiaries receiving only a fraction of PDS allocations, and even that was of a poor quality.

Green Revolution resulted from the above, and led to a paradigm shift in the dietary preferences of India, as it prioritised the dependence on wheat and rice as the primary food-grains. Coarse grains, including different varieties of millets, maize, barley and rye, which constituted the staple food for large section of the population, witnessed a decline in the consumption and production by 21% in rural areas and 11% in urban areas.

Devinder Sharma, a leading agro-economist in India observed that the coarse grains were not only orphaned but led to micronutrient deficiencies. The low quality rice supplied by PDS had very low levels of iron. A study in 2018 (Satpathi and Saha, 2020) found that 500 million people, or more than two-third of the population are affected by Iron (90%), Vitamin A (85%) and protein (50%) deficiencies. The poorer households were disproportionately affected by nutritional insecurities due to less flexibility to diversify the diets, following the changes in national food policy (subsidies favouring rice and wheat over coarse grains). Child malnutrition/under-nutrition became a very sensitive indicator of the overall levels of food security and hunger. Under-nutrition manifests itself in usually three ways, viz., stunting – when height is below the average height for that age, wasting – when weight is low for a given height and underweight – when weight is low for the given age. Stunting is associated with cognitive impairments and wasting results from inadequate food intake which is seen in many states including Odisha. The micronutrient deficiencies were most prevalent among the tribal population followed by the Scheduled Castes. Global Nutrition Report (2017) said that India is at the bottom of the table as India has maximum number of anaemic women of reproductive age in the world (51%). Illiteracy, lack of awareness, family priorities prevent women from taking proper nutrition leading to anaemia. Analysis at district level reveals a correlation between a larger percentage of tribal population and greater malnutrition levels with the relationship being strongest for underweight children.

In order to save the farmers from climate stresses, there is an imperative need for promotion of climate smart agricultural practices among the farmers. Cultivation of millets is considered to be one of the climate smart agricultural practices as millet is known for its exceptional heat tolerance, and less water (Satyavathi et al, 2021).

History of Millets in India

Millet is as old as the history of food in human civilization as its mention is found in many Holy Scriptures. Jones (2016) said that millets first became widespread in North China's heartland about 7500 years ago. From there, these grains spread through nomadic shepherds through Central Asia, Europe, and the South through Thailand and to India.

Millet is usually small-grained and the most important types are pearls, fingers, foxtail millets and so on. In India, various types of millets continued to be an integral part of the diet of the tribal communities in different parts of the subcontinent, until wheat and rice were heavily promoted by the Green Revolution. Given the easy cultivation process of millet, it is ridiculed as the 'lazy farmer's harvest' because the cultivation does not require many technical processes and inputs for a fruitful harvest. Also, millets can withstand biotic and abiotic stresses (Padulosi et al, 2015). Seeds are sent and only harvested after 3 months. Millets is a climate compliant crop (compared to wheat and rice) in terms of marginal growing conditions and nutritional value (Kumar et.al.2018). Similarly, eating millet as food brings social stigma. Despite social disappointment against millet production and consumption, millet is a nutritionally superior food with abundant micronutrients and low glycaemic properties compared to rice and wheat.

FAO reports that over the past two decades, rising incomes, increasing urbanization and government policies have reduced the importance of millet as a staple food in India. Currently, more than 50.0% of millet production is used for alternative uses rather than being consumed solely as a staple food. Among agricultural crops, rice and wheat are primary but they are water intensive crops and may not be sustainable as freshwater resources are depleting. As a new staple food, millet has the potential to become a sustainable alternative to rice and wheat because it takes less time to harvest and requires less water and it can grow in drought prone conditions too (Konapur et al, 2014). It will also help ensure food security for a large population for years to come. Considering the nutritional value associated with millets and its weather resistance, both consumption and production of millet is becoming increasingly important. Despite millet's declining popularity in recent decades, its historical versatility, resilience in harsh environments, nutritional properties and health benefits, long shelf life, and economic potential is getting reemphasized, lately.

Because millet thrives on shallow, low fertile soils with a pH range of acidic 4.5 to basic soils with pH 8.0.17, millet agriculture is very resource-efficient (Gangaiah, 2019). Millets were grown on 36.90 million acres in 1965–1966, the last year before the Green Revolution. Due to dietary changes, lack of availability, low yield, decreased demand, and conversion of irrigated land for rice and wheat farming, the area under millet production decreased to 14.72 million hectares (nearly 60% fall) in 2016–17. (Ministry of Agriculture and Farmers Welfare website, 2018).

Why Women Empowerment Was Taken Up By OMM?

80% of people displaced by climate change are women. (Gender Climate Tracker, 2010); prolonged droughts, severe weather, reduced food production and consumption seriously impact the economic and health conditions of women making them vulnerable. (UNFPA, 2009). Neumayer and Plumper (2007) asserted that between 1981 and 2002, a sample of 141 countries found that more women died in disasters than men in gender-unequal societies. Vulnerability to climate change is exacerbated by inequalities and marginalization related to gender, ethnicity, low income and other social and economic factors. Gender roles make women more vulnerable to environmental degradation and rising temperatures, which can lead to a negative feedback loop of increasing poverty in the future. But research shows that empowering women in these roles can reverse poverty and enable effective solutions to climate change. Solutions to combat climate change will be more effective if they can address this reality. Leadership of indigenous women, in particular, who are at the forefront of environmental conservation possess invaluable knowledge and expertise that can help increase resilience and reduce greenhouse gas emissions. A study by Landesa (landesa.org) in India found that reforestation and canopy growth increased when there was a higher proportion of women in leadership positions, even when planting in smaller, degraded forests. In a review of 17 studies from around the world, the presence of women in conservation and natural resource management led to stricter and more sustainable mining regulations, stronger compliance, greater transparency and accountability, and better conflict resolution asserted Leisher, Temsah et al (2016). In fact, the United Nations reports that community resilience and capacity-building strategies can be more successful when women are involved in the planning process. The SDG also highlights investing in gender equality.

Social Innovation

Social innovation has always been acknowledged for its unique and effective role in responding to welfare crises (The Young Foundation, 2012) and achieving a transformative impact (Avelino et al., 2019). Pieri et al. 2021 reported that social innovation promotes collaborations between citizens, third sector organizations and public actors; therefore, social innovation has been used by policy makers world over. Social Innovation provides a novel solution that is more sustainable, than the current solutions as its

Figure 1. Stakeholders of social innovation
Source: *Tanimoto K, (2012)*

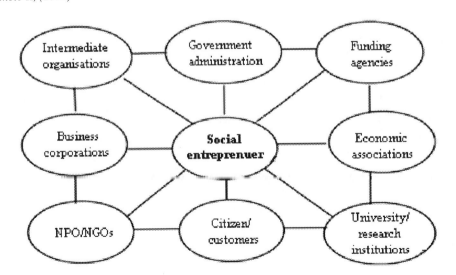

Climate-Resilient Crops and Tribal Women Empowerment

value grows primarily to society rather than to private individuals (Stanford Centre for Social Innovation, 2017). The social entrepreneur has many stakeholders as depicted in Figure 1.

Ümarik, Loogma, and Tafel-Viia (2014) noted that social innovations share a number of traits: the basis of legitimacy; social benefit or to meet social needs; agents of change, or drivers of change; social interaction and learning, which promote the transfer of knowledge; trigger for change, unmet needs or facing a crisis. Based on Rodríguez Herrera & Alvarado, (2008) study that social innovations envisage the merging of diverse actors and social agents, as well as different processes around the recognized problem with demonstrable impact, the present study examines OMM in the light of the above to highlight it as a social innovation.

FINDINGS OF THE STUDY

The Birth of Odisha Millets Mission

As mentioned, a strategic approach begins with an assessment of the current scenario: our SWOT (strengths, weaknesses, opportunities, and threats) (**Table 1**) to achieve sustainable nutrition security for all. OMM was launched in 2016 to address the growing concerns of agricultural uncertainty and nutrition insecurity as a five-year project (2016-2021). In 2018, the government declared the year as "National Year of Millets." (Ministry of Agriculture and Farmers Welfare, March 22, 2018). The first three years of the program period represented the implementation phase of the programme, while the next two years consisted of consolidation, expansion and institutionalization. Different organizations were involved at State, District and Block levels as depicted in **Table 2**.

Based on the above, the government of Odisha decided to revive farming and consumption of millets by launching a special programme for promotion of millets in tribal areas called as the Odisha Millets Mission in 2017. OMM was born after state-level consultations organized by the Office of Planning and Convergence of Odisha on "Extensive millet revitalization in tribal areas of Odisha" for food security and drought mitigation in South Odisha", held at Nabakrushna Chowdhury Centre for Development Studies (NCDS), on 27 January 2016.

A memorandum of understanding (MoU) was signed on February 27, 2017 with the Department of Agriculture and Food Production (DAFP), and the state-level hub agency that oversees and implements the programme, NCDS, and Watershed Support Services and Activities Network (WASSAN), District ATMA (Agricultural Technology Management Agency), NGOs and community based organizations to anchor the programme secretariat as part of the state secretariat.

The purpose of the OMM is to address both the supply and demand side aspects of the use of millet. The program aims to improve millet farming practices, restore household consumption of millet to improve food and nutrition security, establish millet businesses and farmer-producer associations, and support existing food sources such as Integrated Child Development Services (ICDS) for robust integration of millet into nutrition programs, and the public distribution system "PDS". This initiative is unique in that it leverages a variety of stakeholders involved in farm and tribal development, including community-based organizations, grassroots NGOs, and technical advisors. (Satpati, Saha and Basu, 2020)

According to program guidelines, the key project goals include increasing household consumption of millet by approximately 25%, improving household food security, and increasing demand for millet using productivity mechanisms, with women and children, establishing farmer producer organizations,

Table 1. SWOT (strengths, weakness, opportunities, and threats)

STRENGTHS	WEAKNESSES
o Hill areas highly amenable for millet cultivationo Friendly government schemes which can be integrated to form a strategic sustainable solution to food insecurity and nutritional outcomeso The state has adequate infrastructure to handle the food insecurity challenges.	o Erratic Monsoono Monsoon washing away top fertile soil on the hills making rice and wheat cultivation unsustainableo Tribal farmers losing livelihoodo Men migrating to urban areas throwing the domestic and livelihood burden on womeno Food insecurityo Stillborn babies due to nutritional defect
OPPORTUNITIES	THREATS
o Growing demand for millets across urban Indiao Growing potential of millets as a solution to handle nutritional, climatic, and agricultural concerns.o Millets yield and returns is faster than rice and wheat.o Empowering women	o Stunting, Wasting, Underweight issues among children under the age of 5o Rice and Wheat cultivation unsustainable on hill slopes given the resource intensivenesso Unpredictable climate-change effects on the lives and livelihoods of tribal men and women farmers and children

Source: Compiled by the author from various sources

promoting millets in urban areas, establish custom hiring centers for implements, improve agronomic

Table 2. The stakeholder matrix of OMM

S.No.	Stakeholders	Level	Responsibilities
1	Mission on Millets Committee	State	Policy related decisions such as inclusion of millets in PDS.
2	Directorate of Agricultureand Food Production	State	Nodal Agency for the Program. Appoints a Nodal person to work with Programme Secretariat. Periodically monitors the Programme.
3	Program Secretariat(WASSAN) and ResearchSecretariat (NCDS)	State	Programme Management, Monitoring,Development and maintenance of web-based MIS, Coordination with State Departments and District Administration, Capacity Building, Convergence and Research.
4	District ATMA GoverningBoard	District	Facilitates disbursement of funds to CBOs and Facilitating Agencies (FA) against six monthly action plans. Regularly monitors the program on a monthly basis. Administrative head of the Project at the District Level.
5	Program Secretariat(WASSAN)	District	Program Secretariat support District ATMA Governing Board in Monitoring of implementation of the program. Verification of financial and process documentation of CBO.
6	Facilitating Agencies (NGOsat Block Level)	District	Monitors work of Community BasedOrganizations (CBOs) and Community Resource Persons (CRPs) for timely implementation of the project. Build capacities of CBOs, CRPs. FA will be responsible for relevant financial and process documentation of CBOs.
7	Community BasedOrganizations (CBO)	Block	Responsible for implementing of the program with support from Community Resource Persons (CRPs), Facilitating Agencies andProgram Secretariat. FA is responsible for project implementation by the CBOs and all relevant documentation by CBOs.

Source: Satpati, Saha and Basu, (2020)

Climate-Resilient Crops and Tribal Women Empowerment

practices. The idea is to bring back these nutri-cereals, ensuring food and nutrition security, while enabling climate-resilience to farmers. OMM intends to increase millet cultivation from 25,000 hectares to 2 lakh hectares by 2023 and make Odisha state as the millet hub of India.

The project is most likely to fall under the broad auspices of the National Nutrition Mission (POSHAN Abhiyaan), which concentrates on close supervision and coordination of activities aimed at better nutrition through several state-level departments.

Based on the above, OMM stands on five following pillars:

A. Consumption improvement of millet at household level
B. Improvement of millet production and productivity
C. Promotion of Millet Enterprises through women entrepreneurs
D. Value Chain Development of Millets
E. Special inclusion of millet in state nutrition programme such as Integrated Child Development Services (ICDS) and Public Distribution System(PDS)

The uniqueness of OMM lies in the direct benefit transfer model-based incentive structure, which offers farmers a conditional cash transfer if they adopt the suggested practise.

Strategy Involved

OMM was carried out in three phases.

The first phase focused on creating awareness about the benefits of millets among farmers, consumers, and other stakeholders. This includes organizing awareness campaigns, training programs, and capacity building initiatives for farmers, as well as promoting millets through various media channels. The first phase implementation of Odisha Millet Mission was started in seven southern Odisha districts Gajapati, Kalahandi, Kandhamal, Koraput, Malkangiri, Nuapada and Rayagada. Present Nuapada District comprises one sub-division (Nuapada), and five Blocks (Khariar, Sinapalli, Boden, Nuapada and Komna).

Classification of millet farmers on the basis of social category reveals that majority of millet farmers, overall, to the extent of 76.7%t are Scheduled Tribes (STs) followed by other castes (12.0%) and the remaining 1.2% are SCs. Highest incidence of millet farmers to the extent of 98.8% at Komna block are found as tribals. The incidence of tribal millet farmers at Boden and Sinapalli blocks is found at 80.0 and 81.3% respectively. The mean age of millet farmers is overall found at 47.5 years. This implies that experienced farmers are found to have been registered as millet farmers under OMM. With respect to religion, all of the millet farmers are Hindus by religion with less than 1% Christians in all of the blocks covered under OMM. The education levels range from Illiterate, Primary, Upper Primary, HSC, and above HSC.

In order to avail the benefits of OMM project intervention, the farmers in the programme area are required to register themselves with OMM. Incidence of female millet farmers registered under OMM is comparatively higher at Komna block in relation to Boden and Sinapali blocks.

Ragi (finger millet) received focus in first phase of implementation. Ragi area in 2010 compared to 2000 has decreased by 33.77% in Nuapada district. The procurement and distribution flowchart is presented in Fig 1:

During first phase programme intervention of OMM, 531.41 hectares of land in three blocks covering Boden, Komna and Sinappali are taken up for ragi cultivation in Nuapada district. Out of the total

81

Table 3. First phase implementation

Sno	Blocks	Number of farmers covered under first phase OMM by districts, blocks and crop years (No. of farmers)				% Share of the block in district total	% Share of the district in state total
		2017-18	2018-19	2019-20	All Years		
1	Boden	345	402	642	1389	41.3	
2	Komna	92	359	474	925	27.5	5.3
3	Sinapali	184	357	510	1051	31.2	
	Sub total	**621**	**1118**	**1626**	**3365**	100.0	
	All districts	**8636**	**21972**	**32394**	**63002**		100.0

Source: Computed from WASSAN Official data

Figure 2. Ragi procurement and distribution flowchart
Source: *https://milletsodisha.com/*

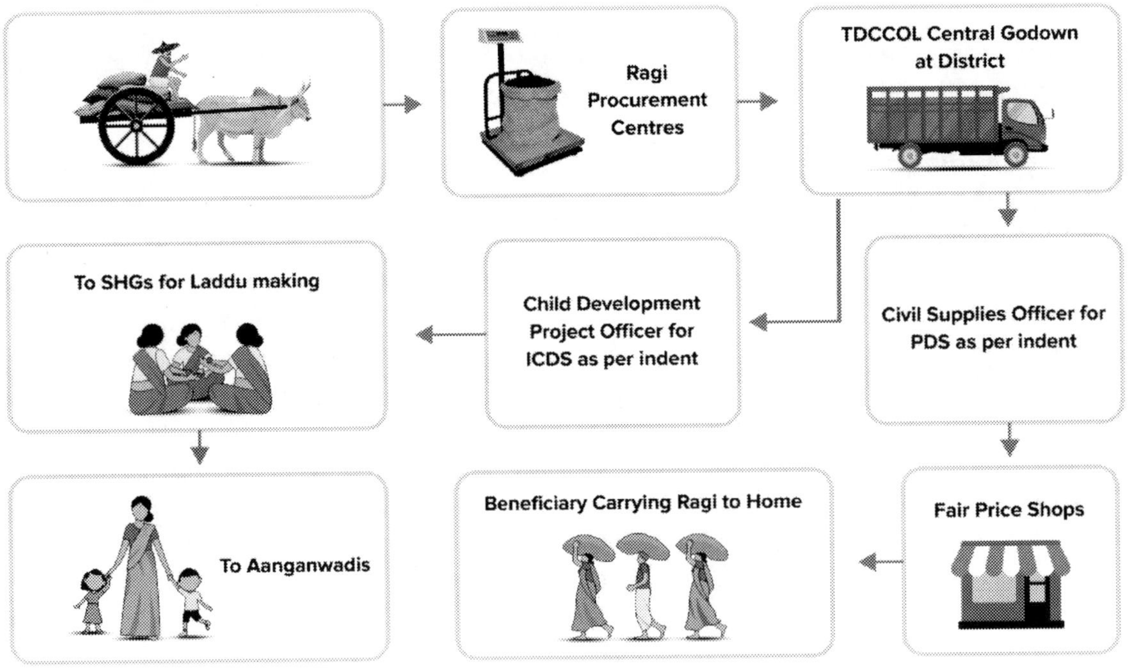

land area taken up for ragi cultivation, the percentage share of Boden, Komna and Sinapalli stand at 51.5, 23.3 and 25.2 percent respectively. Out of overall land area taken up in the state, percentage share of Nuapda district is only 2.4%.

In the second phase, the focus was on improving the production and productivity of millets by providing farmers with access to high-quality seeds, fertilizers, and other inputs. The OMM has set up seed banks and production centres to ensure the availability of quality seeds for farmers. The mission also provides training and technical support to farmers on millet cultivation, including best practices in soil health management, pest and disease management, and post-harvest management.

The third phase focused on creating a market for millets by promoting their consumption among consumers, especially in the urban areas. The OMM has partnered with various stakeholders, including

Figure 3. The program delivery mechanism
Source: *https://milletsodisha.com/*

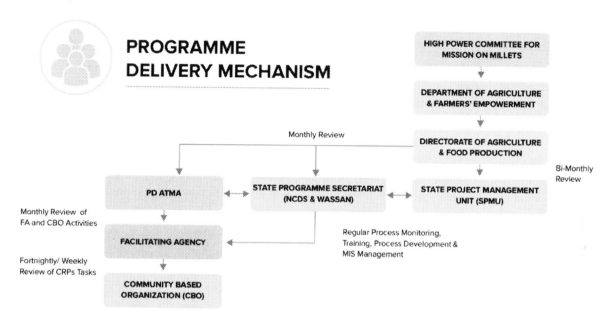

the private sector, to develop value chains for millets, including processing, packaging, and marketing. The mission has also launched the Millets on Wheels program, which involves mobile millet vans that sell millet-based products in urban areas.

The programme also aims at promoting millet processing enterprises at Gram Panchayat and block level to ensure household access for easy processing and value-added millets and millet products. Improvement of millet productivity, profitability from millet cultivation, development of millet-based enterprises with market led value chain activities, promotion of women entrepreneurs for millet-based activities, inclusion of millet in state nutrition programme including public distribution programme are the added objectives for which the special programme on millets is implemented in the state. The highly regarded and discussed OMM can provide important lessons in terms of programme and administrative processes that can be created as a model for other states, in addition to having a long-term impact on the nutritional profile of the state. The Program Delivery Mechanism is shown below:

- The OMM programme was initially launched in the year 2017 in 30 blocks of the 7 tribal populated districts of southern Odisha. In the same year, it achieved 6000 ha of area coverage under improved agronomic practices of millet cultivation, covering 8030 farmers.
- In the following year, 2018-19 the programme was expanded to 55 blocks in 11 districts covering 13000 ha of area coverage and including 29056 farmers. Subsequently, the programme was expanded to 72 blocks in 14 districts of Odisha
- In the year 2019-20, covering 21552 ha and reaching out to 51045 farmers.
- The programme was further expanded to 76 blocks in 14 districts during the financial year 2020-21, covering 47339 ha and including 108731 farmers. During the reporting year, the programme was cumulatively implemented across 5845 villages of 1018 panchayats in 14 districts of Odisha.

How Millets Are Empowering Tribal Women

Being locally available and rich in nutrients, eating millets is beneficial to children's health. It is no wonder that the district administration has included millets in the special nutrition programme for children under the Integrated Child Development Services (ICDS) scheme of the state's Women & Child Development Department. Under Mission Shakti (strength or power) – a programme to empower millions of women through various economic activities – women self-help groups (SHGs) prepare ragi laddu mix with powdered millets, sugar and peanuts for the government-run anganwadi (child care) centres. Under Mission Shakti the state government has been training Self-help groups (SHG) members to prepare snacks and food items with millets and to manage kiosks and cafes to offer millet-based foods. To make millet-based food popular and provide livelihood opportunities to these SHG women, the district administration plans to set up 81 "Millets Shakti" kiosks in different places. The first kiosk was recently opened to a wide reception at Rajgangpur block. Millets are also being supplied through the public distribution system to ration card and food security card holders. For this the government purchased 709 tonnes of ragi cultivated in Sundargarh at a minimum support price of Rs 329.5/tonne.

During the pandemic, when the centres were closed, Anganwadi workers went on a door-to-door drive to distribute the ragi mix. In the first year of implementation, over 60,000 children aged between 3 and 6 years in Sundargarh benefitted from this initiative. Millets are also being supplied through the public distribution system to ration card and food security card holders.

Poshansakhi (a nutrition friend) was an innovative method to improve health and nutritional status of adolescent girls, pregnant women and lactating mothers through an integrated multi-sectoral approach. Odisha Livelihood Mission piloted the *swabhimaan* (self-respect) program with support from UNICEF and Living farms, an NGO. It resulted in picking-up and training community women resource persons as *Poshansakhis* to engage with self-help groups and form clubs and groups of adolescent girls, farmer, and mothers on nutrition-related agricultural techniques and support. A *Maitribaithak* (friendship meeting) is held once in a month wherein all expectant mothers, lactating mothers, adolescent girls, and other women attend to discuss sensitive matters such as sanitation, menstrual hygiene, pregnancy-related matters, and health and nutrition issues. A *Poshansakhi* guides every woman on the above matters as she is a trained nutrition advocate. *Psohansakhis* check the mid upper arm circumference (MUAC) of every pregnant woman; MUAC is a mandatory measurement for every pregnant woman. OMM has helped women improve their nutrition as well as their family's nutrition and helped them deliver babies with appropriate weight. OMM's focus on millet cultivation and consumption has helped lactating mothers, and expectant mothers with nutrition and saved them from miscarriages or still-born babies. Adolescent girls' nutritional status has improved and they would give birth to a healthy future generation.

The revival of millets is helping women farmers in the tribal districts to enhance their livelihood and become financially independent as millets are resilient to drought, salinity, extreme heat, pests and diseases. With high lands, Sundargarh suits millet cropping pattern. A sizable number of women farmers are a part of this growth story. Women are taking the lead in cultivation of native varieties of millets as they grow well in dry zones (need 60% less water than paddy) as rain fed crops and are low-duty crops besides being nutritious. They can be harvested within 70-100 days as against 120-150 days for paddy/wheat and that too chemicals free, thus making millet cultivation economical.

Women are no more limiting themselves to individual farm activities but have formed groups comprising landholders and landless and take up some patches of land on lease and grow different nutrition-rich crops like vegetables and pulses besides millets. This method of collectiveness has come from the tribal

Climate-Resilient Crops and Tribal Women Empowerment

systems of *Panch (togetherness)* wherein a team of tribal men work together for levelling and terracing the sloped land in the hily terrain to make them farming ready. The women collectiveness then enter the scene and engage in sowing, planting, watering, monitoring, reaping, plucking, piling and so on.

OMM has fortified the strength and conscientiousness of women collectives by creating primary processing and post-harvesting units for millets in rural areas. These millet-based enterprises can address malnutrition and unemployment among tribal farmers. The awareness created by local NGOs about cultivating millets through improved farming techniques, and the institutional impetus given by OMM such as assured purchase and higher prices is steadily yielding good results.

DISCUSSION

OMM has fortified the strength and conscientiousness of women collectives by creating primary processing and post-harvesting units for millets in rural areas. These millet-based enterprises can address malnutrition and unemployment among tribal farmers. The awareness created by local NGOs about cultivating millets through improved farming techniques, and the institutional impetus given by OMM such as assured purchase and higher prices is steadily yielding good results.

To substantiate, in Mayurbhanj, the number of women farmers involved in millet cultivation has gone up by 104% since 2019. While women share the responsibility of cultivating paddy with their male counterparts, millet cultivation is taken up solely by women, helping them generate an independent income and become aware of their financial holdings.

Odisha has recorded a 215% increase in gross value of millets produced per farmer household from Rs 3,957 in 2016-17 to Rs 12,486 in 2018-19 (NITI Aayog, 2020), as well as, the area under millet cultivation has increased with an increased yield rate of 120%.

During 2020-21, the state government enhanced OMM funding from Rs 65.54 crore to Rs 536.98 crore, for procurement and distribution of *ragi* in public distribution system and Integrated Child Development Service.

OMM has had a significant impact on the state's agricultural landscape and the lives of its people. By the end of 2020, millet cultivation had increased to over 65,000 hectares, with more than 60,000 farmers adopting millet cultivation. The mission has also led to the development of a range of millet-based products, including snacks, beverages, and bakery products, which are being sold in local markets and online. The OMM has also had a positive impact on the nutrition and health outcomes of the people of Odisha. The increased consumption of millets has led to improved nutrition outcomes, especially among vulnerable populations such as women and children.

Is OMM a social innovation? OMM is a government-led initiative aimed at promoting the cultivation and consumption of millets in the state of Odisha, India. Millets are nutrient-rich grains that are often considered "forgotten foods" due to their decreasing popularity over the years. As for whether the Odisha Millets Mission can be considered a social innovation, it depends on how we define social innovation.

Social innovation can be defined in many ways. OECD refers to social innovation as the design and implementation of new solutions that suggest conceptual, process, product, or organisational change that aims to improve the welfare and well-being of individuals and communities. Soule, Malhotra and Clavier from Stanford define social innovation as the process of developing and implementing effective solutions to difficult and often systemic social and environmental problems in order to support the progress of society. Social innovation is not a prerogative of any organizational form or legal structure. Solutions often require the active collaboration of government, business, and non-profit stakeholders.

Going by Soule, Malhotra and Clavier's (2021) definition, the Odisha Millets Mission can certainly be considered a social innovation. The mission addresses a number of social problems, such as malnutrition, food insecurity, and the declining agricultural sector, by promoting the cultivation and consumption of millets. By doing so, the mission creates value for society by improving public health, reducing poverty, and supporting local farmers.

However, if we define social innovation more narrowly as a new approach that involves the use of technology or business models to solve social problems, then the Odisha Millets Mission may not fit this definition as it is primarily a government-led policy initiative. Nonetheless, it is still an innovative approach to addressing social issues, and its success in promoting the cultivation and consumption of millets could serve as a model for other states and countries facing similar challenges.

CONCLUSION

OMM is a unique initiative which is a comprehensive program for millet cultivation not only focussing on increasing the demand and consumption for millets but also empowered tribal women to improve livelihood and improved agronomic practices leading to substantial improvement of soil quality, water usage efficiency, decreased input costs, and increased yield. This is a model to be emulated in all states.

FUTURE IMPLICATION

Despite the growing demand for millets and OMM's ability to handle the challenges, the need for market linkage is still a weak area. The pricing systems need to be re-looked. Reliable market linkage and procurement of organic farm product from woman farmers by government agencies at a fair price would boost their income and mobilize many more farmers into the fold. A post-pandemic study however, is the need of the hour as the second and third phases of implementation have already rolled-in.

REFERENCES

Asian Disaster Preparedness Center. (2008). *Building on Local Knowledge for Safer Homes*. ADPC. http://www.adpc.net/v2007/IKM/ONLINE%20DOCUMENTS/downloads/2008/3_CaseStudyShelterl.pdf

Avelino, F., Wittmayer, J. M., Pel, B., Weaver, P., Dumitru, A., Haxeltine, A., Kemp, R., Jørgensen, M. S., Bauler, T., Ruijsink, S., & O'Riordan, T. (2019). Transformative Social Innovation and (Dis) empowerment. *Technological Forecasting and Social Change, 145*(August), 195–206. doi:10.1016/j.techfore.2017.05.002

Bhatt, R., Kukal, S. S., Busari, M. A., Arora, S., & Yadav, M. (2016). Sustainability issues on rice–wheat cropping system. *International Soil and Water Conservation Research, 4*(1), 64–74. doi:10.1016/j.iswcr.2015.12.001

Chandra, M., Karkun, A., & Matthew, S. (2021). Hot and Flooded: What the IPCC Report Forecasts for India's Development Future. *The Wire.* https://science.thewire.in/politics/ government/what-the-ipcc-report-forecasts-for-india-development-future/

Cuevas, M. C. (2021). *Meet the Puerto Rican sisterhood reinventing the island's future after Maria.* CNN. https://www.cnn.com/2018/09/19/us/iyw-puerto-rico-women-rebuild-trnd/index.html

M. A. D. R. E. (n.d.). *Resources and Results for Women Worldwide.* Sudan: Women Farmers Unite. https://www.madre.org/page/sudan-women-farmers-unite-41.html

M. A. D. R. E. (n.d.). *Resources and Results for Women Worldwide.* Nicaragua: Harvesting Hope. https://www.madre.org/page/nicaraguaharvesting-hope-34.html

Facing a Changing World: Women, Population and Climate. (2009). UNFPA. https://www.unfpa.org/publications/state-world-population-2009

FAO. IFAD, UNICEF, WFP, & WHO. (2019). The State of Food Security and Nutrition in the World 2019- Safeguarding against economic slowdowns and downturns. FAO.

Gangaiah, B. (2019). Agronomy-Kharif Crops. NISCAIR. http://nsdl.niscair.res.in/jspui/bitstream/123456789/527/1/Millets%20(Sorghum%2c%20Pearl%20Millet%2c%20Finger%20Millet)%20-%20%20Formatted.pdf

Gender Climate Tracker. (2010). *Gender and the Climate Change Agenda. The impacts of climate change on women and public policy.* Gender Climate Tracker. https://genderclimatetracker.org/sites/default/files/Resources/Gender-and-the-climate-change-agenda-212.pdf

Greenpeace. (2009). *Meet Ulamila: Climate Activist in the Pacific.* Green Peace. https://www.greenpeace.org.au/blog/meet-ulamila-climate-activist-in-the-pacific/

Jones, M., Hunt, H., Kneale, C., Lightfoot, E., Lister, D., Liu, X., & Motuzaite-Matuzeviciute, G. (2016). Food Globalisation in prehistory: The agrarian foundations of an interconnected continent. *Journal of the British Academy, 4,* 73–87. doi:10.5871/jba/004.073

Kabeer, M., Peterson, N., & Waldron, D. (2021). Resilient Farmers: Investing to Overcome the Climate Crisis. Acumen and Busara Center for Behavioral Science, 16.

Konapur, A., Gavaravarapu, M. S., Gupta, S. D., & Nair, K. M. (2014). Millets in Meeting Nutrition Security:Issues and Way Forward for India. *The Indian Journal of Nutrition and Dietetics, 51,* 306–321.

Kumar, A., Tomer, V., Kaur, A., Kumar, V., & Gupta, K. (2018). Millets: A solution to agrarian and nutritional challenges. *Agriculture & Food Security, 7*(1), 31. https://agricultureandfoodsecurity.biomedcentral.com/articles/10.1186/s40066-018-0183-3. doi:10.118640066-018-0183-3

Lahangir, S. (2021). Odiya tribes discover the wonders of millets. *Villagesquare.in.* https://www.villagesquare.in/odiya-tribes-discover-the-wonders-of-millets/

Leisher, C., Temsah, G., Booker, F., Day, M., Samberg, L., Prosnitz, D., Agarwal, B., Matthews, E., Roe, D., Russell, D., Sunderland, T., & Wilkie, D. (2016). Does the gender composition of forest and fishery management groups affect resource governance and conservation outcomes? A systematic map. *Environmental Evidence*, 5(1), 6. doi:10.118613750-016-0057-8

Mc Carthy, J. (2020). *Understanding Why Climate Change Impacts Women more than Men?* Global Citizen. https://www.globalcitizen.org/en/content/how-climate-change-affects-women/

Merriam, S. B. (1998). *Qualitative research and case study applications in education.* Jossey-Bass.

Neumayer, E., & Plümper, T. (2007). The Gendered Nature of Natural Disasters: The Impact of Catastrophic Events on the Gender Gap in Life Expectancy, 1981–2002. *Annals of the Association of American Geographers*, 97(3), 551–566. doi:10.1111/j.1467-8306.2007.00563.x

Ota, A. B. (2020). Tribal Atlas of Odisha. Academy of Tribal Languages and Culture & Scheduled Castes & Scheduled Tribes Research and Training Institute ST & SC Development Department, Government of Odisha. Commissioner-cum-Director, SCSTRTI & Member Secretary, ATLC, Bhubaneswar.

Padulosi, S., Mal, B. C., King, O. I., & Gotor, E. (2015). Minor Millets as a Central Element for Sustainably Enhanced Incomes, Empowerment, and Nutrition in Rural India. *Sustainability (Basel)*, 7(7), 1–30. doi:10.3390u7078904

Panigrahi, J. K. (2016). Coastal Ecosystems of Odisha–Health and Nutritional Challenges Consequent to Climate Change. Directorate of Economics and Statistics, Odisha.

Patel, A. M., & Jha, M. K. (2007). *Weapons of the Weak -Field Studies on Claims to Social Justice in Bihar & Orissa.* Mahanirban Calcutta Research Group, Kolkata, India http://www.mcrg.ac.in/pp13.pdf

Patnaik, B. K. (2016). *Impact of Global Warming on Agriculture with Reference to Odisha.* Special Issue on Agriculture and Farmer's Welfare, Directorate of Economics and Statistics.

Satpathi, S., Saha, A., & Basu, S. (2020). *Millets as a Policy Response to the Food and Nutrition Crisis—Special Reference to the Odisha Millets Mission.* BRLF. https://www.brlf.in/brlf2/wp-content/uploads/2020/01/Odisha-Millet-Mission.pdf

Satyavathi, C. T., Ambawat, S., Khandelwal, V., & Srivastava, R. K. (2021). Pearl Millet: A Climate-Resilient Nutricereal for Mitigating Hidden Hunger and Provide Nutritional Security. *Frontiers in Plant Science*, 12, 659938. doi:10.3389/fpls.2021.659938 PMID:34589092

Siggelkow, N. (2007). Persuasion with case studies. *Academy of Management Journal*, 50(1), 20–24. Retrieved July 14, 2023, from https://journals.aom.org/doi/10.5465/amj.2007.24160882#:~:text=If%20one's%20conceptual%20argument%20is,is%20usually%20much%20more%20appealing. doi:10.5465/amj.2007.24160882

Soafer, S. (1999). Qualitative methods. What are they and why use them? *Health Services Research*, 34(5 Pt 2), 1101–1118. PMID:10591275

Tanimoto, K. (2012). The emergent process of social innovation: Multi-stakeholders perspective. *International Journal of Innovative Research and Development*, 4(June), 267. doi:10.1504/IJIRD.2012.047561

Thomas, G. (2011). A Typology for the Case Study in Social Science Following a Review of definition, Disclosure, and Structure. *Qualitative Inquiry, 17*(6), 511–521. doi:10.1177/1077800411409884

Ümarik, M., Loogma, K., & Tafel-Viia, K. (2014). Restructuring vocational schools as social innovation? *Journal of Educational Administration, 52*(1), 97–115. doi:10.1108/JEA-08-2012-0100

UNFCC. (2023). *Five Reasons Why Climate Action Needs Women.* UNFCC. https://unfccc.int/news/five-reasons-why-climate-action-needs women#:~:text=Particularly%20in%20developing%20countries%2C%20the,risk%20to%20their%20personal%20safety

U. N. Women. (2022). Explainer: How gender inequality and climate change are interconnected. UN. https://www.unwomen.org/en/news-stories/explainer/2022/02/explainer-how-gender-inequality-and-climate-change-are-interconnected

World Bank. (2016). Odisha -Poverty, growth and inequality. India state briefs. Washington, D.C.: World Bank Group. https://documents.worldbank.org/curated/en/484521468197097972/Odisha-Poverty-growth-and-inequality

www,milletsodisha.com

Xaxa, V., & Ramanathan, U. (2014). *Report of the high level committee on socioeconomic, health and educational status of tribal communities of India.* Ministry of Tribal Affairs, Government of India. https://cjp.org.in/wp-content/uploads/2019/10/2014-Xaxa-Tribal-Committee-Report.pdf

ENDNOTES

[1] The Global Hunger Index 2019 ranking is based on three leading indicators -- prevalence of wasting and stunting in children under 5 years, under 5 child mortality rates, and the proportion of undernourished in the population. https://www.globalhungerindex.org/pdf/en/2019.pdf

[2] GHI ranks are based on 4 key indicators -- undernourishment, child mortality, child wasting and child stunting. www.economictimes.indiatimes.com/articleshow/66226877.cms?from=mdr&utm_source=contentofinterest&utm_medium=text&utm_campaign=cppst

[3] Food security refers to sufficient availability of food for direct consumption, and also the purchasing power to buy food and access to food that meet the food and nutritional requirements for keeping the body in proper health. Food insecurity is lack of the above. Accessed from: FAO, 2001a: Food Insecurity in the World 2001. https://www.fao.org/3/a-y1500e.pdf. Food and Agriculture Organization of the United Nations, Rome, Italy, 8 pp.

Chapter 5
ENGO Communication Management Towards Climate Change:
Solutions in Port City

Valentina Burkšienė
Klaipeda University, Lithuania

Jaroslav Dvorak
iD https://orcid.org/0000-0003-1052-8741
Klaipeda University, Lithuania

ABSTRACT

This chapter explores environmental NGOs' (ENGOs) communication management towards climate change issues. E-communication was researched in the Lithuanian port city of Klaipeda. ENGOs residing in the city unsuccessfully fight against the polluting port business. The authors argue that communication management could bring success. A case study to research the e-communication of Klaipeda ENGOs was used. Content analysis of Facebook (FB) profiles of ENGOs and semi-structured interviews were chosen as approaches. The research revealed a weak understanding of communication management that impacts common e-communication practices and misunderstanding of this instrument for effective networking and joint actions.

INTRODUCTION

Every society deserves to live in a clean unpolluted area, but in times of rapid climate change, it becomes a real challenge. There are worldwide agreements and various regulations that help to organize modern life with regard to sustainability. Almost two hundred countries have agreed on sustainable development goals and Agenda 2030 is prepared that should help to fight climate change and reach better conditions for every living in the Globe. Society itself from the individual point must protect the environment, but

DOI: 10.4018/978-1-6684-8963-5.ch005

many problems are born by the industry that businesses do in the city area. According to Korten (1990), the power of social movements was ignored for decades with the attention focused on money rather than the quality of social life and development. Nevertheless, people have become more organized and active, the society still stands in the conflicts against environmentally harmful businesses and the political power that supports them. On the one hand authorities (local in particular) should fulfil the requirements of businesses that support their election campaigns, on the other hand, they have won a mandate to serve the public, and thus must protect society's rights (Powel 2013; Palttala et al. 2012).

In this complicated situation, nongovernmental organizations (NGOs) can help to strengthen society's power as an initiative to engage society comes rather from NGOs (bottom-up) than government officials. NGOs strategically focus on public policy problems, implementation activities, and cooperation areas. Environmental issues are the area of focus of environmental NGOs (ENGO), which can achieve positive changes in many global and local ecologic issues.

As such, communication experts highlight the importance of focusing on target audiences for releasing appropriate missions and sending effective messages through various communication channels. The model of governance, however, declares not the simple hearing of society's voice but its involvement in decision-making. Therefore, Hue (2017) proposes a two-way symmetrical communication that engages all parties as equal partners seeking common relations based on mutual understanding and needs. The success of such communication would be determined by clearly formulated strategic priorities, vision and mission of communication activities. It is also important to monitor stakeholders' perceptions and to understand the way of thinking as well as the behaviours of various groups (Palttala et al. 2012). Social media provides opportunities for everyone to message about their goals and achievements, or in other words to communicate to society and officials.

Agreeing on the effectiveness of two-way symmetrical communication management, we perceive that well-managed bottom-up communication can foster ENGO to become more powerful in networking with government and other stakeholders which means more sustainable governance, and argue that the success of such communication depends on its management.

Research context and problem. The future is foreseen as uncertain and challenging due to population growth that influences rapid economic growth and higher greenhouse gas emissions. Moor (2009) argues that climate change is the defining environmental problem of this century. Previous studies revealed a direct interrelation between air pollution and climate change. Orru, Ebi and Forsberg (2017) argue that both factors affect each other (climate affects air quality and vice versa) and at the same time, they both reduce the quality of human life. Both factors also can cause serious health problems. Air pollution is named a serious problem, especially in industrial territories and cities (Orru, Ebi and Forsberg 2017; D'Amato et al. 2015; Michie and Cooper 2015; Moore 2009; Bulkeley and Betsill 2003).

The literature (Fanø 2019; Marin et al. 2017; Hricko 2012; Burskyte, Belous and Stasiskiene 2011; Belous and Gulbinskas 2008; Sharma 2006; Bailey and Solomo, 2004) assumes, specific environmental and air pollution issues in port cities despite their obvious impact on economic well-being. The ports are defined as very complex and growing pollution sources. Diesel emissions with particulate matter, volatile organic compounds and nitrogen oxides are named as the main pollution sources that require firm regulation both globally and locally. Other serious sources of pollution include dredging and dredging sediments, cargo handling equipment, in-port heavy-duty trucks and locomotives as well as particles from trade-related activities (Sharma 2006; Bailey and Solomon 2004). The air pollutants according to D'Amato et al. (2015) can be divided into i) natural (desert sand, sea salt, wildfires and volcanic ash) and ii) manmade (wood or biofuel for heating and cooking, coal and oil excavation, vehicle exhaust, particle pollution). The second one proves the impact of every human on climate change and forces us to agree with Bulkeley and Betsill (2003) that climate change is not only a global but also a local issue with cities being central to the responsibility for air pollution (especially industrial cities with consideration of manmade or industrial pollution).

In this study, we have chosen to research the communication of ENGOs in the only seaport - Klaipeda, Lithuania. Klaipeda is the third-largest city in Lithuania, with a population of about 150 thousand. The city has been facing various environmental issues. Burkšienė et. al., (2020) found out that Klaipeda, together with the capitals of the Baltic States, is characterized by very weak cooperation in the field of environment. In most cases, those environmental issues have been identified or started to be debated in the public sphere not by local politicians but by both active residents and local NGOs. Some of those problems have been lasting since the time of the USSR. After Lithuania regained its independence in the early 1990s, the first green organizations started organizing protests against the companies - polluters, that just have been privatized. In the face of the first shocks of the market economy, however, when corporate bankruptcies began, rising unemployment, in turn, the arguments of green organizations were weakened as society preferred livelihoods instead of another "empty space". Later with the increasing economic well-being, environmental issues have risen again. The latest research (Kotseva-Tikova, Dvorak, 2022) still highlighted the issue of climate change and pollution in this port.

The dominant environmental issues of Klaipeda are related to the Seaport. A brief online press publication review about air pollution in Klaipeda (Kauno diena 2021 ; Klaipėda aš su tavim 2021 ; Rumšienė 2021a, 2021b) revealed that there are two main issues concerning the port business activities: air pollution and the radiating noise. Residents living nearby the port suffer from the excess particulate matter permanently. Until May 2021 there was no specific regulation, therefore only after that date did port authorities equip special means for measuring the concentration of particles in the air within the territory of the port. Still, residents complain about the noise radiating from the port, especially during the night and weekends. The general director of the port states that they are aware of the complaints of residents and have strategic objectives prepared for air pollution diminishing. Issues concerning the noise, by the director, were addressed to the contractors who have permission for work on 7/24 but with no extension of noise limits. Strict deadline terms of ongoing EU projects were also provided as defensive explanations.

Rapidly growing cargo volume and increasing vessel capacities have intensified Klaipeda port activity which has made a significant environmental impact because air and water pollution are directly interrelated to climate change. Therefore, several solutions (engineering, monitoring, etc.) are necessary for balanced and sustainable development not only of the port but even of the port city (Burskyte, Belaus and Stasiskiene 2011; Belous and Gulbinskas 2008). There can be no single solution, however, to such a complex issue, therefore engagement of society is necessary for consensus achievement and moving the port as well as the whole port city toward a sustainable model. Orru et. al., (2017) and Moor (2009) propose that the negative effect of climate change and the improvement of air quality can be decreased by the new attitude to pollutant emissions, climate negotiations and regulatory interventions. Following these proposals, in the public model of governance, all climate regulatory interventions require to be negotiated with society first. Community-based organizations and environmental justice groups should be treated seriously and engaged in the decision-making process. It is supposed that, if society is actively engaged in decision-making, its attitude and behaviour will change and be more friendly towards climate issues.

Several active ENGOs have drawn the attention of residents, city authorities and businesses to the environmental problems (i.e., hard particles in the air, port handling, constant odours in certain areas) influencing climate change. The NGO communication takes place starting with active actions in protest against the city authorities and/or companies and constantly provoking the population to start acting on social networks or electronic media for the benefit of the city, families and children.

Klaipeda city authorities (as they refer in the city strategy for 2021 -2030) recognize the value of communities and NGOs, stating that an active and inclusive community is an important condition for cooperation among residents, businesses, science and government for the creation of governance in the port city development (Klaipeda city municipality 2021). Specific attention in the mentioned document is focused on the city's ENGOs while recognizing them as most active in the formation of civil society and putting a valuable impact in strengthening democracy and well-being. Actually, neither the population is as active as it is supposed to be, nor the local government is fast to engage and participate in the solution of global and local environmental issues. Therefore, we argue that these ENGOs lack specific knowledge and skill in communication management and due to this they do not develop appropriate (i.e., two ways horizontal and vertical) communication that would help ENGOs to achieve more effective networking in the context of climate change. In this study, we research e-communication as the modern communication type and focus on the e-communication of Klaipeda city ENGOs analyzing their experience in communication management. First, we propose the theoretical model of communication management. Empirically we researched e-communication practices of the respected ENGOs related to climate change issues and reveal limitations in their e-communication management.

As there can be more port cities with similar issues in countries of a young democracy, we perceive that the case of Klaipeda port city will reveal some commonalities and propose general solutions for the improvement of e-communication management concerning climate change.

Methodology. Literature analysis was used to highlight important aspects related to climate change in port cities, which could be in the area of communication focus by ENGOs. Analysis of management theories helped to construct a theoretical framework of communication management. This framework helped to evaluate the communication process of researched ENGOs. To understand the e-communication management process a case study of three Klaipeda port city ENGOs was utilized. First, the content analysis was used as a methodological approach, which helped to logically classify and categorize the communication data collected from Facebook. Second, the semi-structured interviews with leaders of respected ENGOs helped to reveal their attitude to communication management, e-communication practice and communication networking with other stakeholders.

THEORETICAL BACKGROUND

Climate Change Mitigation in Port Cities

According to the literature analysis, we argue that several solutions aimed at reducing the port's impact on climate change can be proposed. Such categories are 1) health impact assessment; 2) stakeholder-oriented programs and action plans; 3) smart city; 4) good local governance.

Air toxins impact the health of residents living near ports and major sea transport corridors and can cause several diseases such as asthma, bronchitis, cardiovascular disease, lung cancer, and even premature death. Therefore, the authors (Fanø 2019; Sharma 2006; Bailey and Solomon 2004) mostly emphasize the effectiveness of regulation and regulation techniques (including appropriate assessments) that put the responsibility on decision-makers at all levels in order to seek a "greener" seaport. According to

Dvorak (2015, 132): in democratic governments, impact assessment opens the decision-making process to stakeholders because it is based on consultation and is much more accountable to the citizens. Ports themselves make assessments on air pollution and climate impact, but these are criticized and independent health impact assessments are proposed on the contrary before deciding on potential infrastructure projects for port expansion or changing its developing policies (Hricko 2012). Independent assessments guarantee more transparency and the possibility of citizens' engagement.

Best practice teaches that stakeholders' involvement in the development of relevant programs or action plans is the most appropriate way for seaports to properly address the challenge of climate change. For instance, the program of the Port of Rotterdam (Netherlands) states that port development with the inclusion of port reconstruction will be designed to be climate-proof, and climate-change assessments will be integrated into the port's spatial planning (Becker et al. 2013). The Port of San Diego (California) has included stakeholders' input and responsibility for emergency response, critical utility protection, and stormwater drainage in its plan (Becker et al. 2013).

Some scientists (Deng, Zhao, and Zhou 2017; García Fernández and Peek 2020) believe that as cities increasingly face climate change, they need to use the tools of a smart city. It is recognized that smart city governance based on big data, cloud computing and other next-generation technological tools will replace urban governance and bring more sustainability. As Ejdys (2020) argues, trust in e-technology solutions is a necessary factor for their effective use in the future. The characteristics of a smart city become a critical precondition for the emergence of smart NGOs that are equipped with various information, communication, and mobile technologies and can share information, participate, and cooperate with the government in the innovative delivery of services (Ho 2016; Burkšienė et.al., 2019).

Indeed, that good local governance should matter and be very effective in urban industrial cities if we consider a democratic world. Governance means an active engagement of non–state actors and various stakeholders (from the public, private and voluntary sectors) in the decision-making process while working in networks of close partnership. It is the partnership environment that helps people make sure that there is a common interest between them. Based on these statements we argue that legal regulations of the smart port city authorities and model of governance can make a positive impact on polluting port city businesses.

Environmental NGOS As Catalysts for Climate Change Action

Political and bureaucratic structures are traditionally well organized and follow common rules that help them to fulfil situational demands and to navigate even in pressuring situations (Bloodgood, Tremblay-Boire and Prakash 2014; Bleyen et. al. 2017; Keulemans 2021). On the contrary, the society (in particular, that from a new democracy) has no strong traditions to act jointly in purposive groups or in nongovernmental organizations. Nevertheless, Urry (2015) states that society plays an essential role in actions concerning the changing environment, as people use to fight for or against the changing aspects locally or globally. Burksiene and Kazdailiene (2023) summarizing the research of other authors found out that the problem of climate change is not a matter of direct environmental protection, but it is closely related to the goals of sustainable development, which in turn requires fast, effective solutions by national and local authorities.

The situational theory states that the public differs from each other, but they can unite to feel solidarity in common issues and be able to react similarly. These individuals can remain institutionally unorganized without knowing each other or they can set up a formal organization that becomes a permanently acting citizens' initiative or a non-profit organization (NGO) that is primarily seen as a social development agency aiming to prevent any further issues (Hue 2017; Raupp and Hoffjann 2012). NGOs are appreciated as significant units in contemporary society because of the higher degree of institutionalization in comparison to the sporadic movements and because of the transparency and trust that support their work (Herranz de la Casa, Alvarez-Villa and Mercado-Sáez 2018). Any NGO is also understood as a platform to convey public reactions, ideals, desires or arguments which are expressed in their mission, goals and priorities. Korten (1990) appeals to the management skills that are essential for the voluntary organization supporting people's movements. Such an organization should energize a self-managing network and be skilful in the strategic positioning of its resources, ideas and values.

According to the theory of action, 21st-century society, reacting to common development failure, engages in volunteering actions that focus on support of global social vision such as the Global Agenda (Korten 1990). The most significant global issue concerns sustainable development that has emerged from the problem of changing climate and was first declared in the report Our Common Future by the World Commission on Environment Development (1987) or also well known as the Brundtland report (Bulkeley and Betsill 2003). Scientific communities and ENGOs are understood to be powerful in the decision-making process concerning environmental politics. They advocate a participatory, bottom-up approach to managing the environmental problems in the conception of a sustainable city (Bulkeley and Betsill 2003). ENGOs are simply entitled as Greens or Green Movements (Telesiene and Kriauciunaite 2008). Different studies analyze these movements globally and locally and rank them as being of more or fewer radical forms (Tumulyte 2012).

Konstantinaviciute (2003) reductively presented the main stakeholders in the analysis of climate change mitigation policies in Lithuania with ENGOs mentioned as one of them. According to the author, together with the media and trade unions, these ENGOs would have to represent the public interest, inform, educate and consult society. Balunde, Perlaviciute and Steg (2019) suppose that Lithuanian ENGOs need to educate about environmental issues and promote civic environmental activism. On the other hand, Vavtar (2014) in his research proved that the effectiveness of environmental lobbying by Lithuanian ENGOs is very low. Consequently, the author proposes for ENGOs to find ways to increase environmental activism not only by informing but also by educating.

Some articles propose communication as effective means for proper information dissemination, promotion of environmental ideas and education (Herranz de la Casa, Alvarez-Villa and Mercado-Saez 2018; Rajhans 2018; Hey 2017; Palttala et al.2012; Raupp and Hoffjann 2012; Liu 2012). Examples of multifaceted communication illustrate the impact of ENGOs on climate change decision-making. As Corell and Betsill (2001) found, the following indicators of communicative success can be identified: 1) presentation of written information; 2) presentation of oral information; 3) provision of advice; 4) the possibility to set the agenda; 5) the ability to incorporate the text into the agreement. These identifications (provision of information, setting the agenda) represent one-way communication and lead to criticisms of the ability of ENGOs to use electronic communications (representing two-way vertical and horizontal communication) that would impact the development of governance (Nulman and Özkula 2016) and increase ENGO's power when acting against climate change.

Conception of Communication Management

Rajhans (2018) warns that communication can be problematic because the communicating NGO deals with very different people whose attitudes and behavior are unpredictable and can change over time.

Also, individuals and groups are not always able to define their interests, and the media and powerful interest groups can fool them. Therefore, different types of dialogues for different stakeholders are useful and can change and define the real interests. In order to achieve acceptance, communication should be adapted according to the needs and perceptions of particular stakeholders (Palttala et al. 2012).

As NGOs need to work in close relation with communities and governments to achieve their goals and fulfil their missions, they should plan communication both vertically and horizontally (Andres 2011) and create dual or two-way symmetrical communication that results in productive feedback (Hue 2017; Mihai 2017). By using one way or descending communication NGOs can lose close contact with reality and miss real understanding of specific problems they seek to solve. On the contrary, purposive two-way communication allows them to send and receive the right messages from the right sources at the right time.

Professional communication management is important because of the complexity of communication. Raupp and Hoffjann (2012) explain that communication management is beneficial when the problem cannot be solved employing routine action. Rajhans (2018), Herranz de la Casa, Alvarez-Villa and Mercado-Saez (2018) and Palttala et al. (2012) argue the importance of strategic planning, vision, mission, objectives, functions, monitoring and measuring the progress as valuable components of communication management. Muszynska et. al., (2015) proposes that it is important to know the roles and working methods of other stakeholders or society to strengthen cooperation ability. All these components of managed communication create several alternatives of action to discuss and evaluate them for the final decision to be agreed upon (consensus achieved). Unfortunately, small NGOs mainly work voluntarily and do not give much attention to proper communication (Muszynska et. al., 2015). The development of communication strategy is weak, because much attention is paid to the technical aspects of communication (Liu 2012)

Comfort and Hester (2019), Ruehl and Ingenhoff (2015), Pavlovic et al (2014) state that NGOs are very interested in developing social media for communication while creating differentiated and personalized networks for communication with other users. Various researchers propose different communication forms and types: traditional (mass media, school curricula, major media events); digital (recorded media, social networks) etc. (Korten, 1990; Palttala et al. 2012; Ruehl and Ingenhoff 2015). According to Hue (2017) and Pavlovic, Lalic and Djuraskovic (2014), it is very beneficial for social movements (and thus for NGOs as well) to use electronic means for effective and efficient communication. We know relatively little, however, about how ENGOs turn these forms and types of communication into actions. Advanced technologies allow the rapid creation of social networks. Virtual social space provides various platforms for online communication. Therefore, citizen groups receive the channels to speak out on issues relevant to them. Continuous extension of technological features proposed many applications and social media with the most users like Facebook, Twitter and video portal YouTube that can increase public acceptance (Ruehl and Ingenhoff 2015). Social media is a relevant channel to increase the transparency of actions and empowers NGOs to raise their voice to receive sufficient public support for confronting issues, but unfortunately, this modern tool is mostly used for one-channel messaging rather than for dialogues and discussions.

Although the success largely depends on the NGOs' communication competencies, unfortunately, Hue (2017), Muszynska et. al., (2015), and Palttala et al. (2012) found that there is not much scholarly work on the communication concept of NGOs. Reasonably we also lack research in the field of communication of ENGOs. Nonetheless, we can state that communication practices are associated with most of the success dimensions: strategic priorities, vision and mission of communication activities, understanding

Table 1. Dimensions of successful communication of ENGO

Dimension	Description
Volume	Reflects how frequently messages of NGOs are shared by any other social media users and other NGOs in particular. More shares mean higher volume
Topic/ valence	Reflects how long the message stays on topic with the support of NGOs position, because counter-public opinions can emerge that are difficult to be controlled and can even upend NGO's goals
Participants	Refers to: (i) NGOs target sympathetic audience by responding to them and engagement; (ii) attracting the supportive news media; (iii) counter messaging individuals or their groups Refers to: who is speaking, in terms of posting NGOs' content, and whom their messages are directed, in terms of who is "tagged" in the post

and monitoring perceptions of society, and two-way communication in networks that are managed by ENGOs. These dimensions will be used as criteria in the research.

Agreeing on the above dimensions of communication management we decided to additionally analyse articles directly related to ENGOs with the perception of more dimensions for successful communication to be found. Comfort and Hester (2019) propose that the success of digital communication of ENGOs in social networks largely depends on three dimensions: volume, topic–valence, and participants (Table 1).

Reviewed studies confirm that communication of ENGO's vision is really important. And it needs to be *formulated and expressed via all available communication channels* (with the inclusion of social and mass media). To increase communication effectiveness, ENGOs should be *creative and skilful when using untraditional ways for problem solutions* and changing the behaviors towards more climate-friendly. These dimensions as well as the *three dimensions proposed by Comfort et al* (ibid) complement the package of successful communication. Relying on them we propose the list of dimensions of ENGO's successful e-communication management:

1) clearly formulated strategic priorities,
2) vision and mission of communication activities,
3) monitoring perceptions of all stakeholders and society,
4) understanding the way of thinking and behaviors of various groups,
5) two-way communication channels (vertical and horizontal),
6) creativity in unordinary problem solutions,
7) knowledge and skills on how to engage and change behaviors of society members to be more climate-friendly.

Decision to apply communication management can be likened to a project which normally has to pass three phases starting from the primary idea and ending with its permanent implementation, leading to a more effective outcome (Gido, Clements and Baker 2018). There is only one difference from the project: absence of the closing phase because communication obviously should be developed in the permanent cyclic management mode. Following this logic, we will have the cyclic framework consisting of seven steps that are coherent to the specific phase (Figure 1).

Agreeing with Comfort and Hester (2019) we argue that practical employment of these e-communication management steps (dimensions) can lead to higher communication volume, stronger valence and support of top message and more engaged participants acting for climate change.

METHODOLOGY

Modern social research is becoming more localized and episodic due to the huge flows of information. Naturally, the experience of NGO communication management as a research approach is constantly constructed and reconstructed: the present research interprets known concepts, constructs new climate change response categories for seaport cases, reveals new opportunities for NGO communication and language reflection. This study covers the following conceptual explications of ENGO climate change communication: 1) ENGOs have difficulty surviving in market conditions, lack operational experience, weak advocacy and communication skills; 2) many ENGOs have a simulation nature: their activities are registered but frozen, so no significant actions are provided to the public; 3) the activities of most ENGOs are chaotic: poorly strategic and networked, and therefore episodic.

Data in Klaipeda municipality's strategy for 2021 - 2030 refers to a total of 388 NGOs in the city. But our revision of a sample of the NGOs by the types revealed that some of them are mentioned repeatedly. This means, that there are no right numbers of NGOs (with environmental NGOs among them) announced statistically.

According to the website of the Lithuanian Ministry of Environment[1] there are three voluntary environmental organizations in Klaipeda, but we found that one of them is not active and replaced it with-the Association of Klaipeda Communities that also acts in the environmental field (Table 2).

In the frame of the current research the method of case study was used to evaluate social media (FB) activities and findings from interviews of three Klaipeda city ENGOs. Data on the FB of the respected ENGOs were analyzed in the period from 1 January to 15 July, 2021.

Relying on dimensions and their definitions in Table 1, we developed research criteria as categories and subcategories (Table 3) for FB analysis of respected ENGOs, which helped to reveal and assess their communication.

Table 2. Klaipeda environmental non-government organizations

ENGO	Brief characteristic and link to NGOs information
Association "Klaipėdiečių iniciatyva už demokratiją ir ekologiją" (KIDE) (Klaipeda residents initiative for the democracy and ecology)	https://www.facebook.com/KIDEKLAIPEDA Mission - uncontaminated chemical and acoustic pollution of the living environment. Activities: advocacy; organization of research; participation in the processes of formation and implementation of environmental and public health safety policy; education.
Association "Klaipėdos žalieji" (Klaipeda Greens)	https://www.facebook.com/profile.php?id=100008449032535 A broad-based environmental organization providing proposals on landscaping, water, soil, air and noise pollution, soil, waste issues.
"Klaipedos bendruomeniu asociacija" (the Association of Klaipeda Communities)	https://www.facebook.com/Klaipedosbendruomeniuasociacija The aim is to unite the residents into a unified network of communities and speed up the application of the principles of participatory governance in the local government system, thus ensuring sustainable urban development, participation of involved citizens in municipal decision-making and their growing satisfaction with life in the city.

Table 3. Categories and subcategories for successful e - communication of ENGO

Category	Subcategory and explanation
1. Volume	1.1. Numbers of sharing of NGOs messages by any user (calculation of shares of every message; more shares more volume)
2. Topic/ valence	2.1. Days for the message supporting NGOs position staying on top 2.2. Number of messages with counter public opinions for a particular message (these messages are not under the NGO control and can upend its goals)
3. Participants	3.1. Numbers of sympathetic audience of NGO (NGOs responds to them and invitations to actions) 3.2. Numbers of supportive news media attracted (reactions or shares by news media on the messages of NGO) 3.3. Who is posting NGOs' content, and who are target audiences? (Numbers of direct tags and who is "tagged" in the post)

Content analysis has proven to be a suitable approach in similar studies (see Comfort and Hester 2019; Herranz de la Casa Alvarez-Villa and Mercado-Saez 2018; Rajhans 2018; Pavlovic, Lalic and Djuraskovic 2014; Burksiene, Dvorak, 2022) and was used in our case as suggested by Krippendorff (2013) and Mayring (2014). This approach helped us to analyze the information available on FB sites of respected NGOs and to logically classify and interpret all data collected.

A semi-structured interview was the second approach used in our research in order to test how the proposed framework of successful e-communication management (Figure 1) is applied by respected ENGOs of Klaipeda port city. The questions were formulated not to directly reveal the opinions in "question-answer" form. The questions served as guidelines and helped to develop more comprehensive answers. Content analysis of interviews was used for searching statements and phrases related to every step of the model.

The structure of questions reflected every phase:

Figure 1. Model of ENGOs effective e-communication management

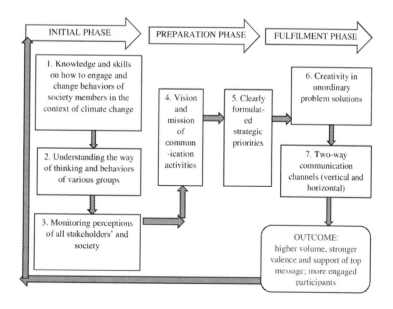

Table 4. Data of e - communication of Klaipeda ENGOs

Subcategory	NGO1	NGO2	NGO3
Total number of messages	64	147	136
1.1. Numbers of sharing of NGOs messages by any user (calculation of shares of every message; more shares more volume)	212	169	172
2.2. Number of messages with counter public opinions for a particular message (these messages are not under the NGO control and can upend its goals)	2	9	4
3.1. Numbers of sympathetic audience of NGO (NGOs responds to them and invitations to actions)	4	50	26

Initial phase: e - communication practices and understanding of the value of such communication for achieving the goals (communication channels, ways and message design; purpose and perception of and from communication, etc.).

Preparation phase: management of strategic communication (importance of mission, vision and priority goals and interrelation to the climate change; target audiences).

Implementation phase: Features of governance (interaction with other NGOs, scientific institutions, government and different stakeholders; forms and types of interaction).

FINDINGS

Analysis of FB data of three ENGOs revealed that the criterion 2.1 (Days dedicated to supporting the NGOs message position) could not be calculated because the activity of comments, shares and counter opinions is quite low. Also new messages appear very often and overlap previous ones (i.e., NGO2 sometimes announces up to 4 new messages a day). No single media was attracted by any ENGO. On the contrary, most messages of all ENGOs are based on the sharing of information from different online media sources. No tags were used in messages by any ENGO. And all messages are published by only one person - the administrator (on the name of ENGO) with rare extra comments of NGO 1 and NGO 3 leaders (as individuals) who, we guess, are indeed the same administrators.

Data interpretation of message shares can be challenging because in many cases ENGOs shared their messages themselves too and it was the only share at all. Most activism in e - communication is developed by Klaipeda Greens (NGO2) with the largest counter public as well as sympathetic audience (see Table 4). But dominating messages of Greens contribute to protection or even saving trees, parks and forests. Only ten messages were dedicated to air pollution issues and one to noise pollution. Greens also do not put much attention to the port problems.

In general, we argue that Klaipeda ENGOs do not apply e - communication professionally (do not manage it) and thus do not reach the real attitude and support from other stakeholders, public. Their voice is too weak in order to engage other stakeholders, society as well as local or national governments. The weak communication is not interesting for national or even local mass media too.

Interview findings are described following the phases of the model.

Initial phase. All respected ENGOs agree on the importance of e - communication All three respective ENGOs agree on the importance and usefulness of e-communication (emailing, phoning, messaging,

ENGO Communication Management Towards Climate Change

etc.) and use it daily. Covid - 19 even fostered the development of more e - tools (Zoom in particular) and employed them intensively.

All of them use the FB platform for communication, and no one has built a personalized web page yet. The informants indicate lack of time, human and financial resources as the reasons. The issue of limited resources and time can manifest in the way messages are designed. Only one informant ingeniously observed the practices of designing their messages:

<...>Yes, I personally think this is very important and I urge my colleagues from time to time. We take the time to talk to someone like a message because we have had a super network of participants since I recently created it<...>.

The informant has even prepared the rules for a super active group of members who can administer messages by specific fields (engagement, events, international relations, etc.). It is worth mentioning however, that nobody uses regular e-newsletter, neither circulates cartoons nor jokes and links to the campaign website by email, which is commonly used in the cases of the UK and Australian ENGOs (Hall and Taplin 2007).

Nevertheless, the ENGOs in question do not think about wider political influence in their messaging because they do not form coalitions with other NGOs. That was confirmed by informants when stating unwillingness to have any serious cooperation or coalition with other (E)NGOs in the city. In contrast, the informants stated that everyone wants to be a leader and act separately. Some anger was felt when talking about other colleagues. In alternative ENGOs classifications, such organizations are described as confronters and supporters of the politics of blame (Alcock 2008). Leadership manifests itself, however, through who the municipal administration will talk to about environmental issues, but not through the fragmented informing about alternative energy sources in the city, asking the population to reduce electricity consumption, and striving for lower emissions.

All respected ENGOs briefly mentioned the intention to educate society by provoking them with the messages or simply messaging scientific messages in simplified and easier to understand ways. But these actions are directed only to the audience that are already subscribers and followers on FB. Nothing was told about education related to climate change or increasing the numbers of the audience. In addition, it is important for education whether the residents trust ENGOs, because only then will they be able to carry out their mission (Lee, Johnson and Prakash 2012). Of course, we do not answer this question in the current study, but the small number of followers on the social network may indicate that the ENGOs in question lack the trust of the residents.

Preparation phase. Interviews with the leaders of three ENGOs revealed that vision and mission are not treated as important. No one could express it in words. Mission and vision can be perceived from the definitions of the priorities or goals that can be related to climate change but are not directly expressed. Environment protection is mentioned instead. This shows that currently Klaipeda ENGOs are failing to contribute to a deeper understanding of climate change and will not translate this understanding into a need for a government response.

The qualitative data proves that the main audience of all respected ENGOs' consists of residents and communities, local government and businesses. The informants claimed, however, that their main emphasis has been put only on municipality authorities and environmental bureaucrats. Two ENGOs complained what was already supposed from the FB research, that they wish to reach the interest of national mass media (*it depends on mass media favor to be published or not*), but fall in trouble with this and Greens having that problem on the local level too.

Implementation phase. The attitude of respected ENGOs to the cooperation with academics is not equal but seems not to be seriously treated. Two informants seek some help and support from academics (i.e., Klaipeda University (KU) or researchers - practitioners) when the third informant in contrast seeks to propose their help and support to KU. Two respective ENGOs communicate research findings in their FB accounts in an attempt to ensure their legitimacy in the eyes of a specific audience and to maintain the impartiality of researchers.

And no one is thinking about real cooperation with the local government bodies. *Even though there are some positive and good politicians it is too dangerous to go in contact with them*. The informants rather expressed their complaints towards the municipality and dissatisfaction of provision of finances and other support for NGOs. Contacts with the municipality are in the written form of complaints, resolutions, proposals or even conflict solutions in the court. All respective ENGOs have got support from attorneys and lawyers. Their main goal is to be more effectively engaged in decision making of environmental and infrastructural issues. All researched ENGOs also wish to be involved in performance of Klaipeda NGO Board as every of them think that they are skilful enough to make positive changes both in the Board and in its decisions. ENGO 2 proudly expressed their higher activity and better success (in comparison to other ENGOs) when fighting and winning in the court. This can indirectly prove the Greens being more aggressive (Tumulyte 2012).

One informant mentioned about their actions toward legal cooperation with business enterprises:

<...> *Undoubtedly, as much time as we have left from the main activities, we dedicate to making contacts with business, encouraging it to operate responsibly. And one of the biggest our achievements is - formalization of our contacts with the Klaipeda Confederation of Industrialists. as they also want to cooperate*<...>.

The other informant precisely told how they received support from Klaipeda port authority for their cognitive trip to Ventspils (Latvia) and about plans to apply to them for help rather than to the municipality in the future.

Interviews revealed that all three ENGOs do not follow e - communication steps as it is proposed in the theoretical framework. They use just the one-way e - communication channel mostly for informing the audience, consulting or activating. Conflicts with governmental institutions, complaints and a negative attitude towards real cooperation limit searching for creative solutions in unordinary issues.

Communication, in general, has a vertical bottom-up (one way) expression with too little attention paid for developing two way vertical and horizontal networking. ENGOs' e-communication on FB is not used to emphasize climate change policy and problems neither locally nor globally. Aspects covering the use of renewable energy sources, reduction of automobile emissions, circular economy and energy imports from the Astravets nuclear power plant in Belarus are not considered. Although much attention is paid to the problem of port and industrial pollution, the issues of Klaipeda Liquefied Natural Gas Terminal are not articulated as climate change problems. In spite of their communication being related to air pollution or tree felling, the ENGOs do not strive to educate and change the behaviors of the audience concerning climate change. Although according to Dolšak (2013) the precise goal of ENGOs is to collect, systematize and disseminate knowledge on that global issue.

CONCLUSION

In sum, it can be stated that air pollution and climate change are interrelated and port cities play a specific role in this case. Industrial cities are named as being mostly responsible for air pollution with man-made pollution dominated. Ports are very complex in their business activities and port cities are among the growing pollutants having a variety of pollution sources despite their valuable impact on the city's economic well-being. Air pollution in the port cities negatively affects living conditions and impacts the health of residents.

ENGOs of Klaipeda can be treated as examples of e-communication in the post-soviet country with a tradition of competition rather than inter cooperation. Their e-communication does not create two-way communication channels and strong networking. Therefore, ENGOs are not powerful enough to engage bigger numbers of both residents and stakeholders in their activities. ENGOs need to be more creative in message designing to reach the minds of society and government. For becoming real *bottom-up power, ENGOs* should gain and develop lobbying and PR skills that would help to find other specific forms to communicate with the government and municipality. Strengthening the cooperation and collaboration with the academy in the field of climate change issues could develop the right messages to appropriate groups. Knowledge of global issues of sustainability (including climate change) is necessary for a contemporary ENGO to be interesting for social and mass media.

Further complex research is needed in the field as well as comparative analysis among various ENGOs. Our research can be treated as pilot research due to a shortlist of researched ENGOs. Despite the big number of NGOs announced in the Klaipeda strategy, we found only three of them acting in the field of environmental protection. And this is the main limitation. The other limitation is related to the pandemic lockdown that influenced the interview process. Due to restrictions for direct contact, we had to do interviews remotely and were not able to unwind them in full scale.

REFERENCES

Alcock, F. (2008). Conflicts and coalitions within and across the ENGO community. *Global Environmental Politics, 8*(4), 66–91. doi:10.1162/glep.2008.8.4.66

Andres, S. (2011). Communication, The Essence of Management of A Nonprofit Organization. *Annals of Eftimie Murgu University Resita, Fascicle II. Economic Studies, 1*(1), 121–130.

Bailey, D., & Solomon, G. (2004). Pollution prevention at ports: Clearing the air. *Environmental Impact Assessment Review, 24*(7-8), 749–774. doi:10.1016/j.eiar.2004.06.005

Balunde, A., Perlaviciute, G., & Steg, L. (2019). The relationship between people's environmental considerations and pro-environmental behavior in Lithuania. *Frontiers in Psychology, 10*, 2319. doi:10.3389/fpsyg.2019.02319 PMID:31681111

Becker, A. H., Acciaro, M., Asariotis, R., Cabrera, E., Cretegny, L., Crist, P., & Velegrakis, A. F. (2013). A note on climate change adaptation for seaports: A challenge for global ports, a challenge for global society. *Climatic Change, 120*(4), 683–695. doi:10.100710584-013-0843-z

Belous, O., & Gulbinskas, S. (2008). Klaipėda deep-water seaport development. In Conflict resolution in coastal zone management (Environmental Education, Communication and Sustainability). Peter Lang Pub Inc.

Bleyen, P., Klimovský, D., Bouckaert, G., & Reichard, C. (2017). Linking budgeting to results? Evidence about performance budgets in European municipalities based on a comparative analytical model. *Public Management Review*, *19*(7), 932–953. doi:10.1080/14719037.2016.1243837

Bloodgood, E. A., Tremblay-Boire, J., & Prakash, A. (2014). National styles of NGO regulation. *Nonprofit and Voluntary Sector Quarterly*, *43*(4), 716–736. doi:10.1177/0899764013481111

Bulkeley, H. B., & Betsill, M. M. (2003). *Cities and Climate Change. Urban sustainability and global environmental governance*. Routledge.

Burkšienė, V., & Dvorak, J. (2022). Local NGO e-communication on environmental issues. In *The Routledge Handbook of Nonprofit Communication* (pp. 269–278). Routledge. doi:10.4324/9781003170563-32

Burksiene, V., Dvorak, J., & Burbulytė-Tsiskarishvili, G. (2020). City Diplomacy in Young Democracies: The Case of the Baltics. In A. Sohaela & E. Sevin (Eds.), *City Diplomacy* (pp. 305–330). Palgrave Macmillan Series in Global Public Diplomacy. Palgrave Macmillan. doi:10.1007/978-3-030-45615-3_14

Burkšiene, V., Dvorak, J., & Duda, M. (2019). Upstream Social Marketing for Implementing Mobile Government. *Societies (Basel, Switzerland)*, *9*(3), 54. doi:10.3390oc9030054

Burskyte, V., Belous, O., & Stasiskiene, Z. (2011). Sustainable development of deep-water seaport: The case of Lithuania. *Environmental Science and Pollution Research International*, *18*(5), 716–726. doi:10.100711356-010-0415-y PMID:21104330

Comfort, S. E., & Hester, J. B. (2019). Three Dimensions of Social Media Messaging Success by Environmental NGOs. *Environmental Communication*, *13*(3), 281–286. doi:10.1080/17524032.2019.1579746

Corell, E., & Betsill, M. M. (2001). A comparative look at NGO influence in international environmental negotiations: Desertification and climate change. *Global Environmental Politics*, *1*(4), 86–107. doi:10.1162/152638001317146381

D'amato, G., Vitale, C., De Martino, A., Viegi, G., Lanza, M., Molino, A., & D'amato, M. (2015). Effects on asthma and respiratory allergy of Climate change and air pollution. *Multidisciplinary Respiratory Medicine*, *10*(1), 1–8. PMID:26697186

de la Casa, J. M. H., Álvarez-Villa, À., & Mercado-Sáez, M. T. (2018). Communication and effectiveness of the protest: Anti-fracking movements in Spain. *Zer: Revista de estudios de comunicación= Komunikazio ikasketen aldizkaria, 23*(45).

Deng, D., Zhao, Y., & Zhou, X. (2017). Smart city planning under the climate change condition. *IOP Conference Series. Earth and Environmental Science*, *81*(1), 012091. doi:10.1088/1755 1315/81/1/012091

Dolšak, N. (2013). Climate change policies in the transitional economies of Europe and Eurasia: The role of NGOs. *Voluntas*, *24*(2), 382–402. doi:10.100711266-012-9260-6

Dvorak, J. (2015). The Lithuanian Government's Policy of Regulatory of Regulatory Impact Assessment. *Management and Business Administration. Central Europe*, *23*(2), 129–146. doi:10.7206/mba.ce.2084-3356.145

Ejdys, J. (2020). Trust-Based Determinants of Future Intention to Use Technology. *Foresight and STI Governance*, *14*(1), 60–68. doi:10.17323/2500-2597.2020.1.60.68

Fanø, J. J. (2019). Enforcement of the 2020 sulphur limit for marine fuels: Restrictions and possibilities for port States to impose fines under UNCLOS. *Review of European, Comparative & International Environmental Law*, *28*(3), 278–288. doi:10.1111/reel.12306

García Fernández, C., & Peek, D. (2020). Smart and sustainable? Positioning adaptation to climate change in the European smart city. *Smart Cities*, *3*(2), 511–526. doi:10.3390martcities3020027

Gido, J., Clements, J., & Baker, R. (2018). *Successful Project Management*. Cengage Learning.

Hall, N. L., & Taplin, R. (2007). Solar festivals and climate bills: Comparing NGO climate change campaigns in the UK and Australia. *Voluntas: International journal of voluntary and nonprofit organizations*, *18*(4), 317–338. https://doi:org/ doi:10.1007/s11266-007-9050-8

Ho, E. (2017). Smart subjects for a Smart Nation? Governing (smart) mentalities in Singapore. *Urban Studies (Edinburgh, Scotland)*, *54*(13), 3101–3118. doi:10.1177/0042098016664305

Hricko, A. (2012). Progress & Pollution Port Cities Prepare for The Panama Canal Expansion. *Environmental Health Perspectives*, *120*(12), 470–473. doi:10.1289/ehp.120-a470 PMID:23211315

Hue, D. T. (2017). Fourth Generation NGOs: Communication Strategies in Social Campaigning and Resource Mobilization. *Journal of Nonprofit & Public Sector Marketing*, *29*(2), 119–147. doi:10.1080/10495142.2017.1293583

Keulemans, S. (2021). Rule-following identity at the frontline: Exploring the roles of general self-efficacy, gender, and attitude towards clients. *Public Administration*. doi:10.1111/padm.12721

Klaipeda City Municipality. (2021). *Klaipėdos miesto savivaldybės 2021 - 2030 metų strateginis plėtros planas*. Klaipėda [Strategy of Klaipeda city municipality 2021-2030]. https://www.klaipeda.lt/lt/planavimo-dokumentai/klaipedos-miesto-savivaldybes-2021-2030-metu-strateginis-pletros-planas/8827

Konstantinaviciute, I. (2003). Climate change mitigation policies in Lithuania. *Energy & Environment*, *14*(5), 725–736. https://www.jstor.org/stable/43734595. doi:10.1260/095830503322663429

Korten, D. C. (1990). *Getting to the 21st Century: Voluntary Action and the Global Agenda*. Kumanian Press.

Kotseva-Tikova, M., & Dvorak, J. (2022). Climate Policy and Plans for Recovery in Bulgaria and Lithuania. *SSRN*, *22*(2), 79–99. doi:10.2139srn.4294988

Krippendorff, K. (2013). *Content Analysis. An Introduction to Its Methodology* (3rd ed.). Sage Publications.

Lee, T., Johnson, E., & Prakash, A. (2012). Media independence and trust in NGOs: The case of postcommunist countries. *Nonprofit and Voluntary Sector Quarterly*, *41*(1), 8–35. doi:10.1177/0899764010384444

Liu, B. F. (2012). Toward a better understanding of nonprofit communication management. *Journal of Communication Management (London)*, *16*(4), 388–404. doi:10.1108/13632541211279012

Marín, J. C., Raga, G. B., Arévalo, J., Baumgardner, D., Córdova, A. M., Pozo, D., Calvo, A., Castro, A., Fraile, R., & Sorribas, M. (2017). Properties of particulate pollution in the port city of Valparaiso, Chile. *Atmospheric Environment*, *171*, 301–316. doi:10.1016/j.atmosenv.2017.09.044

Mayring, P. (2014). *Qualitative content analysis: theoretical foundation, basic procedures and software solution*. Austria: Klagenfurt. https://nbn-resolving.org/urn:nbn:de:0168-ssoar-395173

Michie, J. & Cooper, C.L. (2015). *Why the Social Sciences Matter*. NY: Palgrave Macmillan

Mihai, R. L. (2017). Corporate Communication Management. A Management Approach. *Valahian Journal of Economic Studies*, *8*(22), 103–110. doi:10.1515/vjes-2017-0023

Moore, F. C. (2009). Climate Change and Air Pollution: Exploring the Synergies and Potential for Mitigation in Industrializing Countries. *Sustainability (Basel)*, *1*(1), 43–54. doi:10.3390u1010043

Muszynska, K., Dermol, K., Trunk, V., Đakovic, A., & Smrkolj, G. (2015, May). Communication management in project teams–practices and patterns. In *Joint International Conference* (pp. 1359-1366).

Nulman, E., & Özkula, S. M. (2016). Environmental nongovernmental organizations' digital media practices toward environmental sustainability and implications for informational governance. *Current Opinion in Environmental Sustainability*, *18*, 10–16. doi:10.1016/j.cosust.2015.04.004

Orru, H., Ebi, K. L., & Forsberg, B. (2017). The Interplay of Climate Change and Air Pollution on Health. *Current Environmental Health Reports*, *4*(4), 504–513. doi:10.100740572-017-0168-6 PMID:29080073

Palttala, P., Boano, C., Lund, R., & Vos, M. (2012). Communication gaps in disaster management: Perceptions by experts from governmental and non-governmental organizations. *Journal of Contingencies and Crisis Management*, *20*(1), 2–12. doi:10.1111/j.1468-5973.2011.00656.x

Pavlovic, J., Lalic, D., & Djuraskovic, D. (2014). Communication of Non – Governmental Organizations via Facebook Social Network. *Inzinerine Ekonomika-Engineering Economics*, *25*(2), 186–193. doi:10.5755/j01.ee.25.2.3594

Powell, S. (2017). *Sustainability in the Public Sector: An Essential Briefing for Stakeholders*. Routledge. doi:10.4324/9781351275729

Rajhans, K. (2018). Effective Communication Management: A Key to Stakeholder Relationship Management in Project-Based Organizations. *The IUP Journal of Soft Skills*, *XII*(4), 47–66. https://ssrn.com/abstract=3398050

Raupp, J., & Hoffjann, O. (2012). Understanding strategy in communication management. *Journal of Communication Management (London)*, *16*(2), 146–161. doi:10.1108/13632541211217579

Sharma, D. C. (2006). Ports in a Storm. *Environmental Health Perspectives*, *114*(4), 222–231. doi:10.1289/ehp.114-a222 PMID:16581529

Telešienė, A., & Kriaučiūnaitė, N. (2008). Trends of Nongovernmental Organizations' Environmental Activism in Lithuania. *Public Policy and Administration*, *1*(25), 94–103.

Tumulytė., I. (2012). Darnaus vystymosi komunikacija. Pilietinės iniciatyvos aplinkosaugos komunikacijoje. [Communication of sustainable development. Citizen initiatives in environmental communication]. *Informacijos mokslai, 62,* 7–17.

United Nations. (2019). *The Sustainable Development Goals Report 2019.* UN. https://www.un-ilibrary.org/content/books/9789210478878

Urry, J. (2015). Climate Change and Society. In J. Michie & C. L. Cooper (Eds.), *Why the Social Sciences Matter* (pp. 45–59). Palgrave Macmillan. doi:10.1057/9781137269928_4

Vavtar, L. (2014). Environmental lobby effectiveness–the case of Lithuania and the United Kingdom. *Socialinių mokslų studijos, 6*(2), 313–330. https://repository.mruni.eu/handle/007/13317

World Commission on Environment and Development. (1987). *Our common future.* Oxford University Press.

ENDNOTE

[1]   Ministry of Environment of the Republic of Lithuania (2021). Environmental NGOs. https://am.lrv.lt/lt/nuorodos/aplinkosaugos-nvo

Chapter 6

People–Centered Urban Governance in Latin America and the Caribbean:
Sociocybernetics, Climate Justice, and Adaptation

Shar-Lee E. Amori

https://orcid.org/0000-0002-9617-435X

McGill University, Canada

ABSTRACT

This chapter delves into the complexity of sustainable urbanization, climate justice, social inclusion, and participatory governance. Grounded in a one-year descriptive ethnographic study and meta-synthesis, the analysis deconstructs the disparities between the urban rich and poor in Jamaica, Panama, Trinidad and Tobago, and Columbia, across five key development domains- wellbeing, education, security, infrastructure, and governance. Through sociocybernetics, the decision-making processes in urban ecosystems are interrogated, revealing unique challenges faced by the urban poor, trapped in a cycle of recovery, versus the mitigation-oriented urban rich. The analysis extends to the role of urban citizens, designers and integrators, governance structures, levels of social inclusion, resource allocation, and their amalgamated implications for socio-climate justice. It evaluates the international policy arena, translation of global mandates into local development plans, and the need for hyperlocal strategies that encourage a more people-centred planning approach for sustainable urbanization.

PEOPLE-CENTRED URBAN GOVERNANCE IN LATIN AMERICA AND THE CARIBBEAN: SOCIOCYBERNETICS, CLIMATE JUSTICE & ADAPTATION

Global climate change has heightened the need for sustainable urban development, in the Global South regions, particularly in Latin America and the Caribbean (LAC), a region marked by pronounced socio-

DOI: 10.4018/978-1-6684-8963-5.ch006

People-Centered Urban Governance in Latin America and the Caribbean

economic disparities and a complex colonial history. LAC, the most urbanized region in the developing world, has experienced rapid urbanization, moving from 50%- 80% between 1940 and 2014, necessitating people-centred urban planning that emphasizes urban plurality, interdisciplinary innovation, climate justice, and inclusive governance (Inter-American Development Bank, 2015). This chapter, while not offering a comprehensive design framework, contributes insights for developing one using descriptive ethnography and meta-synthesis centred on socio-climate justice and participatory governance. It examines the theoretical and practical aspects of sustainable urban planning, aligning with the United Nation's Sustainable Development Goals (SDGs), with a focus on the lived experiences of the urban rich and urban poor in Bogotá, Columbia, Kingston, Jamaica, Panama City, Panama, Port-of-Spain, Trinidad, and Tobago. These countries offer Global South perspectives on people-centred urban planning strategies that adeptly balance spatiality and humanity for sustainable urban futures.

Sustainable Development Goal (SDG) 11, which aims to "make cities and human settlements inclusive, safe, resilient, and sustainable," and SDG 13, which calls for "urgent action to combat climate change and its impacts" are directly relevant to this discussion. When these SDGs are translated into local policies and applied at the macro, meso, and micro levels, the complexity of sustainable urbanization, which intersects with other SDGs, becomes evident. It therefore demands multifaceted governance, considering the plurality of urban ecosystems in LAC from their colonial birth to the current neoliberalist ideologies. Within this context, sociocybernetics- a field that examines the dynamics of social systems and the interactions between human agency and structural forces (Geyer & Van der Zouwen, 2014)- brings another level of human-centeredness to spatio-temporal urbanism. It acknowledges that while urban planning itself is fundamentally a spatial endeavour, it does not exist in a vacuum indifferent to the diverse human usability and security needs for which it must be fit for purpose. Therefore, three core themes are interrogated here: (1) the intersectionality of colonialism, class, citizenship, and climate change); (2) the dynamics of climate vulnerability, resilience, and justice through the lens of sociocybernetics and; (3) the role of governance and social inclusion in fostering urban resilience.

COLONIALISM, CLASS, CITIZENSHIP, AND CLIMATE CHANGE

While the global expansion of capitalism was made possible through various political, economic, and cultural manoeuvres- colonialism, class-compromise of social democracy; spatiotemporal fixes of footloose capital...The challenge posed by the concrete reality of natural limits in the Anthropocene re-signifies the centrality of distributional conflicts. (Arsel, 2023, p. 70)

Colonialism, class, and climate change are interconnected issues that have undeniably shaped LAC, institutionalizing inequality that persists today. Defined by cultural, political, and economic domination, colonialism involved the exploitation of natural resources, labour, and land for the benefit of the colonizing country (Horvath, 1972). It has had far-reaching impacts on the environment, with the poorest and most marginalized communities disproportionately affected. The Intergovernmental Panel on Climate Change (IPCC) acknowledged these interrelated themes stating that "present development challenges causing high vulnerability are influenced by historical and ongoing patterns of inequity, such as colonialism, especially for many Indigenous peoples and local communities" (Pörtner et al., 2022, p. 12). This period in the Plantationocene reflects the environmental and social changes of plantation economies, influencing today's labour and consumption systems. Evidence of this exists in the development histories of Colombia, Jamaica, Panama, Trinidad, and Tobago, which began with the arrival of European colonizers

109

in the late 15th century and the subsequent resource extraction to fuel their growing economies, which many scholars consider the genesis of climate change (Barker, 2012; Whyte, 2018; Varanasi, 2022).

Colombia, with an urban population of 75.5% as of 2018 (UN-DESA, 2022), has an urban development history that began with the establishment of Spanish settlers in the early 16th century, who founded cities such as Bogotá, Cali, and Cartagena to serve as administrative, economic, and cultural hubs throughout the 297-year (1525-1822) colonial period. Neighbouring Panama, from the time of the 320-year Spanish rule (1501-1821), was a strategic location, acting as a bridge between Central and South America. Panama City, established in 1519, became a key port and trade centre, facilitating the movement of goods between the Atlantic and Pacific Oceans. The construction of the Panama Canal in the early 20th century, and the era of U.S. Imperialism exemplified through its ownership (1904-1999), further contributed to the city's urban development, as the Canal Zone and surrounding areas saw rapid economic expansion and displacement, with Panama's urban population at 68.8% as of 2021 (UN-DESA, 2022). Both Jamaica and Trinidad were colonized first by the Spanish, and then by the British. However, the British conquest of Jamaica, in 1655, marked a turning point in its trajectory with the creation of the City of Kingston in 1692 as a strategic port and trade centre. British rule lasted 307 years (1655–1962), nearly twice as long as the Spanish 161 years (1494-1655). As such, the island's urban anatomy, housing 56.7% of its population in 2021 (UN-DESA, 2022), has a decidedly British mark with remnants of the Spanish colonial era. Similarly, Trinidad and Tobago, with a 53.2% urban population in 2019 (UN-DESA, 2022), was finally ceded to the British in 1802, who concentrated urban development around Port-of-Spain, its main administrative and trade centre under British rule (1797-1962).

Out of these colonial histories came cultural imperialism, the erasure of existing languages and cultures, and the imposition of European beliefs, values, and patterns of behaviour (Hovarth, 1972; Ryan, 1999; Tsaaior, 2011). The ensuing acculturation made the European way of life the dominant cultural narrative imposed through religion, language, education, social and economic structures, architecture, and urban planning. A by-product of this was urban planning imperialism—the systemic institutionalization of foreign planning ideas and models onto the region without much consideration for the local context (Angotti and Irazábal, 2017; Duer & Vegliò, 2019). Bogotá, Kingston, Panama City, and Port-of-Spain all carry these colonial imprints, and their spatial dynamics reflect the broader strategies of colonialism as a tool to control, reinforce class hierarchies, and ensure the efficient extraction of resources. Reinforced socio-spatial demarcations saw European settlers occupying the infrastructurally superior central areas, while Indigenous and African populations were relegated to the peripheries. This institutionalized spatial segregation in the Plantationocene led to the associated socio-economic inequalities and climate vulnerabilities. Khan (2022) aptly explained the similarities between this and gentrification:

Both gentrification and colonialism require an economically empowered few to oversee an operation to economically and politically displace one group, to be replaced by another, while the former achieving financial gain and political power. In many instances, such processes are camouflaged or obfuscated through commercial activity and permissions within legally sanctioned frameworks, making a re-evaluation of the greater complexities behind gentrification a matter of necessity.

For this reason, a growing body of literature calls modern gentrification "new urban colonialism," with a focus on its internal aspects as forms of citizen-citizen displacement, rather than the conquest and subjugation of the colonial era (Alexandri, 2014).

Socio-Spatial Segregation, Social Inequality, and Climate Change

Colonial socio-spatial segregation was steeped in ethnic discrimination, and positioned Europeans at the top of the food chain, followed by mestizos (those of mixed European and Indigenous descent). Indigenous people and Africans held the lower strata, facing constrained opportunities and restricted access to resources (Runk, 2012). Land ownership, a critical resource for building wealth and household socio-economic stability (Christophers, 2016; Stiglitz, 2012), was only allowed to the elites, and even post-slavery was unequally distributed, further catalysing a vicious cycle of poverty and social injustice. Already marginalized populations were, by virtue of this, forced into precarious living conditions in areas prone to natural hazards, After the abolition of slavery, these colonial-era land policies led to the emergence of informal settlements as a form of self-help that persists today. Those now living in these types of communities in Bogotá, Kingston, Panama City, and Port-of-Spain, for this chapter (pseudonyms used), fit broadly into six major and interconnected categories:

1. Generational: With ties to informal occupation that can be traced back for generations.
2. Economic: Rural-urban/urban-urban migration for socio-economic reasons.
3. Environmental: Indigenous migration due to rising sea levels, decreased resilience, and a rising Indigenous population without adequate infrastructure (Bogotá, Panama City). Wigundun, a 20-year-old who lived on one of the smaller islands in the Indigenous Guna Yala territory of Panama, explained that:

I had to leave the island I was originally born on, because of the rising sea level. I went to live with relatives in another, but the space was small, and I said let me come to the mainland and work. I have a son, so I needed to work.

4. Deprivation Settling/Migration: Resulting from a lack of reliable needs-based infrastructure, and closely linked to conservation as living close to natural resources, such as rivers, addresses basic needs (peri-urban areas of Kingston and Bogotá).
5. Conservational: Settlers are motivated by the desire to protect natural environmental resources from capitalistic exploitation (Bogotá, Panama City, Kingston).
6. Political: Urban informal settlements were supported by political hopefuls as a means of vote maximizing, and in Colombia, where there is displacement due to internal conflict (Port-of-Spain, Kingston).

Dexter, a senior citizen living in a garrison community in Kingston said:

This was one of those slums, but during the 80s, with all that political war, it kind of got cemented because it is an important political place and plenty of votes come from here. I have lived here since I was a boy and through the 80s. I still don't have any title to the land, but the government will not move us. They know if they do, they lose the votes.

Access to land is crucial for socioeconomic mobility, as studies show that countries with more egalitarian land distribution tend to have higher levels of economic stability and investments in education and social infrastructure (Asher & Novosad, 2020; Stiglitz, 2012). The opposite is true for the urban poor in LAC, who often generationally must prioritize immediate survival over long-term advancement (Runk, 2012; Stiglitz, 2012). Urban areas, with their high population densities and disparities, demand a focus on land and housing insecurity, as equitable land and housing distribution can reduce income inequality, foster social integration, and boost economic vitality, leading to balanced urban development that can prevent the concentration of poverty and wealth in specific neighbourhoods and increase resilience (SDGs 10, 11, and 13). However, achieving this requires overcoming systemic barriers that institutionalize socio-economic disparities.

For example, Guzman and Bocarejo (2017) noted that while Bogotá's dense urban layout, with a central activity hub, supports large-scale public transport, their spatial equity analysis showed it disproportionately disadvantages low-income households in terms of residential and employment locations, mobility, extended travel times, higher costs, and lower accessibility. This disparity is primarily due to the inferior public transportation quality for low-income areas, stemming from spatial segregation and historical land occupation patterns that unduly benefit the city's higher-income inhabitants. They concluded that a household's location within the city significantly influences its access to opportunities. When asked about this, Carlos a resident in Cuidad Bolivar, Bogotá's largest informal settlement, explained:

The government knows of the transport issues we have, but this far from the business centre is not a priority. You either drive or make do with what is available. And because a lot of us do not own this land then we cannot demand they put [in] better transport networks.

At the heart of this issue is the need to deal with land ownership and use disparities, and governments in these cities have implemented various programs and laws to improve socio-legal tenure and resource access.

In Jamaica, while the Land Administration and Management Programme (LAMP) and Registration of Titles, Cadastral Mapping, and Tenure Clarification (Special Provisions) Act (2005) do not explicitly target descendants of colonial displacement, they work in tandem to alleviate land insecurity through land titling and ownership regularization, which is particularly important for informal settlers. Panama's Law 72 (2008) does better at focusing on historical displacement and allows collective land titling for Indigenous communities in a significant step towards reversing historical injustices. Law 80 (2009) gives added support by regularizing state land allocation for poor families' housing. In Trinidad, the State Lands Act, supported by the Land Adjudication Act (2017), similarly aims to update and clarify land ownership records thereby enhancing land administration efficiency. However, Colombia's Victims and Land Restitution Law 1448 (2011) confronts both rural and urban land displacement in a more contemporary way, with a focus on the return and formal recognition of the lands of those displaced by the 50-year Colombian civil conflict in a period known as "La Violencia," (Columbia Events, 2021). Despite these efforts, challenges persist in each of these countries, including informal settlements, political land administration issues, bureaucratic delays, and socio-political barriers. (Angotti & Irazábal, 2017; Berney, 2011; Mullings et al., 2018).

Urban Political Economy and Ideas of Citizenship: Jamaica and Panama Case Studies

Jamaica

The political economy of urban planning in post-independence LAC serves as a microcosm of the enduring colonial legacies intertwined with contemporary political dynamics. In Jamaica, the political economy landscape has been dominated by the two-party political system, a legacy of constitutional decolonization built on the exported Westminster-Whitehall British governance model (Mills, 1997). The capitalist-oriented Jamaica Labour Party (JLP) and the democratic socialist People's National Party (PNP) emerged in the post-colonial period rife with political tensions, social injustices, and inequality, to which Dexter referred. Within this political dichotomy, patronage became deeply entrenched and historically steered resource distribution in elitist ways, reinforcing socio-spatial class divides and displacement (Clarke, 2006; Mills, 1997; Stone, 1973). In more recent decades, the construction boom has been devoid of green infrastructure considerations, further pushing Kingston towards its predicted 2023 climate departure (Mora et al., 2013). Urban poor, with increased vulnerabilities, lamented that these consequences of anthropogenic climate change and high cost of living have made Kingston unliveable for those who are not in the "uptown" (urban rich) economic bracket.

Curtin (1955) described this as the "Two Jamaicas," a historical class dichotomy that permeates every social justice, resource allocation, and development discourse in Jamaica (Bennett, 2021; Curtin, 1955; Salmon, 2020). This dichotomous concept is infused with colonialist and colourist connotations to highlight how laws and policies still fail to effectively challenge socio-economic disparities from the colonial era while framing elitism as continued oppression. Campbell asserted that:

Unequal citizenship began in Jamaica as a natural outgrowth of colonialism and the coming together of Africa and Europe under chattel slavery...Two analytical frameworks are of interest here: (1) the legal as well as racial/colour basis on which notions of who was qualified to be a citizen were based and, (2) the way in which the social order, based on white supremacy and black inferiority, held this together. (Campbell, 2020, p. 58)

Vision 2030 Jamaica Sector Plans (2008), translated into Jamaica's Urban Development Plan, is a notable policy response to some of these issues, providing a long-term development framework centred on inclusiveness, equity, and sustainable urban development (Planning Institute of Jamaica, 2009). However, this is not always the outcome.

In 2022, the Jamaica National Housing Trust (NHT), which represents a statutory tax deduction with the benefit of financial support to acquire housing after a minimum of three years of contributions, came under fire for building an apartment complex outside the income bracket of many contributors (Bennett, 2021). As its mandate is to build affordable housing, this venture put it at odds with Jamaicans who felt the capitalist-oriented JLP government was again catering to elitism. Debates centred on which citizens were more valued, highlighting the socio-legal dimensions of citizenship based on class. Shantay, formally employed, but residing in an informal settlement on the fringes of Kingston, echoed the general sentiment stating, "If your NHT benefits cannot buy an apartment built by the NHT, then it should not be involved in such developments because it excludes the poor man." Objectively, this calls into question the exclusion of the elite class in the pursuit of housing equity, who, by law, must also pay NHT contributions. Should the urban rich be excluded from these opportunities that their taxes are used for in favour of the urban poor? Its developers argued that no law prohibits high-income projects financed by the NHT. A point for policy clarification here would be the difference between the principles of equity and equality in resource allocation.

Panama

Panama, to deal with Indigenous marginalization and safeguard their rights and culture, established autonomous Indigenous territories known as "Comarcas," in 1938 (Bogotá.gov.co., 2021; Dirlik, 2003). However, the mestizo elitist system, deeply entrenched in the political and economic structures, has continued to shape policies favouring their interests (Dirlik, 2003; Moreno, 1993). The building of the Panama Canal Zone further compounded this problem as it fuelled gentrification and gave rise to informal settlements and high-density lower-income housing (Maurer & Rauch, 2019; Moreno, 1993). Disjointed planning approaches involving the central government, local authorities, and the private sector further led to inefficient and inequitable urban outcomes (Moreno, 1993; Runk, 2012; Sigler & Wachsmuth, 2016). Current programmes such as the Inter-American Development Bank's (IDB) Emerging and Sustainable Cities Program (ESC), aimed at supporting national and subnational governments in the development and execution of city action plans, seek to rectify these issues. However, implementation remains a challenge, with a perceived bias towards wealthier neighbourhoods in the distribution of climate-related investments. To address this, a new project called, "Aligning the financial flows of the Panamanian financial sector with the climate change objectives of the Paris Agreement" was developed and led by the United Nations Environment Programme in Latin America and financed by the Green Climate Fund and the European Union (UNEP, 2023).

Similar to the *Two Jamaicas* debate, perhaps a bit Orwellian in nature, these consequences of colonialism and socio-spatial segregation have spurred discussions in Panama about what citizenship means when urban governance is perceived as treating some communities as "more equal than others." The colonially birthed "othering" is a complex phenomenon that is rooted in broader socio-economic and cultural discrimination (Orwell, 1945, p. 103; Runk, 2012; Tsaaior, 2011; Whyte, 2017, 2018). To centre this, construction of the Panama Canal significantly gentrified the surrounding urban spaces and the privileged status of U.S. citizens, and their Panamanian allies gave them preference to prime land, which some older residents of the city say relegated them to a "lesser" citizenship status (Forbath, 1999; Maurer & Rauch, 2019). The project also attracted a diverse labour force from across the Caribbean and neighbouring Latin American countries, which led to the emergence of racially segregated neighbourhoods, such as the Afro-Antillean settlements of La Boca and Silver City (Angotti & Irazábal, 2017; Moreno, 1993).

Socio-spatial segregation is evident in Kingston's juxtaposition of informal settlements and affluent neighbourhoods, and in Panama City, where luxury high-rises mask underlying disparities and tell another side of the city's story. This divergence in the ways these urban spaces are experienced by different groups reinforces the feedback loops that shape policy priorities with Stiglitz (2012), arguing that with their proximity to political power, the urban rich are better positioned to shape policies in their favour. Furthermore, how these spaces are developed has implications for socio-climate resilience, and the urban rich, though not immune, are better positioned (financially and spatially) to rebound and/or confront their vulnerabilities. In contrast, many existing solutions to dealing with the opposite reality of the urban poor are prescriptive and lack "place-based perspectives" for community adaptation (Barca,

2008; Dirlik, 2003; Schoburgh, 2017). Given the socio-climate justice concerns this presents, a greater understanding of how these present at various levels of analysis, and vary within and across urban eco-systems, is needed. Sociocybernetics becomes useful here.

CLIMATE VULNERABILITIES AND THE SOCIOCYBERNETICS OF RESILIENCE IN URBAN PLANNING

Sociocybernetics is an independent chapter of science on societal complexity, and the self-organisation (autopoiesis) of societal systems, applying the concepts of chaos theory and the mathematical complexity related to it, as well as the transdisciplinary analysis of the manifestation of such complexity. (Geyer & Van der Zouwen, 2014, pp. 4–6)

Pioneered by thinkers like Talcott Parsons (2005), Niklas Luhmann (2012), and Stafford Beer (1984), sociocybernetics, draws from cybernetics, systems theory, and sociology, emphasizing the interconnectedness, communication, and control processes within social systems (Beer, 1984; Parsons, 2005; Luhmann, 2012). An important distinction is that while systems theory provides a broad analytical framework for diverse types of systems, sociocybernetics distinctively applies cybernetic principles to social structures and their inherent feedback loops, self-regulation, and adaptability (Geyer & Van der Zouwen, 2014). These feedback mechanisms can either be positive, amplifying changes in a system (social justice) or negative, counteracting changes to maintain equilibrium (status quo). From this perspective, cities are seen as complex, adaptive systems, characterized by nonlinearity and emergent behaviour.

Employed effectively, sociocybernetics has the potential to identify areas of improvement and strategies for optimal resource allocation, while aiding in the design of more resilient, sustainable, and liveable cities. These efforts contribute to sustainable urbanization, supporting the goal of SDG 16 for inclusive and participatory decision-making, and intersecting with objectives in SDG 3 (Health and Well-being), SDG 6 (Clean Water and Sanitation), SDG 10 (Reducing Inequality), and SDG 15 (Sustainable Terrestrial Ecosystems). This, therefore, reframes urban planning from the "top-down" approach to development to a "bottom-up" strategy that acknowledges the nuanced realities and needs of urban dwellers, and reaffirms the importance of place-based activism for socio-climate resilience and inclusive governance.

A Sociocybernetic Approach to Climate Resilience

Port-of-Spain, a city at sea level and developed in the flood plains of both the Maraval and East Dry rivers, has seen increased flood events, prompting the government in recent years to develop the 2020 Port-of-Spain Flood Alleviation Project (MOWT, 2020). The project is aimed at focusing on the flood and accessibility issues in the area and pulls on targets outlined in Goals 1 and 3 of Vision 2030: The 2015 National Development Strategy of Trinidad and Tobago. These focus on managing environmental systems and improving urban infrastructure. An urban planner stated the major issue is that:

For a few decades, the city's drainage system has not been developed to match the needs of rapid urbanization, and with significant urban expansion, including informal settlements that have expanded on the outskirts of the city,[;] the reduced permeability of the ground has negatively impacted the natural ability of the area to absorb water. In the rainy seasons and high tides, flooding occurs, aggravated by increased frequency and intensity of extreme weather events.

For the urban poor in the peri-urban areas of Port-of-Spain, this threatens housing, food and water security, reduces climate resilience, and disrupts local systems of production. Similarly, water scarcity during droughts in Jamaica, Panama, and Colombia restricts access to potable water, essential for health and well-being. Where this is experienced in Kingston and Bogotá, informal settlers along riverbanks on the city fringes do so for deprivation and conservation reasons, motivated by infrastructural deficits. In Panama City, however, hotels dominate the coastline, and the expansion of the canal has pushed development further inland. With river sources being scarce in the city, the urban poor make do with low-income high-density structures constituting overcrowded public housing, interspersed throughout the city (Rondinelli, 1990; Sigler & Wachsmuth, 2016).

In each of these cases, sociocybernetics offers a valuable framework for understanding vulnerabilities and planning resilience across five development domains: well-being, infrastructure, security, governance, and education. By viewing these vulnerabilities as interconnected components of a broader socio-ecological system, where changes in one sector can have cascading effects, one can develop more comprehensive and effective strategies for socio-climate resilience (Beer, 1984; IRGC, 2018; Islam & Winkel, 2017; Kelman, 2020; Parsons, 2005). Admirably, this holistic perspective is mirrored in the Vision 2030 national development plans and climate policies of all four countries. However, there are significant gaps in execution at meso and micro levels of development, particularly in their lack of responsiveness to how socioeconomic and climate vulnerability are experienced by the urban rich versus the urban poor. Feedback loops reveal the resilience of the urban rich, with better mitigation capacities, and the vulnerability of the urban poor, trapped in a cycle of vulnerability and recovery.

Across all four cities, there is consensus among urban poor citizens that disaster response is primarily aimed at saving lives, but rarely at bettering their circumstances. Presented here is the challenge of making sure urban development and disaster response do not further entrench inequalities. That concern is also epitomized in the paradox that gentrification presents as an urban renewal process. On one hand, it is touted as a strategy for climate resilience stemming from the influx of wealth and resources into gentrified neighbourhoods, which inevitably leads to infrastructural improvements, such as improved drainage and green spaces, as seen in the Panama Canal zone (Sigler & Wachsmuth, 2016). However, those who benefit are not the marginalized that were originally displaced in the gentrification process, which further amplifies socio-climate vulnerabilities as lower-income residents are further displaced. As Khan (2022) argued, this cycle has its genesis in the urban imperialism of the colonial era, further impeding the achievement of SDG 10 (Reduced Inequalities), SDG 11 and SDG 13.

The first step to confronting these issues is recognizing that building urban resilience requires inclusive governance that is about both infrastructural and environmental planning, and attending to equally urgent socio-economic vulnerabilities. Resilience itself is a systemic property that encompasses the ability of a system to absorb disturbances, adapt, and reorganize while maintaining fundamentally inclusive network structures and functions, as system outputs are looped back as input influencing its future state (Beer, 1984; Cárdenas et al., 2021; Fay, 2005). In this context, the Sendai Framework for Disaster Risk Reduction also stresses the importance of recognizing the context-dependent, nonlinear, and multifaceted nature of systematic risk that includes multiple hazard types and vulnerabilities and moves risk governance beyond singular hazard focus. This reconceptualization requires acknowledgement that systematic risk is not natural but occurs because of human processes, policies, and governance

People-Centered Urban Governance in Latin America and the Caribbean

in context and over time (IRGC, 2018; Kelman, 2020). Hence, for there to be climate justice, feedback loops used to justify policy imperatives must have inclusive inputs from urban citizens, recognizing that they are designers and integrators of the spaces they occupy.

Urban Citizens, Designers, and Integrators as Socio-Climate Actors

Urban citizens, designers, and integrators occupy distinct, yet interconnected, positions within the broader sociocybernetic fabric of urban ecosystems, shaping climate vulnerability and resilience in unique ways (Antonova & Grunt, 2019; Geyer & Van der Zouwen,2014). Citizens can act as designers and integrators, designers as citizens and integrators, and integrators as citizens and designers.

Urban citizens, along the entire socio-economic spectrum, are the lifeblood of cities, and their daily behaviours, choices, and actions make them critical components of the feedback loops that shape the biophysical nature of urban spaces. For instance, through community-based initiatives and collective action, informal settlements in Bogotá and Kingston, with high levels of social cohesion, have relatively effective climate resilience strategies and community adaptation at the micro-levels of analysis. Heavily built around principles of self-help, this represents a sort of hybrid governance, where citizens normatively negotiate power and representation to fill gaps in state representation to address their various human insecurities (Dirlik, 2003; Geyer & Van der Zouwen, 2014; Harrison, 1988; Schoburgh 2017). However, the scope and success of these initiatives are often conditioned by broader socio-political structures and processes. As informal settlements are not encouraged, even where the government takes a *laissez-faire* approach, this knowledge creation is not shared. In Jamaica, for example, across the 755 identified informal settlements in Jamaica (National Housing Policy, 2019), adaptation strategies built on the tenets of sociocybernetics and hyperlocalism exist in abundance. However, knowledge networks are not encouraged, as informal settlements are largely considered a failure in social governance. Although Colombia has a better track record of urban slum upgrading programmes in cities such as Medellin, knowledge exchange is still limited by the extent to which it is politically advantageous and not a public administration nuisance.

Urban designers, on the other hand, shape the spatial dimensions of urban systems through their decisions related to land use, infrastructure, and urban form (Antonova & Grunt, 2019; Agenda U. N., 2016). Critically, they mediate the relationship between urban citizens and the built environment and navigate complex social dynamics and power relations. This requires understanding the degrees of social change/control inherent to these social ecosystems, and the change to which there can be resistance. Recall that the interconnections between these urban roles mean urban planners are not exclusively those professionally trained in the discipline. Urban citizens are inherently designers of the spaces they occupy through their usability demands, e.g., Indigenous, youth, elderly, internal/external migrants, and those living with disabilities or exceptionalities.

In this regard, urban integrators, as connectors and coordinators within the urban system, play a pivotal role in aligning stakeholder actions with development goals, embodying feedback mechanisms crucial for policy learning and socio-climate resilience (Islam & Winkel, 2017; Pahl-Wostl, 2009). A lot of this work also includes adapting international frameworks into local policies and strategic initiatives. However, success depends on their ability to effectively mediate diverging interests and navigate the complexities and contradictions inherent to the urban governance landscape. For example, urban designers, motivated by environmental and conservation factors, might advocate for green infrastructure, while urban integrators, influenced by political and economic factors, might prioritize infrastructure that

117

supports economic growth (Panama City and Kingston). The multiplicity of conflicts this can present, if not properly managed, are endlessly referred to here. Healey's (2003) collaborative planning approach aims to mitigate this by promoting democratic values, public participation, and the creation of socially just urban environments. A failure of which will engender what Stiglitz (2012) asserted leads to unfair policy and development influence from those with more power. This means that for Colombia and Panama, the inclusion of Indigenous voices is essential.

Indigenous Peoples as Urban Citizens, Designers, and Integrators

The "Guna Yala" in Panama and the "Embera" in Colombia, represent a distinctive set of urban citizens, designers, and integrators, with the potential to contribute to urbanism with their unique knowledge, skills, and perspectives on environmental stewardship (McSweeney & Jokisch, 2007; Wickstrom, 2003). However, their integration into urban life is hindered by social marginalization and discrimination, with policies often failing to see to cultural preservation (Alexandri, 2014; Dirlik, 2003). Innanagigili, an Indigenous mother of four, who moved to Panama City, a melting pot of modern North American and European cultures, makes a living selling traditional Guna Yala jewellery along the Cinta Costera. She explained that:

Adapting to the culture has been very difficult and the pressure to assimilate is hard especially since mainlanders inside and outside of the informal settlement I live in[ ] do not understand Guna Yala culture nor think it has any place in their city. It was hard to learn the language and jobs have been hard to find[,] because I do not have educational qualifications fitting their standards and documentation.

Seated next to her, Daniel, who no longer uses his Guna Yala name while job hunting because of the discrimination, explained that a combination of these things has forced them into undesirable and limiting circumstances that demand socio-cultural assimilation and not socio-cultural integration. Implicit here, is the ideas of citizenship that see those better assimilated into the urban culture of Panama City as more Panamanian than those with Indigenous cultures. Within this context, household economic stability escapes many of these Indigenous families for generations and leaves them open to exploitation (Tovar-Restrepo & Irazábal, 2014; Wickstrom, 2003; Whyte, 2017).

There is a missed opportunity here to leverage Indigenous culture and traditional knowledge for urban development, as Indigenous people as urban integrators can help to bridge gaps between traditional and contemporary practices to build urban resilience. In the *Panama Corporates 2030 Plan*, and Bogotá's *El Plan de Ordenamiento Territorial-Bogotá 2022-2035* (POT, 2021), the governments recognize these opportunities; however, effective integration requires confronting structural inequalities and socio-political dynamics that limit Indigenous influence in decision-making (Whyte, 2017). In Colombia, for example, despite POT's strategic planning for urban development, Indigenous populations often find themselves on the periphery of these initiatives, and the lack of culturally appropriate social services and adequate representation in political and decision-making spheres further intensifies their marginalization (Berney, 2011; Columbia Events, 2021; Tovar-Restrepo & Irazábal, 2014).

Within LAC's complex urban ecosystems, climate change, spatial segregation, and social equity are intertwined with historical, social, and environmental factors affecting vulnerability and resilience. This means urban vulnerability in the region is not merely geographic or physical, but shaped by socio-economic disparities, highlighting the need for equity-focused development policies with meaningful participatory inputs.

GOVERNANCE AND SOCIAL INCLUSION

The truth is that development as an idea or practice is contentious, serving the purpose of dominant protagonists of theory and model at a particular time. This might explain the difficulty countries- developing countries, especially- experience in matching priorities to context or realizing goals, since goals and priorities are defined externally and continue to shift. (Schoburgh & Gatchair, 2016, p. 128)

What Schoburgh and Gatchair (2016) refer to is the nexus of international development imperatives and national priorities, and how they unite in any localized multi-actor system. This draws attention to impediments inherent to the democratization of decision-making, equity, and intervention efficacy in plural urban ecosystems. In this way, how international frameworks translate to local policies is influenced by context-unique factors, defined by governance and power dynamics, with direct implications for policy and the extent to which social justice and climate adaptation are engendered in strategic sustainable development.

International Frameworks and Local Government

Within an anarchic international system, frameworks, and conventions, such as the Paris Agreement, New Urban Agenda, Sendai Framework for Risk Reduction, and the Sustainable Development Goals (SDGs), guide the institutionalization of participatory governance for socio-climate justice. However, the translation of these international mandates into national development plans is an intricate process and, perhaps as a fault of their nature, these international agreements also assume a linear progression for local implementation. However, within LAC, several barriers to effective and inclusive governance exist, including institutional resistance, political will, financial resources, and stakeholder capacity (Gunnarsson-Östling & Svenfelt, 2017; Jones, 2010; Schoburgh, 2017).

When the focus is placed on socio-climate resilience, development planning requires long-term commitments and changes that may not align with short-term political cycles. Politicians who are primarily concerned with immediate re-election may not prioritize these if they involve difficult decisions or significant upfront and long-term costs, which can later lead to policy disruption (Jones, 2010; Karmack, 2019; Schoburgh & Gatchair, 2016). This concern is referenced by urban planners across all four cities in direct contravention to the tenets of the politics-administration dichotomy and the qualities of the bureaucratic servant who should be free of political interference and guided by tenets of good governance (Kamarck, 2019; Mills, 1997; Wickstrom, 2003).

Specific strategies to undertake this can involve training programs, workshops, and exchange programs with more developed urban centres. Additionally, stakeholder engagement programs can be established to ensure that all relevant parties, including marginalized communities, have a voice in the urban planning process. These programs facilitate dialogue, gather diverse perspectives, and foster a sense of ownership among stakeholders. To increase political will, public awareness campaigns and advocacy efforts are crucial. These campaigns can educate the public and policymakers about the benefits of participatory governance and the risks of maintaining the status quo. Advocacy efforts can also be directed at reforming policies and regulations to support more inclusive and equitable urban development practices.

There is also considerable emphasis on the role of local institutions and governance mechanisms in a prescriptive oversimplification of these relationships. Local institutions with limited human and structural resources or weak governance structures may struggle to facilitate and promote social inclusion. Schoburgh (2017) explored the role of meso-level actors (municipalities) in local development, with a focus on the Anglophone Caribbean policy systems in Jamaica and Trinidad and Tobago. It is argued that local government is increasingly seen as an appropriate institutional context to pursue both short- and long-range goals of social transformation. However, Schoburgh (2017) raised two central concerns: the institutional and organizational imperatives of a developmental role for local government, and the extent to which these imperatives have been dealt with in reform, and revealed substantive convergence around local development as an outcome of reform, but an important divergence in the approach to achieving this goal. That observation suggests the absence of a cohesive model that links local government more consistently with a local development strategy, as per the benefits of sociocybernetics and hyperlocal integration. Schoburgh asserted that such a strategy must incorporate gender equality, the informal economy, and institutional organizational capacity to form the basis for creating a local context in which all types of resources can be maximized in the potential to drive both economic and social transformation.

Practical steps to overcome this institutional resistance should include capacity-building initiatives to enhance the skills and knowledge of local government officials and stakeholders, along with the creation of dedicated task forces or committees. These bodies would be responsible for dissecting international mandates, understanding their relevance to local contexts, and formulating actionable policies. Additionally, establishing clear timelines for implementation and setting up accountability mechanisms can ensure that these policies are not only adopted but also effectively executed with a commitment to regular monitoring and evaluation to adapt strategies as needed (Schoburgh, 2017).

An argument can be made here for regional organizations, such as the Economic Commission for Latin America and the Caribbean (ECLAC), the Union of South American Nations (UNASUR), and the Caribbean Disaster Emergency Management Agency (CDEMA) to play a more mediatory role between international imperatives and national development plans for inclusive urban planning, social justice, and regional cohesion. They can achieve this by organizing regional dialogues that bring together policymakers, experts, and local stakeholders to discuss and adapt international guidelines to regional contexts. Additionally, these organizations can provide technical assistance to local governments, helping them to develop the necessary skills and capacities to implement these frameworks effectively. Sharing best practices and success stories from within the region can also serve as a valuable resource, offering practical insights and inspiring local adaptations of international norms for socio-climate adaptation and resilience. By playing this intermediary role, regional organizations can ensure that international goals are met while respecting local needs and sharing this knowledge with urban citizens, designers, and integrators for more inclusive and equitable strategies. However, it is important to note that the success of these strategies is dependent on the political, economic, and social context in which they are implemented, where the ultimate responsibility rests with how inclusive and effective governance is

People-Centered Urban Governance in Latin America and the Caribbean

Navigating Citizen Participation With Arnstein's Ladder and Fung's Democracy Cube

A dilemmic convergence across all four cities is that of a lack of consultation beyond tokenism, especially for the urban poor. Guided by sociocybernetics principles and sustainable development imperatives, Fung's (2006) "Democracy Cube" and Arnstein's (1069) "Ladder of Citizen Participation" are both seminal models applicable to resolving this issue.

Arnstein's (1969) "Ladder of Citizen Participation" is a linear model with eight rungs, each symbolizing a level of citizen involvement in decision-making processes. The model illustrates a power dynamic, with the lower rungs representing non-participation (manipulation and therapy), the middle rungs indicating degrees of tokenism (informing, consultation, and placation), and the top rungs embodying citizen power (partnership, delegated power, and citizen control). This ladder suggests a normative progression to citizens' influence on outcomes, implying that higher levels of participation are inherently more democratic and desirable. On the other hand, Fung's "Democracy Cube," a three-dimensional model, offers a more nuanced perspective (Fung, 2006). It categorizes democratic practices along three axes: the level of participation, the mode of communication and decision, and the degree of influence. Unlike Arnstein's ladder, Fung's cube does not inherently rank forms of participation as superior or inferior. Instead, it acknowledges the diversity of democratic practices and suggests that different forms of participation may be suitable in different contexts.

Consider a scenario where Port-of-Spain, given frequent flooding, is planning to redevelop a neighbourhood. Using Arnstein's ladder, one might evaluate the process based on the degree of power given to the residents. If the city merely informs residents about the redevelopment plans (informing), this would be seen as a low level of participation. If the city consults with residents but does not necessarily incorporate their feedback into the final plans (consultation), this would be seen as tokenistic. True citizen power would only be achieved if residents are given a significant say in the decision-making process, through a partnership where residents are given delegated power or full control over the redevelopment plans through hyperlocal co-production arrangement.

Alternatively, using Fung's "Democracy Cube," one might look at not just the degree of power given to residents, but also who is included in the decision-making process and how decisions are made. Meaning, are all residents allowed to participate, or only a select few? Are decisions made through a process of deliberation, where residents can voice their opinions and concerns, or are they simply asked to vote on pre-determined options? And what influence do residents' inputs have on the final decision? Here, Fung's model is particularly useful in recognizing the diversity of stakeholders, the complexity of decision-making processes, and the role of urban integrators, who mediate the intrinsic power dynamics. What to consider here is that any homogenous neighbourhood can present different needs arising from its heterogeneity in socio-cultural realities, and so a "one-size-fits-all" approach to participation is rarely effective.

For example, consider urban Indigenous citizens such as Inanagigili and David, who often find homes in informal settlements with diverse cultures, ethnicities, and demands, all seeking to fit in without the erasure synonymous with cultural imperialism of the colonial era. Additionally, Arnstein's (1969) model serves as a reminder that the process of participation is not just about dialogue and deliberation, but also about power. Even the most inclusive and deliberative process may be meaningless if residents' inputs do not influence the final decision. Sociocybernetics, with its emphasis on feedback loops, self-organization, hybrid (negotiated/normative) governance, and system adaptation, becomes useful here.

121

However, Arnstein (1969), Fung (2006), and the tenets of sociocybernetics also remind us that systems tend to resist change (negative feedback loops), especially when there are power imbalances. If not properly managed, participatory processes can be co-opted by more powerful or vocal groups, exacerbating rather than mitigating social inequalities (Arnstein, 1969; Healey, 2003; Fung, 2006, Inner & Booher, 2010). Special consideration must also be given to communities with negotiated normative hybrid governance, which through auto-poietic processes, have developed their internal governance structures with the nation's government that is seen as illegitimate, particularly in historically underrepresented communities (Campbell, 2020; Geyer & Van der Zouwen, 2014; Harrison, 1988;). This also raises the question of the extent to which these informal hybrid governance systems are to be given more power and autonomy in the co-creation processes.

In 2017, when the redevelopment of the Kingston downtown waterfront area began as part of Jamaica's urban renewal focus, "Dons" from the surrounding communities (some labelled as garrisons) expressed an interest in being a part of the process (Cross, 2017). Given Jamaica's complex relationship with "donmanship," garrison communities coming from the instability of the 1980s political debacle (Harrison, 1988; Clarke, 2006; Campbell, 2005, 2020), and its history of violent crimes associated with gangs in and around the area, the leadership of the Urban Development Corporation (UDC) was quick to state that this would not be a case of endorsement of these informal leadership arrangements by giving them a seat at the table (Cross, 2017). This posed two dilemmas: (1) further exclusion of communities with these normatively accepted forms of leadership, where governance gaps have allowed them to thrive; (2) potential retaliatory disruptions to development initiatives and redevelopment that lead to further displacement because of the exclusion. The redevelopment of Kingston's waterfront proceeded, but from the perspective of these community stakeholders, inclusivity was seemingly conceptualized as an end-user experience.

Urban designers and integrators can assist in changing these inclusion dilemmas through active partnerships that enhance system-level understanding and appreciation of community knowledge as a normal and valued part of civic life. However, power shifts also need to occur at a deeper, institutional level, reflecting changes to the legal and policy frameworks that govern urban planning to ensure that participatory processes are not just optional add-ons, but mandatory elements. It must involve changes in the culture and practices of planning organizations, to ensure that planners see residents not as passive recipients of their expertise, but as active co-producers of urban spaces, which has the potential to challenge stereotypes and prejudices and break down barriers to inclusive participation. Public participation is not a panacea, however, so careful attention must be paid to the equitable design and implementation of these processes. Furthermore, while public participation can help to mitigate "othering," it cannot eradicate it on its own. In the post-independence era, each of these countries must critically examine how they have institutionalized discriminatory and socially unjust practices, and how to resolve them.

OTHER COUNTRY-SPECIFIC RECOMMENDATIONS

Bogotá, Colombia

The Bogotá City Master Plan, also known as *El Plan de Ordenamiento Territorial-Bogotá 2022-2035* (POT, 2021), is laudable for its focus on social inclusion, sustainable mobility, and environmental protection. It aims to reduce or eliminate social inequality by promoting mixed-income neighbourhoods, affordable housing, and improved access to essential services for marginalized communities. Evident in the expansion of the TransMilenio Bus Rapid Transit system and the development of bicycle infrastructure, it also prioritizes inclusive and sustainable mobility, with measures to conserve green spaces, encourage urban agriculture, and mitigate climate change impacts. However, while it acknowledges social inclusion, it lacks specific strategies to comprehensively deal with social inequalities, particularly in access to quality education, healthcare, and job opportunities. The plan could further emphasize climate adaptation and resilience, attending to the challenges and disparities posed by climate change more explicitly in risk governance frameworks.

Kingston, Jamaica

Jamaica's Vision 2030 aims to make "Jamaica the place of choice to live, work, raise families, and do business." Despite its inclusive vision, the program has faced criticism for not adequately challenging the socio-spatial disparities, levels of administrative corruption, and climate vulnerabilities faced by marginalized communities. In keeping with the recommendations by Schoburgh (2012), Jamaica's Vision 2030 should empower local governance mechanisms and localized decision-making bodies to build on community-based knowledge and risk governance for greater adaptation. It should also prioritize "participatory vulnerability assessments" to understand community-specific climate risks and incorporate these findings into mitigation and adaptation strategies. Urban anthropological research is key to this being effective, and so the government should develop more robust and integrated research frameworks and data-sharing capacity. IDB's ESC project has also been implemented in Jamaica's second city, Montego Bay, and so urban designers and integrators in Kingston should seek to learn from the successes and failures of that project.

Panama City, Panama

Panama should focus on formulating policies that respect the rights of Indigenous peoples, particularly regarding land ownership and social integration that acknowledges the value of Indigenous environmental stewardship. The government should also pursue targeted and "sustainable resettlement strategies" that consider the preservation of the Guna Yala's traditions, and culture. Although the 1938 law creating *Comarcas* has served its purpose well, the impacts of climate change, population growth, and socioeconomic needs, demand targeted integration strategies, especially in Panama City, a melting pot of culture, economics, and tourism.

Port-of-Spain, Trinidad

Urban citizens and designers in Trinidad and Tobago have called for the prioritization of community-led flood management initiatives that foster public-private partnerships to invest in sustainable urban infrastructure resilient to recurring floods. Moreover, flood risk maps and early warning systems should be co-developed with local communities, incorporating their lived experiences and insights. There is also merit to developing a "Community Flood Resilience" program, where local communities are involved in co-designing flood management strategies that would leverage the benefits of sociocybernetics feedback loops and hyperlocal strategies to include community-based perspectives for flood defences and locally managed early warning systems.

CONCLUSION

The exploration of climate vulnerability and resilience in urban planning within Colombia, Panama, Jamaica, Trinidad, and Tobago uncovers a complex interplay of factors shaping their urban landscapes. Historical influences, notably colonialism, have significantly impacted the socio-spatial dynamics of these nations, influencing urban development trajectories. Urban planning imperialism, characterized by top-down approaches, has historically marginalized certain communities, leading to socio-spatial segregation and reinforcing inequities. These factors contribute to heightened socio-climate vulnerability, particularly for marginalized urban populations who disproportionately face environmental hazards and climate change impacts.

Central to confronting these challenges is the role of participatory governance in urban planning. The involvement of urban citizens, designers, and integrators is crucial in creating equitable, resilient, and sustainable urban environments. Emphasizing social inclusion and equity, people-centred urban planning speaks to both the symptoms and root causes of climate vulnerability and inequity. Sociocybernetics emerges as a valuable tool in enhancing participatory governance and urban resilience. However, challenges such as socio-political divisions, resource limitations, and gaps between policy and practice, persist. Overcoming these challenges requires strengthening participatory mechanisms, enhancing capacities, and ensuring tangible outcomes from participatory processes.

International frameworks, such as the Sustainable Development Goals (SDGs), alongside national development plans, significantly influence urban planning agendas. These frameworks provide a platform to combine climate resilience and social equity goals. Their effectiveness, however, hinges on their translation into inclusive, participatory, and context-specific policy practices. This analysis highlights the interconnected nature of sustainable development, socio-climate justice, and participatory governance in urban planning. The way forward involves harnessing participatory governance and people-centred planning to create urban spaces resilient to climate change and grounded in equity and inclusivity, ensuring no one is left behind.

REFERENCES

U. N. (2016). *The new urban agenda*. In *The United Nations Conference on housing and sustainable urban development (Habitat III) held in Quito, Ecuador*. UN. https://habitat3.org/the-new-urban-agenda/

Alexandri, G. (2014). Reading between the lines: Gentrification tendencies and issues of urban fear in the midst of Athens' crisis. *Urban Studies (Edinburgh, Scotland)*, *52*(9), 1631–1646. doi:10.1177/0042098014538680

Angotti, T., & Irazábal, C. (2017). Planning Latin American cities: Dependencies and "Best Practices.". *Latin American Perspectives*, *44*(2), 4–17. https://www.jstor.org/stable/26178807. doi:10.1177/0094582X16689556

Antonova, N. L., & Grunt, E. V. (2019, December). Citizens' role in formation of urban environment design. In *IOP Conference Series: Materials Science and Engineering*. IOP Publishing. 10.1088/1757-899X/687/5/055053

Arnstein, S. R. (1969). A ladder of citizen participation. *Journal of the American Institute of Planners, 35*(4), 216–224. doi:10.1080/01944366908977225

Arsel, M. (2023). Climate change and class conflict in the Anthropocene: Sink or swim together? *The Journal of Peasant Studies, 50*(1), 67–95. doi:10.1080/03066150.2022.2113390

Asher, S., & Novosad, P. (2020). Rural roads and local economic development. *The American Economic Review, 110*(3), 797–823. doi:10.1257/aer.20180268

Barca, F. (2008). *An agenda for a reformed cohesion policy: A place-based approach to meeting European Union challenges and expectations.* (No. EERI_RP_2008_06). Economics and Econometrics Research Institute (EERI), Brussels. https://ec.europa.eu/regional_policy/archive/policy/future/barca_en.htm

Barker, D. (2012). Caribbean agriculture in a period of global change: Vulnerabilities and opportunities. *Caribbean Studies (Rio Piedras, San Juan, P.R.), 40*(2), 41–61. https://www.jstor.org/stable/41917603. doi:10.1353/crb.2012.0027

Beer, S. (1984). The viable system model: Its provenance, development, methodology and pathology. *The Journal of the Operational Research Society, 35*(1), 7–25. doi:10.1057/jors.1984.2

Bennett, K. (2021, November 9). *NHT defends sky-high Ruthven Towers prices.* Jamaica WI-The Gleaner. com. https://jamaica-gleaner.com/article/lead-stories/20211109/nht-defends-sky-high-ruthven-towers-prices

Berney, R. (2011). Pedagogical urbanism: Creating citizen space in Bogotá, Colombia. *Planning Theory, 10*(1), 16–34. https://www.jstor.org/stable/26165894. doi:10.1177/1473095210386069

Bogotá.gov.co. (2019). *Bogotá's Master Plan Targets Major Global Development Agendas.* Bogotá.gov. co. https://Bogotá.gov.co/internacional/Bogotás-master-plan-targets-major-global-development-agendas

Campbell, H. (2005). Reflections on the post-colonial Caribbean state in the 21st century. *Social and Economic Studies, 54*(1), 161–187. https://www.jstor.org/stable/27866408

Campbell, Y. (2020). *Citizenship on the margins.* Springer International Publishing. doi:10.1007/978-3-030-27621-8

Cárdenas, M., Bonilla, J. P., & Brusa, F. (2021). *Climate policies in Latin America and the Caribbean: Success stories and challenges in the fight against climate change.* Publications.iadb.org. https://publications.iadb.org/publications/english/viewer/Climate-policies-in-latin-america-and-the-caribbean.pdf

Christophers, B. (2016). For real: Land as capital and commodity. *Transactions of the Institute of British Geographers, 41*(2), 134–148. https://www.jstor.org/stable/45147008. doi:10.1111/tran.12111

Clarke, C. (2006). Politics, violence and drugs in Kingston, Jamaica. *Bulletin of Latin American Research, 25*(3), 420–440. https://www.jstor.org/stable/27733873. doi:10.1111/j.0261-3050.2006.00205.x

Columbia Events. 2021. (2021, December 10). *World Report: Colombia.* Human Rights Watch. https://www.hrw.org/world-report/2022/country-chapters/colombia

Cross, J. (2017, October 6). *Dons want in: Downtown Kingston area leaders call meeting to discuss role in redevelopment*. Jamaica WI-The Gleaner.com. https://jamaica-gleaner.com/article/lead-stories/20171010/dons-want-downtown-kingston-area-leaders-call-meeting-discuss-role

Curtin, P. D. (1955). *Two Jamaicas: The role of ideas in a tropical colony, 1830–1865*. Harvard University Press.

Dirlik, A. (2003). Globalization, indigenism, and the politics of place. *ARIEL: A Review of International English Literature, 34*(1).

Duer, M., & Vegliò, S. (2019). Modern-colonial geographies in Latin America. *Journal of Latin American Geography, 18*(3), 11–29. https://www.jstor.org/stable/48618849. doi:10.1353/lag.2019.0058

Fay, M. (2005). *The Urban Poor in Latin America*. World Bank Publications. doi:10.1596/0-8213-6069-8

Forbath, W. E. (1999). Caste, class, and equal citizenship. *Michigan Law Review, 98*(1), 1–91. doi:10.2307/1290195

Fung, A. (2006). Varieties of participation in complex governance. *Public Administration Review, 66*(s1), 66–75. https://www.jstor.org/stable/4096571. doi:10.1111/j.1540-6210.2006.00667.x

Geyer, R. F., & Van der Zouwen, J. (Eds.). (2014). *Sociocybernetics: An actor-oriented social systems approach* (Vol. 1). Springer.

Gunnarsson-Östling, U., & Svenfelt, Å. (2017). Towards social-ecological justice. In *The Routledge Handbook of Environmental Justice* (p. 160). Routledge.

Guzman, L., & Bocarejo, J. P. (2017). Urban form and spatial urban equity in Bogotá, Colombia. *Transportation Research Procedia, 25*, 4491–4506. doi:10.1016/j.trpro.2017.05.345

Harrison, F. V. (1988). The politics of social outlawry in urban Jamaica. *Urban Anthropology and Studies of Cultural Systems and World Economic Development, 17*(2/3), 259–277. https://www.jstor.org/stable/40553119

Healey, P. (2003). Collaborative planning in perspective. *Planning Theory, 2*(2), 101–123. doi:10.1177/14730952030022002

Horvath, R. J. (1972). A definition of colonialism. *Current Anthropology, 13*(1), 45–57. https://www.jstor.org/stable/2741072. doi:10.1086/201248

Innes, J. E., & Booher, D. E. (2018). *Planning with complexity*. Routledge. doi:10.4324/9781315147949

Inter-American Development Bank. (2015). *Panamá metropolitana: Sostenible, humana y global*. IDB. https://www.iadb.org/en/urban-development-and-housing/emerging-and-sustainable-cities-program

IRGC. (2018, September). *IRGC Guidelines for the Governance of Systemic Risks*. doi:10.5075/epfl-irgc-257279

Islam, S. N., & Winkel, J. (2017). *Climate change and social inequality*. UN Department of Economic and Social Affairs. https://digitallibrary.un.org/record/3859027?ln=en

Jones, E. (2010). Contending with local governance in Jamaica: Bold Programme, Cautionary Tales. *Social and Economic Studies*, 59(4), 67–95. https://www.jstor.org/stable/41803728

Kamarck, E. (2019, September). *The challenging politics of climate change.* Brookings. https://www.brookings.edu/research/the-challenging-politics-of-climate-change/

Kelman, I. (2020). *Disaster by choice.* Oxford University Press.

Khan, M. (2022, March 27). Examining gentrification: A new internal colonialism. *Inverse Journal.* https://www.inversejournal.com/2022/03/27/examining-gentrification-a-new-internal-colonialism-an-academic-essay-by-m-moosa-khan/

Luhmann, N. (2012). *Theory of society* (Vol. 1; R. Barrett, Trans.). Stanford University Press.

Maurer, S., & Rauch, F. (2019). Economic geography aspects of the Panama Canal. *Centre for Economic Performance, 1633.* https://cep.lse.ac.uk/pubs/download/dp1633.pdf

McSweeney, K., & Jokisch, B. (2007). Beyond rainforests: Urbanisation and emigration among lowland Indigenous societies in Latin America. *Bulletin of Latin American Research*, 26(2), 159–180. doi:10.1111/j.1470-9856.2007.00218.x

Mills, G. (1997). *Westminster style democracy: The Jamaican experience.* Grace Kennedy Foundation Lecture. http://gracekennedy.com/lecture/GKF1997Lecture.pdf

Ministry of Works and Transport (MOWT)- Trinidad and Tobago. (2020). *The Port of Spain Flood Alleviation Project.* MOWT. https://www.mowt.gov.tt/Divisions/Programme-For-Upgrading-Roads-Efficiency-(PURE)-Un/Projects/The-Port-of-Spain-Flood-Alleviation-Project

Mora, C., Frazier, A. G., Longman, R. J., Dacks, R. S., Walton, M. M., Tong, E. J., Sanchez, J. J., Kaiser, L. R., Stender, Y. O., Anderson, J. M., Ambrosino, C. M., Fernandez-Silva, I., Giuseffi, L. M., & Giambelluca, T. W. (2013). The projected timing of climate departure from recent variability. *Nature*, 502(7470), 183–187. doi:10.1038/nature12540 PMID:24108050

Moreno, S. H. (1993). Impact of development on the Panama Canal environment. *Journal of Interamerican Studies and World Affairs*, 35(3), 129–150. doi:10.2307/165971

Mullings, J., Dunn, L., Sue Ho, M., Wilks, R., & Archer, C. (2018). Urban renewal and sustainable development in Jamaica: Progress, challenges and new directions. *An Overview of Urban and Regional Planning.* doi:10.5772/intechopen.79075

National Housing Policy (Draft). (2019). Ministry of Economic Growth and Job Creation.

Orwell, G. (1945). *Animal Farm: A fairy story and essays collection.* (75th anniversary). Penguin Publishing Group.

Pahl-Wostl, C. (2009). A conceptual framework for analysing adaptive capacity and multi-level learning processes in resource governance regimes. *Global Environmental Change*, 19(3), 354–365. doi:10.1016/j.gloenvcha.2009.06.001

Parsons, T. (2005). *The social system* (B. Turner, Ed.; 2nd ed.). Routledge., https://voidnetwork.gr/wp-content/uploads/2016/10/The-Social-System-by-Talcott-Parsons.pdf

Planning Institute of Jamaica (PIOJ). (2009). *Urban planning and regional development: sector plan 2009–2030.* PIOJ.

Pörtner, H. O., Roberts, D. C., Tignor, M. M. B., Poloczanska, E., Mintenbeck, K., Alegría, A., Craig, M., Langsdorf, S., Löschke, S., Möller, V., Okem, A., & Rama, B. (Eds.). (2022). *Climate Change 2022: Impacts, Adaptation and Vulnerability.* IPCC Assessment Report. https://www.ipcc.ch/report/ar6/wg2/

Rondinelli, D. A. (1990). Housing the urban poor in developing countries: The magnitude of housing deficiencies and the failure of conventional strategies are worldwide problems. *American Journal of Economics and Sociology, 49*(2), 153–166. https://www.jstor.org/stable/3487429. doi:10.1111/j.1536-7150.1990.tb02269.x

Runk, J. V. (2012). Indigenous land and environmental conflicts in Panama: Neoliberal multiculturalism, changing legislation, and human rights. *Journal of Latin American Geography, 11*(2), 21–47. doi:10.1353/lag.2012.0036

Ryan, D. (1999). Colonialism and hegemony in Latin America: An introduction. *The International History Review, 21*(2), 287–296. https://www.jstor.org/stable/40109004. doi:10.1080/07075332.1999.9640860

Salmon, D. (2020, June. *A Jamaican tale of two cities.* Jamaica WI-The Gleaner. https://jamaica-gleaner.com/article/focus/20200621/david-salmon-jamaican-tale-two-cities

Schoburgh, E. D. (2017). Is a self-help orientation sufficient basis for local [economic] development? *Journal of Human Values, 23*(3), 151–166. doi:10.1177/0971685817713287

Schoburgh, E. D., & Gatchair, S. (2016). Managing development in local government: frameworks and strategies in Jamaica. Developmental Local Governance: A Critical Discourse in 'Alternative Development. doi:10.1057/9781137558367_8

Sigler, T., & Wachsmuth, D. (2016). Transnational gentrification: Globalisation and neighbourhood change in Panama's Casco Antiguo. *Urban Studies (Edinburgh, Scotland), 53*(4), 705–722. https://www.jstor.org/stable/26151056. doi:10.1177/0042098014568070

Stiglitz, J. E. (2012). *The price of inequality: How today's divided society endangers our future.* WW Norton & Company.

Stone, C. (1973). *Class, race and political behaviour in urban Jamaica.* Institute of Social and Economic Research.

Tovar-Restrepo, M., & Irazábal, C. (2014). Indigenous women and violence in Colombia. *Latin American Perspectives, 41*(1), 39–58. doi:10.1177/0094582X13492134

Tsaaior, J. T. (2011). History, (re) memory and cultural self-presencing: The politics of postcolonial becoming in the Caribbean novel. *Journal of Caribbean Literatures, 7*(1), 123–137. https://www.jstor.org/stable/41939271

UN-DESA. (2022). *Total and urban population: UNCTAD Handbook of Statistics 2021.* United Nations Conference on Trade and Development (UNCTAD). https://hbs.unctad.org/total-and-urban-population/

United Nations Environment Programme (UNEP). (2023, March 31). *Panama to strengthen its sustainable finance framework: United Nations Environment – Finance initiative.* UNEP. https://www.unepfi.org/themes/climate-change/panama-to-strengthen-its-sustainable-finance-framework/*Urban law in Colombia.* (2018). UN-Habitat.

Varanasi, A. (2022, September 21). How colonialism spawned and continues to exacerbate the climate crisis. *State of the Planet, 23.* https://news.climate.columbia.edu/2022/09/21/how-colonialism-spawned-and-continues-to-exacerbate-the-climate-crisis/

Vision 2030. (2015). *Vision 2030: The national development strategy of Trinidad and Tobago.* Ministry of Planning and Development. https://www.planning.gov.tt/content/vision-2030

Vision 2030 Jamaica Sector Plans. (2008). Vision 2030. https://www.vision2030.gov.jm/vision-2030-jamaica-sector-plans/

Whyte, K. (2017). Indigenous climate change studies: Indigenizing futures, decolonizing the Anthropocene. *English Language Notes, 55*(1–2), 153–162. doi:10.1215/00138282-55.1-2.153

Whyte, K. (2018). Settler colonialism, ecology, and environmental injustice. *Environment and Society, 9*(1), 125–144. https://www.jstor.org/stable/26879582. doi:10.3167/ares.2018.090109

Wickstrom, S. (2003). The politics of development in Indigenous Panama. *Latin American Perspectives, 30*(4), 43–68. doi:10.1177/0094582X03030004006

ADDITIONAL READING

Kopytoff, B. K. (1976). Jamaican Maroon political organization: The effects of the treaties. *Social and Economic Studies, 25*(2), 87–105. https://www.jstor.org/stable/27861598

Wiener, N. (1973). *Cybernetics or Control and Communication in the Animal and the Machine.* MI. press.

KEY TERMS AND DEFINITIONS

Anthropocene: The current geological epoch in which human activities have had a significant global impact on the earth's ecosystems.

Autopoiesis: The self-creating, self-maintaining, and self-replicating nature of living systems. It describes the ability of a system to (re)produce its components continuously, maintaining its internal organization despite changes in its external environment.

Cultural Imperialism: The dominance of one culture over another, particularly when the dominant culture is imposed onto the less powerful one. This can be seen in the spread of language, religion, social norms, and economic practices.

Dominant Culture Narrative: The stories, values, beliefs, and norms most prevalent and influential within a society, reflecting the perspectives and interests of the society's dominant groups.

Hyperlocalism: The emphasis on small, local initiatives and solutions in dealing with larger issues. In the context of urban planning, it means focusing on community-based strategies and using local knowledge and resources to solve urban challenges.

Plantationocene: The current geological epoch, emphasizing the significant global impact of plantations as a specific form of land and labour use. It highlights the role of colonialism, capitalism, and racial hierarchies in shaping our world and its ecosystems.

Socio-Climate Justice: A concept that links social justice and climate change, recognizing that the impacts of climate change are not evenly distributed and often disproportionately affect marginalized and disadvantaged communities. It advocates for fair and equitable solutions to climate change.

Sociocybernetics: A field of study that applies systems theory and cybernetics to social sciences. It views societies as complex, self-regulating systems and seeks to understand the interactions and feedback loops that drive social dynamics.

Urban Imperialism: A term used to describe the dominance of urban values, norms, and ways of life over rural ones. It can also refer to the expansion of urban areas into rural ones, often driven by economic and political forces.

Chapter 7

Self–Defeat and Psychological Fragility as Predictors of Psychosomatic Disorders:
In Light of Climate Changes Among University Students in Egypt and the Emirates

Islam Hassan Abdel Wareth
Faculty of Education, Alexandria University, Egypt

Marwa Abdel Hamid Tawfiq
Ain Shams University, UAE

ABSTRACT

The current study aimed to reveal the relationship of self-defeat and psychological fragility with psychosomatic disorders among samples of university students in Egypt and the UAE. The study sample consisted of (200) university students in Egypt and the Emirates, and the results revealed a correlation between each of self-defeat and psychological fragility, and between psychosomatic disorders, and also found statistically significant differences between the mean scores of university students on measures of self-defeat and psychological fragility, and psychosomatic disorders according to gender in favor of females, as well as found statistically significant differences between Average scores of university students on measures of self-defeat, psychological fragility, and psychosomatic disorders according to the difference of nationality in favor of the Egyptian nationality.

INTRODUCTION

We all have the right to enjoy the highest attainable standard of physical and mental health. According to the Intergovernmental Panel on Climate Change, the main health impacts of climate change will

DOI: 10.4018/978-1-6684-8963-5.ch007

include increased risks of injury, disease and death from more intense heat waves and fires, among others; increased risk of undernutrition as a result of reduced food production in poor areas; Increased risk of food- and water-borne, and communicable diseases People, especially children, who are exposed to traumatic events such as natural disasters exacerbated by climate change, can suffer from post-traumatic stress disorder.

Climate changes have become a reality and have an impact on environments in general. It affected plants and animals, and its impact extended to humans, and one of its manifestations was that it appeared in poor countries, such as Arab societies, more than its impact on any other regions, including the need for water, which affected agricultural and industrial dates, and the spread of some diseases among individuals, including its impact on food security. For humans, the emergence of famines, the poor environments that have become widespread in the world, and there have been social and psychological problems resulting from these manifestations, of which there are many, many of which we have become in need of survey and awareness studies and the existence of programs and strategies to treat such disorders, especially psychosomatic disorders resulting from these climate changes.

A major shift has occurred in the recent period in many states and countries, and these climatic challenges have affected reality and life in all aspects, especially human beings, as millions of people are already suffering from the catastrophic effects of severe weather disasters that are exacerbated by climate change - starting with drought that lasts for periods long in sub-Saharan Africa, to the destructive tropical cyclones that sweep across Southeast Asia, the Caribbean and the Pacific Ocean. Extreme temperatures caused deadly heat waves in Europe, wildfires in South Korea, Algeria and Croatia, there were severe floods in Pakistan, while severe and prolonged drought in Madagascar left one million people with very limited access to adequate food.

Climate changes are long-term shifts in temperatures and weather patterns. These shifts may be natural and occur, for example, through changes in the solar cycle. However, since the nineteenth century, human activities have become the main cause of climate change, mainly due to burning Fossil fuels, such as coal, oil, and gas. Burning fossil fuels produces greenhouse gas emissions that act like a blanket wrap around the globe, trapping the sun's heat and raising temperatures. Examples of greenhouse gas emissions that cause climate change include carbon dioxide and methane. These gases are produced, for example, by using gasoline to drive cars or coal to heat buildings. Clearing land of weeds, shrubs, and cutting forests can also release carbon dioxide. Landfills are a major source of methane emissions. Energy production and consumption, industry, transportation, buildings, agriculture, and land use are among the major sources of emission (Faraj Ibrahim Ibrahim, 2022, 227).

Self-defeat is a state of feeling helpless, helpless, frustrated, lack of self-confidence, lack of personal effectiveness in life, and general unhappiness. It indicates, in part, at least exposure to traumatic events, difficult circumstances, and stressful life events, especially if it encounters a fragile psychological structure that is subject to breakage as a result of inappropriate socialization methods. Normality, as well as its association with exposure to coercion and coercion, with its negative effects on the personality of the individual, and self-defeat is a state that is achieved when the individual stops giving and his aspiration ceases from natural interaction with the things around him, and his view of the world becomes dark, carrying many tragedies and deprivation, and he imagines That it is impossible to achieve his goals and objectives, as it is a state of self-refraction, with the feelings and perceptions that it carries, that he loses motivation, self-confidence, and intolerance for his mistakes, as well as the tendency to insult and despise oneself and not stop whipping it (Laith Hamza, 2013, 12).

Self-Defeat, Psychological Fragility as Predictors of Psychosomatic Disorders

Self-defeat is a general psychological state with cognitive, emotional and behavioral contents that dominate the afflicted, embodied in a feeling of helplessness and helplessness towards the various events and realities of life in the present and the future, and is associated with feelings of depression, despair and shame, with the person's lack of self-activity and vitality, which leads him to surrender, rest and accept His personal reality without making any effort to change it and being completely dependent on the other at the level of thinking, emotion and action, and the tendency to belittle, humiliate and humiliate the self and consider it a material thing that has no life (Firas Abbas, 2004, 12).

We may sometimes exaggerate any problem that appears in our lives to the point of portraying it as an existential disaster, in a process called in psychology catastrophic pain, as it is an emotional state that an individual experiences when he falls into a problem that makes him believe that what happened to him is greater than his ability to bear, and he feels helpless and collapse. When a problem occurs, he describes the problem with exaggerated terms that are not equal to its real size, but rather negative descriptions that do not exist in his imagination, which only increases his suffering because he feels that it is greater than his ability to bear, although it is not in reality, and this exaggeration makes him drown in the feeling of spiritual wreck. And psychological exhaustion with no way out of it, his feeling of being lost and completely losing the ability to resist, and he surrenders to pain and his life collapses because of this problem (Ismail Arafa 2022, 24).

Radwan Zaqar (2015, 165) believes that the psychological fragility of the individual indicates a weak ability to withstand pressures and frustrations on the one hand, and his weak ability to manage his fidgety or aggressive impulses, bearing in mind that these two capabilities are essential in the process of building the individual, as a person needs throughout his life because He faces varying degrees of frustration, which helps him to form a sound perception of the reality in which he lives, and keeping with frequent and moderate frustrations allows him to choose reality, and develop the individual's stamina, which is necessary for his individual and social existence, and the individual acquires psychological strength and immunity when his endurance is high Which makes him in constant confrontation with himself and his surroundings, and he tends towards change for the better, and when the social conditions and educational patterns lack the frustration necessary for life, this makes the young man acquire psychological fragility that does not enable him to face the various challenges that life requires in general, and then the young man lives in a state From hesitation and even dispersion, which negatively affects his future and the future of his society.

Based on the foregoing, there was an urgent need to carry out the current study, which seeks to examine the correlation between self-defeat, psychological fragility, and psychosomatic disorders among samples from segments of Egyptian society in light of the current climate changes, to find out the factors and reasons that may contribute to predicting psychosomatic disorders among samples of students. University in Egypt and the Emirates.

THE PROBLEM

The current study is trying to investigate two very important variables that have appeared recently as a result of changes, including what is called self-defeat and psychological fragility, and that the individual resulted in him as a result of these changes, which became and became worse in the reality of these changes, which were reflected in his psychological and physical life and that defeat Self-defeat may lead to helplessness, lack of learning, lack of motivation, and it may lead to suicide, as self-defeat is intended as a general psychological state with cognitive, emotional, and behavioral contents that control its owner, and is embodied in a feeling of helplessness and helplessness towards the various events and facts of life in the present and the future, and is associated with feelings of depression. Despair and shame, along with a person's lack of self-activity and vitality, which leads him to surrender, complacency, and accept his personal reality without making any effort to change it and complete dependence on the other at the level of thinking, emotion, and action, and the tendency to belittle, humiliate, and humiliate the self and consider it a material, lifeless thing.

The problems of water poverty, drought, and lack of rain are also considered, along with the presence of population increase and the challenges of all of this on human mental health, a reason for the emergence of some disorders in him, and also the existence of limitations with the presence of land, and the shrinkage of individuals in specific areas, which led to the spread of many diseases that have already affected On the psychological state of the individual, especially the emergence of some psychosomatic disorders, and it was very important to provide methods and means that contribute to the treatment or guidance of these individuals to confront these phenomena and the resulting problems. Psychological and emotional blackmail, self-harm and self-defeat, emotional discharge and those mental illnesses were very important to treat because they affected the physical and physical side of individuals.

Accordingly, the current study sought to try to reveal the relationship of self-defeat and psychological fragility to psychosomatic disorders among samples of university students in Egypt and the UAE. Therefore, the study problem can be formulated in the following main question:

What Is the Relative Contribution of Self-Defeat and Psychological Fragility in Predicting Psychosomatic Disorders Among Samples of University Students in Egypt and the UAE?

From this main question, the following sub-questions emerge:

1. What is the nature of the correlation between self-defeat and psychosomatic disorders among samples of university students in Egypt and the UAE?
2. What is the nature of the correlation between psychological fragility and psychosomatic disorders among samples of university students in Egypt and the UAE?
3. Do the degrees of self-defeat, psychological fragility, and psychosomatic disorders differ among samples of university students in Egypt and the UAE according to gender (male/female)?
4. Do the degrees of self-defeat, psychological fragility, and psychosomatic disorders differ among samples of university students in Egypt and the UAE according to nationalities (Egyptian/Emirati)?
5. What is the extent of the contribution of self-defeat and psychological fragility in predicting psychosomatic disorders among samples of university students in Egypt and the UAE?

Aims of the Study

The aim of the current study is the following:

1. Understanding and interpreting the nature of the relationship between the dimensions of self-defeat and psychosomatic disorders among samples of university students in Egypt and the UAE.

2. Understanding and interpreting the nature of the relationship between the dimensions of psychological vulnerability and psychosomatic disorders among samples of university students in Egypt and the UAE.
3. Detection and interpretation of differences in self-defeat, psychological fragility and psychosomatic disorders according to the variables of gender, specialization and nationality.
4. Predicting psychosomatic disorders through the dimensions of self-defeat and psychological vulnerability among samples of university students in Egypt and the UAE.

The Importance of Studying

The importance of the study at the theoretical and applied levels is as follows:

A- Theoretical importance of the study: It is summarized in the following:
   1. The importance of the study lies in the importance of the study variables, self-defeat, psychological fragility, as predictors of psychosomatic disorders in light of climate changes.
   2. The lack of Arab studies that dealt with self-defeat and psychological fragility as predictors of psychosomatic disorders in light of climate changes.
   3. The current study can provide the researchers with some theoretical foundations that enable them to rely upon in subsequent research that may be complementary to this study or contribute to the expansion of other educational issues.
B- The applied importance of the study: It is summarized in the following:
   1. The current study is expected to contribute to identifying the level of self-defeat and psychological vulnerability as predictors of psychosomatic disorders in light of climate changes.
   2. The results of this study can be useful in developing appropriate plans, training and treatment programs aimed at reducing self-defeat and psychological fragility among individuals exposed to mental disorders.
   3. The current study attempts to enrich the Arab library with a measure of self-defeat, psychological fragility and psychological disorders, and to provide it to researchers in the educational fields.

Terminology of Study

1- Self Defeat:

And the researchers defined it in the current study as: "a state of failure formed by the individual as a result of reactions when he undertakes to do something that represents a necessary need for him, and this failure often results in different negative emotional patterns that may be unintended, which leads him to think about things that cause him impediment and lack of control that lead to defeat in the face of events", and is measured procedurally through the total score obtained by the participants on the self-defeat scale used in the current study (prepared by the two researchers).

2- Psychological fragility:

The researchers defined it in the current study as: "One of the types of crises that affect the psyche of the individual, as a result of several factors and conditions that he may go through, such as feeling constant tension and anxiety, lack of confidence in his abilities to manage crises, and the problems that he faces in practical and daily life is a problem that affects many of us, It makes the individual vulnerable to being controlled by negative feelings and psychological crises, and pushes him to become attached to others, and thinks that he loves them, while the truth is that he compensates for his weakness through them, as it is a personality disorder that needs to be treated so that the situation does not get worse." It is measured procedurally through the total score obtained by the participants. On the psychological fragility scale used in the current study (prepared by the two researchers).

3- Psychosomatic disorders:

In the current study, the researchers defined it as: "a group of physical symptoms that affect one of the body's systems and functions and do not respond to medical-pharmaceutical treatment", and it is measured procedurally through the total score obtained by the participants on the scale of psychosomatic disorders used in the current study (prepared by the researchers).

THEORETICAL FRAMEWORK AND PREVIOUS STUDIES

The First Axis Is Self-Defeat

Psychological defeat is a state of general psychological loss, the individual lives without a goal, he does not have any positive direction towards life, he is the greatest aggression to himself, his surrender, the loss of his self-confidence, the deterioration of his mental and physical health, and his feet to fall victim to psychological death.

The Concept of Self-Defeating

Riyad Al-Assimi (2012, 21) refers to self-defeat as "thinking that drives individuals to behavior that causes them failure, trouble or harm, as if it is a strategy that hinders the self, and individuals always feel a constant fear of not achieving success and slipping towards failure. Individuals with self-defeating behavior hinder themselves to avoid responsibility for their failure.

Campellone, Sanchez & Kring (2016, 1343) defined self-defeat as "a set of negative thoughts about a person's ability to excel in implementing goal-directed behavior that hinder the initiation and participation in the behavior."

And Mahmoud Al-Attar (2019, 391) defined it as "a situation in which the most dangerous thing is that when a person defeats himself, he does not try to defend it, while in the case of others trying to defeat him, he works hard to meet and confront others' attempt to defeat him, and therefore the greatest damage is in the case of defeat." the individual for himself."

Dimensions Of Self Defeat

Muhammad Abu Halawa (2013, 154) refers to the dimensions of self-defeat as follows:

1- **Feeling of shame**: It is defined as an emotion that possesses its owner and pushes him to feel contempt, disgust, humiliation, embarrassment, a sense of inferiority or inferiority, insignificance, uselessness, and the desire to hide from others.

2- **Self-deprecation:** It is defined as a person's feeling of worthlessness and the lack of his ability and capabilities compared to others, with the tendency to underestimate and multiply the self.

3- **Reification:** It is defined as a psychological state in which a person loses his sense of his personal identity and self-reality, and deals with himself as a material thing that has no life in him.

4- **Cognitive Perceptions**: It is defined as a group of cognitive ideas and beliefs that drive a person to feel the dominance of aspects of weakness and shortcomings over him with his inability to resist or face the events and facts of life, in addition to coloring his emotional life with despair and pessimism.

5- **Lack of self-energy:** It is defined as the individual's feeling of psychological and general behavioral dullness and aversion to life and not welcoming it.

6- **Self-flagellation:** It is defined as a person's view of his mistakes as if they are unforgivable, with the illusion that those around him know them well, as well as the belief that these mistakes are a result of weakness in the personality and a deficiency in the general psychological formation compared to others, as is evident in the negative self-talk.

Stages Of Self Defeat

Self-defeat occurs gradually within a framework of a sense of lack of control and failure, and through several stages, which were explained by Yasser Al-Shalabi (2014, 12) as follows:

1. A feeling of frustration accompanied by high energy and commitment at the beginning, and the person may develop some negative attitudes towards work at this stage.
2. The feeling of delusion dominates over the following aspects: the person assumes impatience, fatigue, and negative self-evaluation, which makes it easy for him to succumb to frustration.
3. Low energy and commitment to work, especially when exposed to external pressures and interference at work. At this stage, the person suffers from a lack of appetite and withdrawal behaviors such as taking sedatives, which may turn into addiction.
4. Losing enthusiasm and starting to mock co-workers, as a person feels that his work is without purpose or meaning when compared to his other life tasks, so work becomes his lowest priority.
5. Despair and surrender, as a feeling at this stage has the following symptoms: tremendous failure, self-doubt, emptiness, desire to escape from work, introversion and isolation. These symptoms result in physical and emotional effects that may develop into chronic disability.

Characteristics of Individuals With Self-Defeating Behavior

Sherry, Stoeber & Ramasubbu (2016, 197) indicate that the most prominent characteristics of individuals with self-defeating behavior are as follows:

1. They do not finish their task after starting it, and they tell themselves not to complete the task even if they start the task.

2. They feel helpless and have a feeling that their tasks are difficult, and there are obstacles in front of their activities and tasks, and they put themselves next to failure, so they direct their efforts away from achieving the goal, and soon they find that achieving those goals is difficult.

3. They stay away from any situation or activity that causes them pain, hinders the achievement of goals, and deviates them from the desired results.

4. Negative self-talk, which stems from their feeling that the tasks are beyond their abilities, so they use phrases that express self-defeat such as (I cannot do that, this does not concern me, the task is very difficult, the task will be higher than my capabilities, I will finish it later.

5. They tend to procrastinate and procrastinate when they start a task or a project. They tend not to complete the project or task and leave it before completing it and start another. Procrastination behavior is a cognitive precursor that develops from fear that accomplishing goals or tasks may lead to more hard work. They set goals and expectations. High-level and at the same time they expect not to complete the tasks in a brilliant way, failure is the only truth achieved for them

6. Poor and negative performance in life: This gives them a bad reputation in their relationships and in solving their problems, so they suffer because of poor achievement and because of their negative behavior in performing tasks.

7. Self-flagellation: They practice cruelty with themselves, so they resort to violence with themselves and lack of compassion for themselves when they make mistakes in situations that are not commensurate with their mistake.

Reasons For Self-Defeating

Mostafa Hegazy (2005, 58) refers to a set of determinants of creating a sense of psychological defeat, the most prominent of which are:

- Exposure to continuous emotional rejection as a form of emotional abuse, and this exposure causes psychological pain that pushes the individual to continue feelings of weakness and non-acceptance, and from here he may resort on the subconscious psychological level to submissiveness, excessive compliance, and loss of self-efficacy.

- Persistent failure of the individual and his failure is one of the most important determinants of creating a state of psychological defeat. One sees himself in failure and failure always falls into his hand.

- The disrespect and contempt of others leads to a negative idea of one's self, which is impotence, helplessness, and lack of self-confidence.

- A struggle between his mind and his feelings, he finds himself unable to push his feelings, get rid of his obsessions, overcome his fears or know their source sometimes.

- Anxiety and fear of the future.

- Recognizing the individual as weak.

- What repression, tyranny and oppression do to the psychological structure of those who are constantly exposed to it, is among the determinants that create contempt and psychological defeat.

- Generating low self-esteem and lack of self-confidence in the learner, and educating him on lack of ambition, constructive discussion and positive criticism, which is known as the phenomenon of behavioral desisting and intellectual insensitivity.

- We find that the manifestations of a sense of defeat and psychological refraction indicate the weakness of the general psychological structure of the individual and the lack of skills, which is called psychological resilience in psychology and its lack of psychological and social support.

Prevention of Self Defeat

A - Prevention at the cultural and social level:
1- Adopting the method of unconditional emotional acceptance of children in parental treatment methods.
2- Restructuring the teaching and learning systems at the level of theoretical, constructive and functional premises by adopting teaching and teaching systems based on creativity to form the mature personality.
3- Facilitating the paths of success for children by restructuring the experiences of socialization, teaching and learning in a manner that ensures their sense of competence and self-confidence.
4- The need to drop the mental illusions that drive towards reliance on imitation, and to dismantle the culture of fear in the structure of socialization, education and teaching systems, and to restore respect for the learner.
B- Awareness of emotions and feelings and their regulation:

The first ability is a person's recognition and identification of the emotions and feelings that control him, the ability to control these feelings and emotions (Ali Abdel-Salam, 2005, 13).

C- Controlling self-impulses:

Tolerating ambiguity, they are able to wait, ponder, and reflect on things or events before making a decision.

D- Optimism: What we mean is realistic optimism that distances a person from imaginary fantasies that are not based on reality, and those who show unrealistic, gelatinous optimism.

E- Causal analysis: It means the ability to comprehensively and in-depth thinking about the problems it faces, and they are the people who are able to look at the problems from multiple angles to gain insight into the various factors causing them and the various possible solutions to overcome them.

F- Self-efficacy: A person's confidence in his ability to solve problems and his focus on the proper use of his strengths to enable him to adapt positively to the events and facts of life (Ashraf Mohamed Attia, 2011, 577).

The Second Axis: Psychological Fragility

Psychological fragility is not a mental illness. Rather, it is a method of upbringing, not nature. When comparing the generation of fathers and mothers with the generation of children, we find that the generation of fathers and mothers is better able to bear the hard life, with the absence of luxuries; Such as service and entertainment technology, that they do not complain or complain, unlike the generation of children, who live in the permanent role of the victim, as their feelings are luxurious, and their psychological toughness is weak, which makes them vulnerable to breakage with the first real experience in real life.

The Concept of Psychological Vulnerability

Nour Hatem (2019, 10) indicated that psychological fragility "is one of the types of crises that affect the psyche of an individual, as a result of several factors and conditions that he may go through, such as feeling constant tension and anxiety, and lack of confidence in his abilities to manage crises and the problems that he faces in practical and daily life. It is a problem that affects Many of us, and it makes the individual vulnerable to being controlled by negative feelings and psychological crises, and pushes him to attach to others, and he thinks that he loves them, while the truth is that he compensates for his weakness through them, as it is a personality disorder that needs treatment so that the situation does not worsen further.

Lana Ayman (2021, 20) defined it as "the inability of the individual to adapt to the pressures he is going through, and they affect his daily life, as they are accumulations and frustrations as a result of the environment and the passage of time, so the person feels that he is in his place that does not develop, and affects his other aspects, so he has a feeling of laziness." Lack of achievement, and high anxiety that makes the individual withdraw from the social environment.

Reasons for Psychological Fragility

Ismail Arafa (2022, 31) stated that psychological fragility emerged as a result of this generation's failure to assume responsibilities since its youth, even in the simplest matters, and its constant habit of relying on others to achieve its goals, study its lessons, or end its concerns. Real life makes him more spoiled, and more defeated in the face of pressures. The child's discovery of life and his exposure to problems from a young age develops in him psychological resistance and makes him more solid in facing life. The individual must be exposed to mental challenges and physical pressures in order to be able to continue in life, so the password It is immunity against people's words and against life's problems and vicissitudes, just as the body gains immunity against viruses and microbes, and gets used to the meaning of suffering in this worldly life, and what it entails in terms of legal costs and psychological and physical hardship.

Vulnerable Age Groups

Zawad Dalila (2014, 107-108) explained the vulnerable groups as follows:

1. The category of childhood: the more the individual lives during his childhood painful experiences such as loss or the difficulties of emigration, the more difficulties it has to modify the course of his existence because these experiences create a sense of inability to assume the responsibilities of life as a whole, and in this context, barriers accumulate in the way of the individual because he was unable to develop his personal abilities in order to control the reins.

2. The category of adolescence: a person goes through some stations in his life, asking himself questions related to his existence, his personality, his behavior, his appearance..either he accepts and continues, or he rebels, gets upset and gets angry at himself to the point of being accused and preferring death over continuing, and among these stations is a crisis Puberty, where the body begins to change some of its features, announcing the transition from childhood to adolescence, and its crisis when the first decisions that determine the fate and path of the professional, family, or even social future begin, in addition to the explosion of feelings, emotions, enthusiasm, rejection of

values, rebellion against parents and the present, and anxiety about adulthood The responsibility awaits.

3. The youth category: This is followed by the crisis of reaching the age of thirty, where specialization and profession have finally decided, to add the choice of the marital partner, the building of the family home, and even the definition of the ideological thought that he prefers, then the crisis of reaching the age of forty, as children exist and securing their requirements is a duty and difficult.

4. The middle age category: Then comes the crisis of reaching the age of fifty, as the first signs of physical diseases, fatigue from the profession and disgust from work, and finally the crisis of reaching sixty, when a person tends to be religious in order to draw closer to his Lord, as anxiety about death scares him as he fears separation from his family And his reality.

5. The category of old age: These existential crises go through every individual, and if the roads are blocked in front of him and his self-criticism increases for the previous stage and his role in it in addition to his dissatisfaction with himself, then suicide here is a present idea that passes through the mind. It is important for others to conspire against him and always fail him, and either he commits suicide by succumbing to a disease, or some fate, or he commits suicide and dies.

Factors of Psychological Vulnerability

Seguin & Huan (2000, 38) explained that there is a group of factors that determine the ability to face difficult situations, which are personal, developmental, medial, and biological factors, as they interfere in various stages of life by means of protection, treatment, and emotional envelopment, which originate in the medium. familial and social, some individuals live more difficult experiences than others, they may find outlets for them because they have lived important successes in their lives, in addition to that they have found in their surroundings individuals who have feelings of confidence, stability and protection. To understand the phenomenon of psychological resistance, the context of growth must be taken into account, therefore The elements of individual resistance, whether strong or fragile, can be determined by the attachment relationships that the individual has built during the period of development, in addition to the accumulation of accidents in his life and the extent of the impact of mental illness.

Ismail Arafa (2022, 31, 57) indicated that each person has his own mechanisms by which he adapts to any problem, works to overcome it, and overcome its bad effects, and affects this by several factors, such as psychological flexibility, the solidity of the individual's belief, the height of his motivation, and the nature of his perceptions About life, its problems and solutions to its crises, and among the means to overcome periods of negative psychological state are as follows:

Do not give in to sadness or distress, and distract yourself from thinking about the negative situation.

- Intimacy with friends, family, and close circles of support.
- Fill the emptiness of time and diligence in living, so if work is stressful, then emptiness is corrupting.

Not resorting to psychiatrists until after a long period of time.

- Leave space for the soul to process its own feelings without the need for therapeutic intervention.

Night prayer and secret worship are among the greatest aids in improving our mental health.

Walking and recreation, listening to sermons and faith lessons.
- Documenting social ties, especially in the family and the extended family.
- Revealing the self and revealing its defects, treating its mistakes and recognizing its problems.

The Third Axis: Psychosomatic Disorders

The detection of these psychosomatic disorders is a great scientific discovery that leads us to dive deep into the soul and the body and reflect on its mysteries. We search for the causative and precipitating factors, which are predominantly psychological in nature and are linked to the emotional state of the individual. Examples of these disorders include asthma, allergies, ulcers, irritable bowel syndrome and high blood pressure. And others, these disorders have been known as psychosomatic disorders, and the effect of the self on the body occurs through the voluntary nervous system, and it is mainly a disorder in the control of the body mediated by that system. There are no apparent organic changes in the diseased organs, but the changes are mainly functional.

The Concept of Psychosomatic Disorders

Psychosomatic disorders are defined in the Fifth Diagnostic and Statistical Manual of the American Physicians Association of Mental Disorders (DSM-V) under the heading of physical states (Galenic) as: "functional disorders that arise, at least in part, from psychological or emotional, affective, affective factors" (Mohamed Ghanem, 2011, 49).

As defined by Ahmed Mahmoud, Tariq Ahmed (2018, 643) as: "Organic disorders in which the emotional factor plays an important, strong and fundamental role, usually through the autonomic nervous system."

Muhammad Hassan (2022, 7) believes that it is: "a group of physical symptoms that affect the organs of the body that are under the control of the nervous system (skin - digestive - respiratory - urinary disorders) as a result of the individual's exposure to psychological pressures that led to emotional disorders that they were unable to express."

Psychosomatic Disorders in Psychoanalysis

The application of analytical principles in the psychosomatic field did not come at the hands of Sigmund Freud, but rather at the hands of his followers, but Freud's ideas and works gave a big leap in the "psychosomatic" field. If we go back to the first texts of the father of psychoanalysis, Sigmund Freud, we will find that the psychosomatic movement started from the foundations and principles that he laid down. And the body (Hanan Taleb, 2011, 22)

Psychosomatic Disorders in the DSM-V

Despite the prevalence of the term "psychosomatic disorders" in the health community in its various specialties, it has undergone a major and radical change in terms of naming and even classification. IV disassembled hysteria into several syndromes, made conversion syndrome associated with other somatic disorders, and included psychosomatic diseases under the general term "somatomorphic disorders", which includes seven categories: somatization disorder, hypochondriasis, conversion disorder, pain

Self-Defeat, Psychological Fragility as Predictors of Psychosomatic Disorders

disorder, fear of deformity Somatic disorders, undifferentiated somatomorphic disorders, and unspecified somatomorphic disorders, and we find (DSM-V) has replaced this term with a new term: "disorders of somatic symptoms and related disorders", and defined them as: "the sum of physical or functional complaints, which can or It cannot, depending on the type of disorder, be explained by a physical disease or a specific physiological-pathological mechanism" (Muhammad Ghanem, 2011, 12).

The Relationship Between Emotion and the Occurrence of Psychosomatic Disorders

Nour Al-Huda (2009, 43) referred to dividing the stages of emotion or psychological stress that lead to psychosomatic disorders into three stages:

1. The shock stage: It is a short stage that lasts minutes or hours, and in it the pulse increases in blood pressure, body temperature, acceleration in heart rate, and a general relapse of the tissues, and from here the danger stage begins. Achieving the organic balance of the individual. In cases of severe trauma, we find that the hypothalamus stimulates the pituitary gland to secrete an adrenal stimulating hormone, which in turn urges the adrenal cortex to secrete cortisone, which works to increase the body's resistance, so the patient wakes up and enters the second stage of shock.
2. The stage of resistance: in which the greatest degree of compatibility is achieved, symptoms disappear, and tissues are rebuilt. This stage lasts for days, months, or years, depending on the type and severity of the ordeal, and ends with the disappearance of the ordeal that caused it, or recovery from it, or the adrenal gland suffers from a deficiency in the secretion of the hormone cortisone, which leads to a weakening of the body's resistance Again, the individual enters the third stage.
3. The stress phase: It is sometimes called the breakdown phase, in which the body exhausts its energies and the body's defenses fail, so signs of fatigue appear gradually, and the individual's ability to adapt and rebuild tissues stops, and the individual weakens and weakens, due to the cessation of the sympathetic nervous system from producing energy, and this may be accompanied by symptoms of depression or psychotic behavior. Or physical illness and sometimes death, and at this stage the pituitary gland fails to secrete the hormone cortisone, which leads to giving it orally in severe psychological cases.

Through the foregoing, it is clear that the value of the severe psychological ordeals experienced by some people varies according to the patient's psychological and neurological composition, or according to the different circumstances of his life, his special experiences, his concepts in life, his psychological flexibility to adapt to it, his acceptance of reality, or his personality, as well as depending on his physical health or his illness with chronic diseases that have already been mentioned. exhausted by his harmonic energy

Characteristics of Psychosomatic Disorders

Ibtisam Muhammad (2010, 50) believes that there are several characteristics of psychosomatic disorders:

1. There is a physiological basis for the disorders.

2. It includes organs and viscera that are affected by the autonomic system, and thus are not subject to voluntary control.
3. Structural changes may be life threatening.
4. Be more controlling and demanding on the affected member.

Zainab Choucair (29, 2002) reported that psychosomatic disorders are distinguished from other disorders as follows:

- The presence of an emotional disorder as a causative factor.
- Some cases are associated with a certain personality type.
- The incidence of these diseases differs significantly between the sexes.
- Different symptoms may be present in the same individual.
- There is often a family history of the same or similar disease.
- The individual tends to take different stages.

Factors and Causes Leading to Psychosomatic Disorders

The emergence of psychosomatic disorders is due to psychological factors in origin that have physical forms and symptoms. The most important factors and causes leading to the emergence of psychosomatic disorder can be clarified. As Faisal Khair (2007, 86) believes, there are several causes and factors for psychosomatic disorders, including:

Factors related to heredity: They mean factors of genetic predisposition and the effect of these factors on the fetus before it is born, conditions of pregnancy and childbirth, mother's diseases, nutritional conditions and blood group (RH)....etc. Sontage and Lister show that the life of the fetus inside the womb is affected. In the emotional life of the mother, we find him irritable after birth, as happens to children of cocaine-addicted mothers in terms of screaming and movements. Medical studies indicate that these factors affect the child's nervous and glandular system and weaken his endurance, so he is usually predisposed to developing disorders.

Disruption of the relationship between the child and the parents: especially the mother, as there is a lack of love, affection, care, loss of security and tranquility, family quarrels...etc. All of these matters are reflected directly and indirectly on the child and affect his organic, psychological, intellectual and social growth and maturity.

Emotional factors: to which the individual is exposed in his life, such as frustration, painful childhood experiences, aggression, suppression of anger, inability to express feelings and desires, etc., all of which generate severe psychological pressures that lead to despair, collapse, and the emergence of disorders.

Clinical Signs of Psychosomatic Diseases According to the DSM-V

The diagnosis of psychosomatic diseases according to the DSM V is based on the following clinical signs:

- The presence of physical symptoms such as pain, discomfort, sensory disturbances such as paresthesia of the extremities, feeling hot or cold...etc.
- These symptoms are accompanied by complaints, usually of continuous intensity.
- These strikes usually require a variety of medical consultations.

- The origin of these symptoms is due to clinically significant suffering, or to a change in social, occupational, or other important areas of the individual's life.
- In the special case of medical injuries affected by psychological factors, the reason for the occurrence or exacerbation of the medical disease is emotion or pressure (mourning, dismissal from work, divorce... etc. (Nour Al-Huda, 2009, 31).

Diagnostic Criteria for Psychosomatic Diseases According to the DSM-V

In its diagnosis, the DSM-V focuses more on the presence of positive signs represented in the presence of physical symptoms associated with maladaptive thoughts, feelings, and behaviors in response to these symptoms, rather than on the absence of a medical explanation (Ismail Eid, 2015, 39).

HYPOTHESES OF THE STUDY

1. There is a correlational and statistically significant relationship between self-defeat and psychosomatic disorders among samples of university students in Egypt and the UAE.
2. There is a correlation and statistical significance between psychological fragility and psychosomatic disorders among samples of university students in Egypt and the UAE.
3. There are statistically significant differences in the degrees of self-defeat, psychological fragility, and psychosomatic disorders among samples of university students in Egypt and the UAE, according to gender (male/female).
4. There are statistically significant differences in the degrees of self-defeat, psychological fragility, and psychosomatic disorders among samples of university students in Egypt and the UAE, according to nationalities (Egyptian/Emirati).
5. Do self-defeat and psychological fragility contribute statistically in predicting psychosomatic disorders among samples of university students in Egypt and the UAE

The Study Method and Its Variables

The current study relied on the use of the comparative descriptive approach, in line with the nature and objectives of the current study, which seeks to investigate the correlation between the variables of the study (self-defeat - psychological fragility - psychosomatic disorders) among samples of university students in Egypt and the Emirates, and to know the differences between them according to gender. And nationality, as well as determining the relative contribution of self-defeat and psychological fragility in predicting psychosomatic disorders in the study sample.

The Study Sample

The current study sample is divided into:

A- A sample of calculating the psychometric characteristics of the study tools and measures:

It is (100) people from the Universities of Alexandria in Egypt, and Al-Ain University in the Emirates, including (50) Egyptian students and (50) Emirati students, and their ages range between (18-22), in order to apply the study tools to them to calculate their psychometric characteristics (internal consistency, honesty, stability) to ensure its validity and use in the current study.

B- The main study sample (participants):

The basic sample of the study includes at the beginning of (200) students from the Universities of Alexandria in Egypt, and Al-Ain University in the Emirates (100 Egyptian students, 100 Emirati students), "120 males-80 females), their ages ranged between (18-22).

Study Tools

The current study used the following tools:

- Self-defeat scale (prepared by the two researchers).
- Psychological fragility scale (prepared / researchers).
- Scale of psychosomatic disorders (prepared / researchers).

STUDY RESULTS AND DISCUSSION

1- The results of the first hypothesis:

Which states: "There is a correlation and statistical significance between self-defeat and psychosomatic disorders among samples of university students in Egypt and the UAE."

To test this hypothesis, the two researchers used Pearson's correlation coefficient to calculate the correlation coefficients between the average degrees of self-defeat and the average degrees of psychosomatic disorders among the study sample. Table (1) shows the results reached:

It is clear from Table (1), the following:

The existence of a positive correlation between self-defeat and psychosomatic disorders, and this indicates that the greater the self-defeat, the greater the psychosomatic disorders, which indicates that the hypothesis is completely fulfilled.

2- The results of the second hypothesis:

Table 1. Correlation coefficients between the average scores of self defeat and psychosomatic disorders (n = 200)

self-defeat/ psychosomatic disorders	value of "r"
Total Degree	0.712

The tabular value of "r" at the level of 0.01 = 0.181 and at the level of 0.05 = 0.138

Self-Defeat, Psychological Fragility as Predictors of Psychosomatic Disorders

Table 2. Correlation coefficients between the average scores of psychological fragility and psychosomatic disorders (n = 200)

Psychological fragility / psychosomatic disorders	value of "r"
Total Degree	0.727

The tabular value of "r" at the level of 0.01 = 0.181 and at the level of 0.05 = 0.138

Which states: "There is a correlation and statistical significance between psychological fragility and psychosomatic disorders among samples of university students in Egypt and the UAE."

To test this hypothesis, the two researchers used Pearson's correlation coefficient to calculate the correlation coefficients between the average degrees of psychological vulnerability and the average degrees of psychosomatic disorders among the study sample. Table (2) shows the results reached:

It is clear from Table (2), the following:

The existence of a positive correlation between psychological fragility and psychosomatic disorders, and this indicates that the greater the psychological fragility, the greater the psychosomatic disorders, which indicates that the hypothesis is fully achieved.

3- The results of the third hypothesis:

Which states: "There are statistically significant differences in the degrees of self-defeat, psychological fragility, and psychosomatic disorders among samples of university students in Egypt and the UAE, according to gender (male/female).

To test this hypothesis, the researchers used the "T" test to denote the differences between the averages of two independent samples to denote the differences between the average scores of the students of the Faculty of Education on the measures of self-defeat, psychological fragility and psychosomatic disorders according to gender (male/ female). This is illustrated by Table (3):

Table (3): There are statistically significant differences between the mean scores of the study sample on the measures of self-defeat, psychological fragility, and psychosomatic disorders, according to gender, and all of them are in favor of females.

Table 3. Results of using the "T" test to indicate the differences between the mean scores of the study sample's mean scores on the measures of self-defeat, psychological fragility and psychosomatic disorders according to gender (male / female) (n = 200)

The variable	gender	n	mean	S. D	degrees of freedom	"t"	significance
self-defeat	M	120	108.27	8.39	198	10.77	0.01
	F	80	122.18	9.62			
psychological fragility	M	120	102.38	10.27	198	12.21	0.01
	F	80	119.97	9.38			
psychosomatic disorders	M	120	139.43	11.68	198	7.86	0.01
	F	80	152.14	10.31			

Tabular t value at the significance level of 0.01 = 2.617 and at the level of 0.05 = 1.98

Figure 1. Differences in measures of self-defeat, psychological fragility and psychosomatic disorders according to the gender variable (male - female)

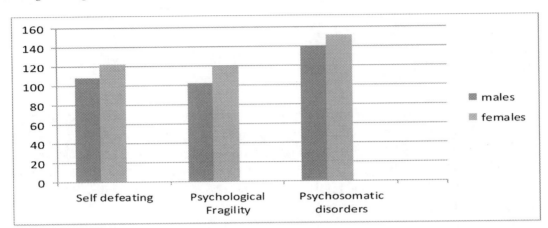

The following figure shows the significance of gender differences on the scale of self-defeat, psychological fragility, and psychosomatic disorders.

4- The results of the fourth hypothesis:

Which states: "There are statistically significant differences in the degrees of self-defeat, psychological fragility, and psychosomatic disorders among samples of university students in Egypt and the UAE, according to nationalities (Egyptian / Emirati).

To test this hypothesis, the researchers used the "T" test to denote the differences between the averages of two independent samples to denote the differences between the average scores of the students of the Faculty of Education on the measures of self-defeat, psychological fragility and psychosomatic disorders according to nationality (Egyptian / Emirati). This is illustrated by Table (4):

Table (4): There are statistically significant differences between the mean scores of the study sample on the measures of self-defeat, psychological fragility, and psychosomatic disorders, according to nationality, and all of them are in favor of the Egyptian nationality.

The following figure shows the significance of the differences in nationality on the scale of self-defeat, psychological fragility and psychosomatic disorders.

5- The results of the fifth hypothesis:

Which states: "Self-defeat and psychological fragility make a statistically significant contribution to predicting psychosomatic disorders among samples of university students in Egypt and the UAE."

To test this hypothesis, the researchers used the Stepwise Regression multiple regression analysis test, and calculated the coefficient of determination or the square of the correlation coefficient to find out the contribution of self-defeat and psychological vulnerability in predicting psychosomatic disorders. Through the analysis, it became clear that there is one model as shown in Table (5.(

We conclude from Table (5) the following:

Table 4. Results of using the "T" test to indicate the differences between the mean scores of the study sample's mean scores on the measures of self-defeat, psychological fragility and psychosomatic disorders according to nationality (Egyptian / Emirati) (n = 200)

The variable	nationality	n	mean	S.D	degrees of freedom	"t"	significance
self-defeat	Egyptian	100	123.38	8.39	99	8.58	0.01
	Emirati	100	111.26	9.62			
psychological fragility	Egyptian	100	135.51	10.27	99	7.94	0.01
	Emirati	100	122.42	9.38			
psychosomatic disorders	Egyptian	100	148.36	11.68	99	7.54	0.01
	Emirati	100	137.25	10.31			

Tabular t value at the significance level of 0.01 = 2.617 and at the level of 0.05 = 1.98

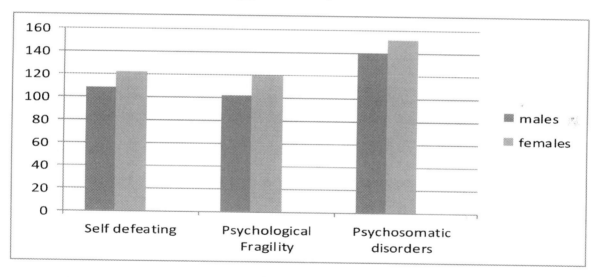

Figure 2. Differences in measures of self-defeat, psychological fragility, and psychosomatic disorders according to the nationality variable (Egyptian-Emirati)

Table 5. Gradual regression of the effect of self-defeat and psychological fragility on psychosomatic disorders among the study sample (n = 200)

The variable	"F"	R	R2	Contribution percentage	equation constant	B	"T"
model	568.145**	0.831	0.771		38.224		
self-defeat				64		9.532	20.112**
model	524.227**	0.798	0.762		36.817		
psychosomatic disorders				63		8.475	22.257**

** Significant at level (0.01)

- The calculated "F" value increased, as it reached (568.145) for self-defeat at the level of significance (0.01), and the percentage of the contribution of the independent variables in this model was (0.771%). Which means that there is a statistically significant effect of self-defeating on psychosomatic disorders.
- The high calculated "F" value, which amounted to (524.277) for psychological fragility at the level of significance (0.01), and the percentage of the contribution of the independent variables in this model was (0.762%). Which means that there is a statistically significant effect of psychological fragility on psychosomatic disorders.

The results of Table (5) indicate that the calculated (T) value is statistically significant at the level (0.01); This indicates that there is a relationship between self-defeat and psychological fragility, and between psychosomatic disorders in the study sample.

Thus, it is clear from the foregoing that there is a statistically significant effect of self-defeat and psychological fragility on the psychosomatic disorders of the study sample.

The regression equation between self-defeat and psychosomatic disorders can be expressed as follows:

Psychosomatic disorders = 38.224 + (9.532 x self-defeat(

The regression equation between psychological fragility and psychosomatic disorders can also be expressed as follows:

Psychosomatic disorders = 36.817 + (8.475 x psychological fragility)

Accordingly, the fifth hypothesis was validated, as it was possible to predict psychosomatic disorders through self-defeat and psychological fragility, among the study sample.

STUDY RECOMMENDATIONS AND SUGGESTIONS

In the light of the current study procedures, the results reached by the two researchers, the explanations provided, and the difficulties they experienced during the application of the study procedures, they suggest some of the following educational recommendations:

1. Achieving sustainable economic growth.
2. Building resilience and the ability to adapt to climate change by mitigating its negative effects.
3. Improving the infrastructure for financing climate activities.
4. Promoting scientific research, technology transfer, knowledge management and awareness to combat climate change.
5. Raising awareness of these changes among the various groups of society.
6. Participation in scientific counseling programs that help to confront self-defeat and psychological fragility caused by climate changes, and to circulate these conferences to all educational and university institutions.

REFERENCES

Abbas, F. (2004). *Self-destruction among the perpetrators of traffic accidents.* [Master's thesis, College of Arts, University of Baghdad].

Abd al-Salam Ali, A. (2005). *Social support and its practical applications in our daily lives.* Egyptian Renaissance Bookshop.

Abu Halawa, M. S. (2013). The factorial structure and discriminatory analysis of psychological defeat in the light of some psychological variables among university students. A proposed model. Arab studies in education and psychology. *Association of Arab Educators.*, *37*(3), 128–171.

Al-Assimi, R. (2012). Contradictions of self-perception and its relationship to social anxiety and depression among Damascus University students. *Damascus University Journal.*, *28*(3), 1–69.

Al-Attar, M. M. (2019). Positive self-talk and its relationship to psychological flow and psychological defeat among students of the Faculty of Education. The Psychological Journal of Psychological Studies. *The Egyptian Association for Psychological Studies.*, *29*(102), 388–432.

al-Haloul, I. E. (2015). Security anxiety and its relationship to psychosomatic symptoms among Palestinian police personnel in the Gaza Strip, The International Specialized Educational Journal, Al-Aqsa University. *Palestine*, *7*(4), 34–58.

Al-Jamous, N. A.-H. (2009). *Psychosomatic disorders.* Al-Bazuri.

Al-Zrad, F. K. (2007). *Mental-physical illnesses. I (2).* Dar Al-Nafees.

Ali, L. H. (2013). *The Self-Defeated Personality and its Relationship to Psychological and Social Status.* [Master Thesis, College of Arts, University of Baghdad].

Attia, A. M. (2011). Academic resilience and its relationship to self-esteem among a sample of open education students. *Journal of Psychological Studies.*, *21*(4), 571–621.

Campellone, T. R., Sanchez, A. H., & Kring, A. M. (2016). Defeatist performance beliefs, negative symptoms, and functional outcome in schizophrenia: A meta-analytic review. *Schizophrenia Bulletin*, *42*(6), 1343–1352. doi:10.1093chbulbw026 PMID:26980144

Ghanem, M. H. (2011). *Psychosomatic disorders.* Dar Gharib.

Hegazy, M. (2005). Social backwardness: an introduction to the psychology of the oppressed person. 9th Edition. Morocco: Casablanca.

Ibrahim, F. I. (2022). Climate Change and Food Security in Egypt. *Scientific Journal of Economics and Trade*, *52*(1), 221–262.

Ibtisam, M. A. A.-A. (2010). *Blood groups and some psychological disorders in a sample of male and female students with learning difficulties and normal in the primary stage in Makkah Al-Mukarramah.* [PhD thesis, Umm Al-Qura University, Saudi Arabia].

Mahmoud, L. A. (2021). *Psychological fragility and its relationship to defense mechanisms among unmarried women.* [Master's thesis, College of Education, Jordan].

Mahmoud, M. H. (2022). *An indicative program based on clinical indications for the T.A.T Subject Comprehension Test to reduce psychosomatic disorders among mothers of children with cerebral palsy (a clinical-psychometric study),* [PhD thesis, Faculty of Education, Alexandria University].

Sherry, S. B., Stoeber, J., & Ramasubbu, C. (2016). Perfectionism explains variance in self-defeating behaviors beyond self-criticism: Evidence from a cross-national sample. *Personality and Individual Differences, 95,* 196–199. doi:10.1016/j.paid.2016.02.059

Taleb, H. (2014). *Introduction to the psychosomatic approach.* University of Ouargla.

Zagkar, R. (2015). Algerian youth between the fragility of psychological formation and the challenges of citizenship. [Unpublished research, Ammar Thaliji University, Algeria].

Chapter 8

Smart Sustainable Cities to Treat the Economic and Climate Repercussions in of Global Climate

Ehab Atalah
Independent Researcher, Egypt

ABSTRACT

Sustainable smart cities aim to provide a number of opportunities to achieve sustainable development and face challenges related to climate change, that is the great challenge facing the future of life on the planet. Information and communication technology is one of the basics in the field of monitoring climate change, mitigating its effects and adapting to it, such as the field of early warning system and smart applications. That transforms the way services are provided in areas including energy, waste, and water management to reduce carbon footprint and address environmental challenges, the role of modern technology, the internet of things (IOT) and artificial intelligence (AI) applications.

INTRODUCTION

Climate change is a very clear and present danger. Though the term usually conjures up images of melting ice sheets and dying forests, our cities are also particularly vulnerable to it too. Currently, 50% of the world's population lives in cities. That percentage is expected to rise to 70% by 2050. Cities are already struggling to mitigate the problems caused by increased population density, and these problems are expected to intensify as other variables such as climate change and extreme weather take hold. To prepare for the future, cities must improve their climate change resilience in order to improve the quality of life of its citizens. With the threat of unmanaged overpopulation, extreme vulnerability to the effects of climate variables, and economic factors, cities would appear to be especially susceptible to the planet's uncertain future. However, cities may be the battleground where mankind turns the tide against climate change (Aleixandre-Tudo et al., 2019). Cities face a number of difficult problems that range from water

DOI: 10.4018/978-1-6684-8963-5.ch008

Copyright © 2024, IGI Global. Copying or distributing in print or electronic forms without written permission of IGI Global is prohibited.

shortages, pollution, and flooding to congestion, heat stress, and waste disposal, but they also provide a unique opportunity to drastically reduce carbon emissions and champion practical sustainability. Historically, cities have been designed with efficiency in mind: they're built on trade routes or watercourses for the easy transportation of people and goods, they serve as economic centers that drive trade, and most importantly, they're meeting places for the sharing and development of ideas. Using cities as innovation testbeds, it's possible for urban areas to develop into greener, more sustainable, more economically-viable, climate-resilient cities (Blunier et al., 1998).

The advantages of these smart technologies are the contribution to preserving the environment, because they simply depend on clean energy sources, and there is no electronic device or smart phone that works with non-renewable energy such as petroleum. climatic; Rather, it helps to improve the quality of human life in general, and reduce waste in natural resources, through the use of artificial intelligence systems that are able to predict fires and natural disasters, and Internet of Things systems that are used to improve resource management and reduce energy waste, and to adopt smart city models that are capable of to create more sustainable societies and preserve the environment (Bonacina, 1947).

THE MOST IMPORTANT FACTS RELATED TO THE PHENOMENON OF GLOBAL CLIMATE CHANGE

There is almost unanimity in the international scientific community on a number of facts related to the phenomenon of global climate change, and the most prominent of these facts are the following:

1- The Earth's average temperature is rising at an unprecedented rate.

2- Human activities, represented by the use of fossil fuels (coal, oil and natural gas), and deforestation are the main driver of this rapid global warming and climate change, as these activities have led to a significant increase in the amount of greenhouse gases, especially carbon dioxide, in Atmosphere causing the planet to warm.

3- Ongoing global warming is expected to have catastrophic environmental consequences if the global temperature continues to rise at the current pace. Scientists of the Intergovernmental Panel on Climate Change (IPCC), a United Nations body established in 1988, have warned of severe impacts expected when the global temperature rises by 1.5°C, for example (Brazier, 2015):

A- Extreme heat waves: This leads to an increase in diseases and deaths as a result of sunstrokes and others.

B- Drought and floods: This will make agriculture more difficult, and crop yields will decline, thus causing food shortages.

C - The melting of ice in the Arctic and the rise in sea levels: This will lead to the inundation of many coastal areas in the coming decades. Small island states will also be at risk of being wiped out.

D- Ocean change: Up to 90% of coral reefs will be destroyed, and the oceans will become more acidic. Consequently, world fisheries will become much less productive.

E- Extinction of more insects, plants and vertebrates: A recent report issued by the United Nations Intergovernmental Panel on Climate Change (IPCC) indicates that about one million species of plants and animals are threatened with extinction due to climate change (Bronnimann et al., 2014).

Of course, the negative effects of global climate change will be much more severe if the Earth's temperature rises to more than 2 degrees Celsius. In this context, many experts warn that the Paris agreement to confront global climate change is not sufficient to prevent the global average temperature from rising by 1.5 degrees Celsius, as the pledges of countries, according to this agreement, are not ambitious enough, in addition to that it will not be implemented. Act fast enough to limit global temperature rise to 1.5°C by the end of the century. In this context, the Climate Change Tracker forecasts indicate that the current policies applied in different countries of the world will lead to a rise of about 2.7 degrees Celsius (4.9 degrees Fahrenheit) by the year 2100 (ClimateHomeNews, 2019).

Third - the controversy over the countries responsible for climate change

Developing countries believe that developed countries bear a "historical and moral responsibility" in confronting global climate change; Because it is the one that has released most of the greenhouse gases, during the process of its economic growth since the industrial revolution at the end of the nineteenth century. Thus, developing countries adhere to the need for developed countries to bear the greatest economic burden in facing global climate change.

The Global Earth's Temperature Will Rise About 4 Degrees Celsius By 2100

Over the past years, experts and scientists have redefined what could be the state of the planet as a result of global warming and the safe thermal status of the Earth in order to be supportive and ensure safe life on the surface of the planet, as climate experts assert that global warming should increase by two degrees Celsius before the year 2100. (Gastner & Newman, 2004) Whereas, if temperatures increase above that rate, severe climatic consequences will occur. Despite the lack of commitment to promises, the Paris Agreement is an important step in combating global warming. During the 21st Climate Summit, the international community formulated a clear goal of limiting global warming to less than 2 degrees Celsius compared to the pre-industrial level, and it was enshrined for the first time in an international treaty.

Since then, the Paris Agreement has become a prominent example of environmental activism, and many countries, cities, and companies are now going even further by calling for carbon neutrality by 2050. The Paris Agreement has had a positive impact. It is clear that the impact on the real economy is not enough, but the European Union was the cause of a clear move in China, which was followed by the United States," said former chief negotiator of the Paris Agreement, Lawrence Tubiana, during a Senate hearing on January 27, 2021., to highlight progress made since 2015 to become net zero emissions the metric (German Watch, 2019).

So, the three largest emitters of greenhouse gases have adopted this new target. What surprised everyone was China's announcement in September 2020, during the United Nations General Assembly, of its intention to achieve carbon neutrality before 2060. For its part, the United States - which left the Paris agreement with Donald Trump before returning to it with Joe Biden - announced in April 2021 Heading towards carbon neutrality by 2050. The European Union did the same thing last June when it established this goal in its climate law (Groneberg-Kloft et al., 2009).

Courts are also taking this new controversy seriously, as they increasingly have to adjudicate "disputes the climate". According to a United Nations Environment Program report released on January 26, 2021, the number of lawsuits related to climate change has increased significantly over the past four years. It has doubled since the last report in 2017 and now stands at 1,550 in 38 countries. As of July 1, 2020, about 1,200 of these cases have been brought to court in the United States and 350 in the rest of the world.

As more countries declare their goal of achieving carbon neutrality by mid-century, the reality of their concrete actions to reduce greenhouse gas emissions contrast sharply with their stated ambitions. "There is an inconsistency between the emission levels projected in current policies, those envisaged by the Nationally Determined Contributions by 2030 and, most importantly, those required to achieve net-zero emissions targets by 2050," the UNEP 2020 report underlines (IPCC, 2019).

Thus, in 2030, to reach the 2°C target of the Paris Agreement, annual emissions must be 15 billion tons of carbon dioxide equivalent less than those projected under the current unconditional NDCs. To reach the 1.5°C target, it would need to cut 32 billion tons of carbon dioxide equivalent.

The UN report on Nationally Determined Contributions, released September 17, confirms the Climate Action Tracker's finding, noting that roadmaps submitted by the 191 parties to the Paris Agreement include a "significant rise" in global greenhouse gas emissions in 2030. Compared to 2010, by about 16%. For her part, the Executive Secretary of the United Nations Climate Organization, Patricia Espinosa, considered that this increase "is a source of great concern." "It stands in stark contrast to the calls made by scientists for rapid, sustainable and large-scale emissions reductions in order to avoid the most serious climate consequences and suffering, especially for the most vulnerable, around the world," she explained.

In addition, many of the road maps put forward by developing countries contain conditional commitments that can only be implemented with external financial assistance (Klingelhofer et al., 2019). The global debate continues about the cost of confronting climate repercussions without material incentives or international sanctions, which causes a sharp decline in the climate issue.

The Idea of a "Smart City"

It has received worldwide attention. Also known as a "wired", "networked" or "ubiquitous" city, a "smart city" is the latest in a long line of catchphrases, referring to the development of technology-based urban systems to drive efficient city management and economic growth. These could be anything from citywide public Wi-Fi systems to smart water meter provisioning in individual homes. Any feature that uses information and communication technologies to make a city more efficient or accessible is said to fall under the umbrella of a 'smart city' (Kopp et al., 2017). Most technologists and engineers are busy investigating how to build smart cities and what features should be given to them. But it is also important to ask who can live in it, and what it means to be a citizen of a smart city (Kopp et al., 2017). At these could be anything from citywide public Wi-Fi systems to smart water meter provisioning in individual homes. Any feature that uses information and communication technologies to make a city more efficient or accessible (Kulp & Strauss, 2019).

Evolution of Sustainable Thinking for Cities

Research and thinking about sustainable cities began in the 1980s, but the term sustainability entered the global conversation in the 1990s, introduced by the World Commission on Environment and Development. In particular, the critical role that environmental and social dimensions of human economic activities play in creating a better world was demonstrated during the 1992 Rio Earth Summit .Influencing Those Discussions, a report on the agenda, was authored by Spotlight International Salt Federation The Nature Conservancy (IUCN), World Wide Fund for Nature (WWF) and United Nations Environment Program (UNEP) highlight how humans are building landscapes at the expense of the environment - and urge a focus on sustainable development (Lockyer, 1910).

With the development of the sustainability discourse, definitions and characteristics of "sustainable cities" are beginning to take shape. In the late 1990s, David Satterthwaite, a leading expert in the field, put forward the characteristics of a "successful" city. He argued that the city needs to ensure a healthy

Smart Sustainable Cities to Treat Economic, Climate Repercussions

working and living environment, and provide infrastructure for basic services such as clean water, sanitation and waste management. It also argued that - in keeping with the basic principles of sustainable development - the city needs to exist in balance with its ecosystems, for example by ensuring balanced water tables and low environmental pollution (Magri & Solari, 1996).

The definition of sustainable cities has continued to grow after the 1990s, incorporating ideas about how resources can be used now without compromising their availability in the future. Some have suggested that all cities must meet the needs of their residents for any city to be truly sustainable. However, the definition of what makes an urban area remains elusive - cities have many different characteristics. Areas can be classified as urban on the basis of population thresholds (above a certain number of people); human density (the number of people per square meter); employment ratios (workers in the agricultural sector compared to service industries); Or the presence of typical urban services such as water pipes, electricity, educational and health facilities (Menzel et al., 2006).

The evangelists of science, technology, and engineering—who proclaim the smart city as the answer to all urban ills—have overshadowed social science critiques about the human problems it creates.

These problems are particularly evident in purpose-built smart cities such as Dholera in India, where farmers have been stripped of their land in order to build the city; in Masdar City in the United Arab Emirates, which sacrificed its carbon-neutral properties after the global financial crisis; and in Songdo, South Korea, which until now has been a ghost town.

All of these cities have backtracked on their massive pledges to address issues accompanying migration, urban population growth, and climate change (Mesolell et al., 1969). The reason could be the economic challenge for this issue .

On the other hand, there are modified smart cities that focus on attracting investment to business districts and urban neighborhoods. They add smart features like e-waste recycling, e-rickshaws, smart water meters and more to the existing infrastructure. Unfortunately, this approach creates winners and losers, depending on who accesses and drives these developments. Most often, the "losers" are those whose interests are not protected by smart city policies (Mills et al., 2010).

A policy nexus for sustainable development

The imperative need for sustainable development remains a key planetary consensus. Sustainable development provides a long-term vision that unites peoples and nations in addressing important questions: how do we want the world around us to function and develop? What are the key priorities? How do we balance and reconcile different priorities? What future do we want? Sustainability calls for development that meets the needs of the present without compromising the ability of future generations to also meet their needs. It calls for the integration of economic development, social equity, and environmental protection. It is the kind of development that puts people at the center and that is just, equitable and inclusive (UN DESA and UNDP, 2012; ND-GAIN, 2019).

The United Nations 2030 Agenda for Sustainable Development provides an ambitious and comprehensive plan of action. At the heart of it are 17 Sustainable Development Goals (SDGs), which represent a blueprint for policymaking and international cooperation. SDGs recognize that a sustainable future depends on how successfully multiple global challenges will be addressed at once: ending poverty and other deprivations, improving health, education and wellbeing, reducing inequality and spurring economic growth, preserving the environment, and tackling climate change. The integrated character of the 2030 Agenda draws attention to linkages and complementarities among many traditionally dispersed policy areas. This publication takes the 2030 Agenda forward in a "cities-based approach" to sustainable development. This approach aims to recognize the central and integrating role that cities play in addressing, initiating and governing sustainability (OECD, 2019).

As the dominant form of the spatial organization of society, cities are driving economic, social, and environmental transformations. Consequently, cities can be, and already are, engines for directing these transformations towards greater sustainability.

INNOVATION MEANING

Innovation involves the generation and application of new knowledge that contributes in a new way to creativity, Pathways to a more sustainable future. Apart from new technology or new product, it may include new innovation Governance and organization practices, new ways of structuring partnerships, or new ways of running a community relation. In short, the concept of innovation goes beyond scientific research, high-tech startups, and even Private sector activities supporting the trend as a whole. It is about experimenting and finding better ways for humans interact and thrive (OneYoungWorld, 2020).

Financing Climate Goals: The Critical Challenge

A century ago, one of the most important challenges for the survival of the human race was how to face the population increase in light of limited resources, but innovation and creativity were able to provide technology that met urgent human needs.

During the UN Climate Change Convention's 15th Climate Summit in Copenhagen in 2009, developed countries committed to a goal of collectively mobilizing $100 billion annually by 2020 for climate action in developing countries. Twelve years have passed, but the $100 billion mark has not yet been reached.

According to a September 2021 OECD report, financing for climate goals for developing countries reached $79.6 billion in 2019, a shortfall of more than $20 billion. Three-quarters of this funding was allocated to mitigating the effects of climate change - that is, related to reducing greenhouse gas emissions, such as developing renewable energy - although the Paris Agreement specifically stipulated the need for a balance between financing mitigation and adaptation - that is, everything that allows to address the consequences Climate change, such as building dams to prevent floods (OneYoungWorld, 2020).

Also worrying: 71% of this funding is given in the form of loans, according to the NGO Oxfam, which stresses that the amount of climate finance in grants hasn't changed much, from about $11 billion in 2015-2016 to $12.5 billion. in 2017-2018.

Alternatives and Challenges of "COP 27"

The current apparent failure to achieve the goal of the "Paris Agreement" to confront global climate change, which is to prevent a rise in the earth's temperature by one and a half degrees Celsius, has led to the emergence of other alternatives that could contribute with the Paris Agreement to confronting global climate change. However, these alternatives face challenges that constitute an obstacle to effectively confronting global climate change, which are at the same time challenges facing the "COP 27" conference, and ways to address them are supposed to be discussed (Parmesan & Yohe, 2003).

The Relationship Between Smart Cities as a Technological Trend and the Climate Change

The relationship between the environment and technology is a historical one. Man has always tried to employ ideas, creativity and innovation in facing the difficulties of the surrounding environment, in order to make it more suitable for living and well-being. As much as technology has helped facilitate human life, it has had negative effects in increasing pollution rates. With regard to the first industrial revolution, it was based on the transition from coal to the steam engine in the eighteenth century. As for the second industrial revolution, which came at the end of the nineteenth century, it was based on the invention of electricity, and its impact on the expansion of productive manufacturing processes with the expansion of exchange in the markets (Plass, 1956).

While the third industrial revolution was marked by the launch of the process of transforming the production movement into "automation", and the development in computer technology and the Internet, which appeared in the sixties of the twentieth century (Pounds, 2006).

The "Fourth Industrial Revolution" came to express the process of merging physical or material sciences with digital and biological systems in manufacturing processes through electronically controlled machines and smart machines connected to the Internet such as: Internet of Things, 3D printing, artificial intelligence, robotics, and others in the form of applications. Interfered in all areas of life and work. The following can clarify the opinions that contribute to explaining the relationship between technology and the environment, which are divided into two main directions (Ritchie & Roser, 2019).

The first trend: It believes that there is a negative impact of technology of the smart cities on the environment and climate. Although this trend acknowledges its role in economic and social well-being, it focuses on the concerns surrounding the harmful effects of each of the technical devices and programs on the environment, such as: high levels of energy consumption, greenhouse gas emissions, and electronic waste, especially since electronic devices are relatively short-lived. It causes environmental damage during disposal (Root et al., 2003).

On the other hand, the emergence and development of cryptocurrencies and their dependence on mining made them energy-hungry. For example, the encrypted "Bitcoin" currency produces more than 22-29 million metric tons of carbon dioxide emissions every year, which is close to the size of emissions for entire countries; Where the volume of cryptocurrency consumption exceeds what is consumed by a country like Finland or the Czech Republic. Bitcoin ranks 35th globally in electricity consumption, and data centers contribute about 2% of global greenhouse gas emissions (Sangam, & Savitha,, 2019).

The second trend: It is the most widespread; It focuses on the role of technology in confronting climate change, and its most important application is "green technology", which involves the latest changes in information technology applications in line with standards for environmental preservation and sustainable development, the possibility of integrating information technology into environmental management efficiently, and the role of information technology in spreading Environmental awareness through digital platforms, the possibility of launching initiatives to protect against environmental pollutants, and awareness of the repercussions of climate change on the environment. The rapid development of the Internet of Things and Artificial Intelligence (AI) can contribute to providing solutions to reduce these harmful effects.

According to the Global e-Sustainability Initiative (GeSI), information and communication technologies have the potential to reduce global greenhouse gas emissions by 20% by 2030 by helping businesses and consumers to use smart, save energy. AI improves environmental governance, safety and environmental risk reduction with a focus on information management for decision-making, use of deep learning applications, big data analytics for water and air quality management, adoption of cloud computing applications, and energy-efficient data centers. In addition to the role of digital transformation in government jobs in reducing dependence on the use of paper, employing social networking applications to share information, and adopting sustainable initiatives that enhance environmental and climate awareness. It also prevents or at least minimizes the impact of misinformation on the public perception of climate change, in addition to strengthening the role of civil society and all stakeholders in confronting the phenomenon (Stott, et al., 2004).

One of the most important applications for adapting to climate change is "smart agriculture". This is done by shifting from traditional systems of agriculture to modern systems in which modern technologies play a crucial role in helping to meet the growing food needs. It relies on introducing technology in the agricultural field on the one hand, and adopting what is known as climate-smart agriculture on the other hand. This is done by employing applications of the Fourth Industrial Revolution such as: artificial intelligence, robotics, and the Internet of Things (The World Bank, 2019a).

Second: Green Technology as a Trend of Sustainable Smart Cities and the Pursuit of Governance

During the second half of the seventies of the last century, the idea of "low and no-waste technologies" appeared, and it tried to search for industries based on "clean technology" and at the same time not contributing to the production of waste harmful to the environment. Another idea called "cleaner production" emerged in the mid-eighties of the last century, in order to seek to form the idea of "waste-free technology". Industry, and the weak ability to inject new investments that would bring about changes in the industry at the level of processes, methods and production methods (The World Bank, 2019b).

In light of the tremendous development in the field of science and technology and its relationship to markets, two trends have emerged: the first, linking economic growth and technological development, and the second, the preservation of that relationship depends on the ability to address the resulting harmful effects.

Although many impartial studies have not been published on the negative health effects of technological applications, this does not necessarily mean that harm is absent, which prompted an attempt to seek to reduce the possibility of harm in the short and long term.

The term "green technology" or "clean" Green Technology (GT) came as a technical application to protect the environment, and contribute to the development of technical solutions in order to reduce carbon emissions and global warming, and its relationship to security, stability and development. Green technology represented an important shift in the application of public policies to technological uses, making it a platform for achieving optimal utilization of resources, achieving greater work efficiency, a high degree of service improvement, and achieving goals at the lowest cost, most quality, and sustainability. "Green technology" plays an effective and strong role in the emergence of the "green economy" as a new approach to efficient dealings in technology applications in the field of industry, which reflects positively on the national economy and addresses environmental and health damages to humans and the surrounding environment (Badyina & Golubchikov, 2016).

The trend towards "green technology" enhances the ability to search for and rely on alternative sources of energy that are friendly to the environment and to humans at the same time. Trends of "humanization of technology" have begun to surface in order to reduce the negative impacts and maximize human capa-

Smart Sustainable Cities to Treat Economic, Climate Repercussions

bilities in the face of the "machine" encroachment on Human feelings and values, the search for spiritual aspects away from the silent material control of the technological revolution, and work to develop and take into account the products, equipment and systems used to preserve the environment and natural resources, which reduces and limits the negative impact of human activities (Bulkeley et al., 2016).

The applications of "green technology" also contribute to confronting the effects of climate change, and reducing or limiting the growth of emissions, not only at the local level, but also at the global level, which prompts the use of energy digitization in reducing waste and generating it from renewable sources. and making digital applications environmentally friendly (Carbon Neutral Cities Alliance, 2019).

The Circular Carbon Economy as a Trend of Smart Sustainable Cities: A Modern Approach to Addressing Climate Change

In addition to interest in green financing that is considerate of the environment and green and low-emissions industries, countries have recently begun to think about other new ways to meet this great and serious challenge. Reducing and recycling emissions. Its focus is on the energy sector and its carbon emissions. It also targets emissions rather than targeting energy. It seeks to achieve a carbon balance or net zero emissions by targeting carbon emissions from gray economic sectors, including the energy sector and manufacturing industries (Charles, 2015).

The concept of a circular carbon economy was first developed by William as an extension of the circular economy concept with its application to the energy sector and CO2 emissions, and in 2017 Mayer proposed a detailed circular carbon economy model that would allow Germany to replace crude oil with household waste and coal as a feedstock. For the chemical industry, in addition to green hydrogen generated from renewable sources, this would allow to completely close the carbon cycle in the long term. After that, a group of innovators in the United States of America published a white paper on (cyclical carbon innovation) to mobilize capital for technologies that seek to generate economic value from reduction or use.

During its presidency of the G20 in 2020, the Kingdom of Saudi Arabia launched the concept of a circular carbon economy as a framework for action to reduce emissions to a level consistent with the goals of the Paris Agreement. KAPSARC) and was launched in 2020 to be a specialized tool used by stakeholders and governments concerned with various climate change policies to assess progress and make decisions in the relationship between energy, emissions and the economy. Its ultimate goal is neutral or zero carbon dioxide emissions. This concept aims to provide a framework Comprehensive, technology-biased and cost-effective (CitiesAct, 2009).

The main goal of the circular carbon economy is to reduce carbon dioxide and greenhouse gas emissions through the four pillars that include reducing emissions through carbon efficiency, reusing carbon by converting it into materials such as polymers and concrete, and removing the dioxide and storing it in natural pools. The four pillars (reduce, reuse, recycle and remove) of the circular carbon economy act as categories of mitigation options (CPH2025 Climate Plan & City of Moscow, 2020a).

The circular carbon economy approach emphasizes the incentives and economic benefits associated with managing carbon, and this includes converting it into products that can be used, or sequestering carbon for long periods of time, and focusing on the most cost-effective solutions. These aspects are important for countries that possess large quantities of hydrogen and have industries with significant consumption-intensive link.

161

Researchers need to understand how smart cities affect citizens' rights, freedom of expression, and participation in democratic politics. These concerns must be placed at the forefront of national smart city agendas.

Smart cities must find ways to encourage more grassroots efforts to engage with marginalized citizens. A good example is the mapping operations of slum children, which have forced policy makers in India to recognize their rights to basic urban services.

We need policies that allow us to closely measure our progress, consider short-term setbacks and create a comprehensive database of smart cities of the future. Many of these policies already exist at the international level. UN Livelihood and Entrepreneurship Rights, Indigenous Peoples' Rights, UN-Habitat Network on Safe Earth Rights for All, UNESCO Convention for the Safeguarding of Intangible Cultural Heritage and UN Guidelines for Power Sharing – all of these advocates the use of social rights. Comprehensive urban development processes.

If smart city policies are to efficiently drive city management and urban governance, they need to fundamentally change women's empowerment and participation, not put an ax on deeper issues of inequality. The state of home life will tell us a lot about the overall effectiveness of smart city policies. Smart city policymakers must think of new ways to engage with both women and men at home, to create and measure positive change. There is still a need for clear incentives for the circular economy with deterrent penalties for polluting activities at the international standard.

We Are Still Fighting to Protect Rights on the Internet

Most private sector organizations that collect and store citizens' data are not legally obligated to protect their rights. For example, violent threats of misogyny and racism are allowed to go unchecked on Facebook and Twitter. Just recently, a member of the Bengali gay community was brutally murdered — an event publicly celebrated in some radically conservative Facebook groups (City of Ottawa, 2016).

Activists in India are under constant threat on social media for criticizing government policies. It is hard to imagine how a smart city can function, when its citizens are subject to violations of their rights to privacy and freedom of expression.

Targets for achieving Goal 11 - Sustainable Cities and Communities - include reducing the negative impacts of natural disasters, ensuring universal access to green spaces and addressing the environmental impact of cities (City of Toronto, 2020).

Here are some of the innovative ways cities are rising to the challenge in Singapore, which has been dubbed the "Garden City" due to its abundance of green spaces.

Cities are on the front lines of increased physical risks associated with climate change. They are home to more than half of the world's population, and by 2050, this number is expected to rise to 68 percent. 2 Urban areas are often located in places with special climate risks, such as coasts floodplains and islands. Moreover, modern urban infrastructure and operating systems are closely related. A failure in one part of the network can affect another, compounding the damage. Flooded roads, for example, can damage public transport links. Severe storms and extreme heat can lead to power outages that disrupt technology systems critical to homes, hospitals, and industries.

Because different cities face different climate risks and have varying levels of vulnerability, adaptation options that are effective in most of them may not be feasible in others. To manage this complexity, cities can focus on actions that play to their strengths (in resources, physical features and assets, and judicial oversight) and offer a high return in reducing risk. Identifying such high-impact adaptations

Smart Sustainable Cities to Treat Economic, Climate Repercussions

can be daunting, given the evolving nature of the climate threat and the dizzying array of adaptation options available.

This report, in partnership with C40 Cities Climate Leadership, a network of large cities committed to tackling climate change, seeks to help leaders set priorities and choose courses of action. Outlines a starting list of 15 high-potential actions that can work in many types of cities. The actions were selected on the basis of three main sources: C40 Cities Climate Leadership and McKinsey analysis, consultations with adaptation experts and city leaders, and a comprehensive review of the literature.

There are two parts to the report. The first identifies the fifteen procedures. Four of them build systemic resilience, which means they fortify all kinds of cities. The other 11 are risk specific, meaning they target specific physical climate risks. Some of the fifteen measures, such as building barriers to protect coastal areas and rehabilitating infrastructure, are complex and expensive. Others, like planting trees next to streets and starting behavior change programs to conserve water, are not. Examples from around the world, in both developed and developing economies, show what is possible.

The second part of the report describes, in broad terms, how cities can implement the measures. We suggest that they start by identifying the most relevant risks and understanding the risks they pose to their communities. On this basis, cities can then conduct detailed analyzes of the impact of risk reduction, costs and feasibility of various measures (False Creek Neighborhood Energy Utility, 2020).

Several important topics emerge from the research. First, nature-based solutions – such as planting trees next to streets, river catchment management, and sustainable urban drainage solutions – are among the most attractive actions because of their impact on risk reduction and their feasibility. Nature-based solutions often provide benefits beyond adaptation in areas such as decarbonization, economic growth, and health.

Second, cities can invest in measures that systematically increase resilience, as well as adapt to specific and immediate risks. Systematic flexibility includes increasing awareness of physical climate risks, integrating risk awareness and preparedness into city operations, improving emergency response, and strengthening financial and insurance programs (City of Vienna, 2013).

Third, there is an important equity component to adapting to climate risks. Vulnerable populations, such as children, the elderly, low-income communities, certain minority groups, people with disabilities, and women may be more vulnerable to climate-related damage. For example, continued rapid urbanization leads to an increase in the population of informal settlements. 7 They often lack the resources and adaptability to withstand major events, such as floods and extreme heat.

Climate risks directly affect people (health, livelihood, and ability to work), assets (businesses, homes, and hospitals), and services (energy supply and food). This report can serve as a starting point to help cities develop their adaptation agendas. Leaders will need to get deep into the action as they create their strategies. Local knowledge is critical to success (City of Vienna, 2014).

At the same time, climate adaptation is a competing priority, and urban resources are limited. By identifying the most effective and feasible actions, cities can focus on getting them done well and build the momentum to do more. This report is a call to action - focused action. We hope this will help cities play an important role in making faster and more assured progress towards a healthy and sustainable future (Clément, 2018).

Egypt and the Smart Cities Technology

Egypt has started implementing a number of smart cities, starting in the Administrative Capital, El Alamein, New Mansoura, New Damietta and New Rashid, according to international specifications, Egypt has paid attention to the establishment of sustainable smart cities, which are known as environmentally friendly technology cities, as Egypt is considered in the face of climate changes and the temptation occupies a large part of the efforts of the Egyptian state in developing cities and ports and protecting the beaches, and Egypt resorts to building smart cities that help control Relative climate and avoiding its effects through governance systems in construction, spaces, isolation and engineering designs that are consistent with health environmental standards and the manufacture of automated meteorological forecasting devices based on artificial intelligence programs and access to comprehensive awareness programs for all individuals and companies through broadcasting by smart means to educate citizens and answer their questions and inquiries With regard to climate change, the matter is not limited to these things, but the mechanism relies on digital modernization in all fields, including agriculture in terms of applications to inquire about the most appropriate planting and irrigation dates, responding to inquiries regarding the irrigation date, harvesting and pricing of the crop, avoiding the over-irrigations system and processes in specific wrong time, as well as programs (SIA, 2018) for forecasting natural disasters, which Smart cities play a major role in reducing waste and losses, as well as the environmental protection program and governance systems. Egypt has begun to enter the era of sustainable smart cities through the establishment of the new administrative capital, which is one of the most recent environmentally friendly cities, which represents a living model for modern smart cities in the world, as it avoids the city deals with the problems of household solid waste and the problems of high temperatures through insulating roofs. Companies are committed to using them and many applications related to green spaces and sports (Clemente & Salvati, 2017).

For the economy, where the report discusses the impact of global warming on the global economy, and although it is not the first economic report on Climate Change, however, is the largest and most widely known report of its kind, outlining a range of government policies to manage adaptation (Cohen, 2017).

With climate changes, it can be mentioned in the light of the following points:

- Providing a high level of information on climate change, to raise the efficiency of markets, and to help audit forecasts at regional levels.
- Develop plans and standards for land use to encourage and direct investments in construction, and develop infrastructure construction standards.
- Develop long-term policies to protect resources and areas that are highly vulnerable to climate change, such as: protecting natural resources and beaches, and preserving (Connolly et al., 2021).

Application of climate-smart agriculture practices and expansion of forests: Studies show that activities resulting from agricultural expansion.

It causes the release of carbon dioxide gas into the air, and to reduce these gases many methods can be adopted, the most important of which are:

- The use of adaptation mechanisms that resist climate change, through specific activities such as the use of crops that are resistant to drought or drought (Dawson, 2017) salinity, more efficient use of water resources, and improvements in pest management. Changes in agricultural patterns may include reducing.

The use of fertilizers and the development of rice production management.

- Agriculture can contribute positively to mitigating carbon dioxide emissions by absorbing it. Estimates indicate that The contribution of croplands to carbon sequestration over the next 20 to 30 years ranges from about 450 to about 610 million (Dawson, 2017) tons of carbon each year, through better land management practices such as improved soil fertilization, water management, and erosion control.

With this emphasis on the need to limit future climate change, and to help the most vulnerable adapt to what can be avoided,

The world must continue the path, and determine the nature of the policies that will help to reach the results it seeks.

Hence, successful coping strategies require action at different levels, by thinking in the long term and taking into account (DeVerteuil & Golubchikov, 2016).

Consider climate change risks at the regional (interstate), national and international level by assessing climate change vulnerability, technologies.

42 Adaptation, capacity assessment, local practices and government actions to address climate change.

T. Investing in renewable energies for a sustainable future: Recently, global fears of continued repercussions have escalated.

Cities Can Play an Important Role in Climate-Smart Development

Today, 55 percent of the world's population lives in urban areas. This number is expected to rise to 68 percent by 2050 - adding an additional 2.5 billion people to cities. Over the next 35 years, more than 1.2 billion people in total, or a third of the world's urban population, are expected to live in Asian cities alone (Luxembourg: Publications Once of the European Union, 2019).

Building cities that "work" - inclusive, safe, resilient and sustainable - requires extensive policy coordination and investment choices. Once a city is built, its physical form and patterns of land use can be locked in for generations, resulting in an unsustainable sprawl. Many cities are already at the forefront of innovative climate solutions, ranging from emissions reduction policies through transit-oriented development and energy efficiency efforts to building more resilient urban infrastructure that can better withstand the effects of climate change.

Asia's cities already consume 80% of the region's energy and produce 75% of its carbon emissions. Asian cities are poised to contribute more than half of the increase in global emissions over the next 20 years if no action is taken (European Commission, 2015).

Smart Cities Performance Experiences in Confronting Climate Change

According to the EU, ambitious Horizon 2020 projects must meet "EU mitigation targets and adapt to national and/or local energy, air quality and climate targets." Initiatives for official reports. Some cities are massively adopting smart city projects, i.e. they are "quickly outgrowing the policies that govern their development." In addition, it can be difficult to define a global framework for smart cities due to the unique characteristics of each city and, therefore, must be designed according to the priorities and vision of a unique city (European Commission, 2019). Cutting-edge technologies such as 5G, artificial intelligence (AI), big data, cloud and the Internet of Things (IoT) will be the catalyst for a rapidly evolving concept of a smart city, where everything from security cameras to sensors at waste collection points will be connected to the Internet. In the business report, Hewlett Packard Enterprise (HPE) highlights some of the IoT use cases that relate to the environment to paint an idea of what's on the horizon within the intelligent environment dimension. Smart water management is able to protect a city's water supply with real-time decision making, while also helping to prevent water waste by using data to detect leaks, temperatures, and water pressure. Building intelligence is useful for monitoring energy use and improving energy efficiency. Furthermore, integrating data from weather, traffic, and environmental sensors has the potential to manage air quality and provide more accurate weather forecasts (European Innovation Partnership on Smart Cities and Communities (EIP-SCC), 2019). The big data generated by these sensors can make prediction more optimized for concrete actions in adaptive policy making. Improving air quality and reducing exposure to heatwaves are potential outcomes of more accurate, real-time environmental data, while telemedicine, tele education, and mobile public health monitoring could make a city more resilient to climate impacts. Digging deeper into the specific applications of smart technologies in climate change adaptation, the European Environment Agency (EEA) states that "smart spatial and infrastructure designs reduce urban heat island effect, air pollution and flooding of streets and homes" (European Environment Agency, 2019).

Some Models of Smart Cities to Face the Climate Change

Barcelona

The Barcelona Ciutat Digital Plan is the city's strategy for the smart city, which sets its goals beyond simply integrating technology into the city but ambitiously intends to address long-term urban challenges, such as climate change and natural resource scarcity. Furthermore, Barcelona strives for a more collaborative and sustainable economy in order to reduce social inequalities, while ensuring leadership in innovation. However, climate action is not named as a specific category, as the key prisms of the strategy are (1) governance, (2) city services, (3) the digital socio-economic fabric and local innovation system, and (4) citizens. In this context, the main challenges facing Barcelona are urban flooding, extreme heat, reduced water availability, and coastal erosion (Salvador, 2017). The city's historic vulnerability to urban flooding created a need for continuous adaptation of the urban drainage system, which led to the implementation of a network of storm water reservoirs. These reservoirs are remotely controlled by ICT systems, providing a more efficient and less dependable drainage system that saves 30 percent on operating costs and reduces the chance of flooding in urban areas by up to 75 percent. Since 2015, Barcelona has implemented smart energy monitoring in 23 buildings. With 31 more charts. The city has also installed 240 pneumatic waste collection points, while 40 percent of Barcelona's parks have an automated irrigation system. (Sinaee, 2016) In addition, the Barcelona City Council has pioneered the development of the Sentilo platform, an open-source software available to citizens for public experiences, which collects. The data is from a network of about 1,800 sensors in use in the city, 50 of which, specifically, measure environmental indicators such as air quality, humidity and temperature (March & Ribera-Fumaz, 2016).

Rotterdam

Rotterdam does not yet have an official smart city strategy, but the city council has instructed PBLQ, a consulting agency, to help develop their vision for a smart city in Rotterdam. According to the report, Rotterdam has already been named a smart city on sustainability because it actively cooperates with national cities and other European cities on EU projects involving sustainability, heat and energy management. However, the report does not consider climate action or sustainability as a separate category, as it does political, economic, social and cultural dimensions. Instead, sustainability is woven into some examples of initiatives, such as the inter-city sectoral cooperation projects in the Rotterdam metropolitan area, Cleantech Delta and Food Delta, which are designated in the economic category (City of Rotterdam, 2013).

Instead, sustainability is incorporated into some examples of initiatives, such as the inter-city sectoral cooperation projects in the Rotterdam metropolitan area, Cleantech Delta and Food Delta, which are planned in the economic category. Besides the Dutch's constant struggle against the sea, Rotterdam also aims to make the city less vulnerable to drought, heat stress, and heavy rains. In order to better withstand heavy rainfall, Rotterdam has developed an innovative water arena, Bentem Square, equipped with a precipitation forecasting system to better protect the city from heavy rainfall and sewage overload, while also serving as a sports and recreation area. It is located in the ZOHO district, which is being converted into the first climate-proofed area in Rotterdam. Other smart initiatives in the ZOHO district include a rainwater storage system that stores and reuses rainwater simultaneously, managed by a smart controller, and the Polder Roof, a green rooftop parking garage that reuses and stores excess rainwater from neighboring buildings in a controlled way (De Urbanisten, 2016).

Vienna

In Vienna, the Smart City Wien Framework Strategy outlines the city's long-term smart ambitions. The strategy highlights that Vienna has already done a lot in the field of climate protection and that the smart city framework builds on existing approaches to environmental and climate policy. It focuses on available resources while ensuring that collaboration between all actors will facilitate joint focus on overarching goals (City of Vienna, 2014). Promoting Vienna as an ecological model city and aiming to conserve the highest possible number of resources are among the objectives set out in the framework, coexisting with other smart city objectives in the three main pillars of innovation, governance and quality of life. In terms of climatic risks, Vienna has already experienced a 2°C increase in average temperatures over the past four decades, causing intense periods of heavy rainfall and almost twice as many heat waves in the period 1976-2005, compared to the period 1961-1990 (Anthopoulos & Roblek, 2019). In cooperation with the European Union and national stakeholders, such as the Austrian Climate and Energy Fund, Vienna has begun financing and implementing several smart city initiatives, such as equipping city traffic lights with environmental and weather sensors, improving water and energy use in school buildings and integrating renewable energy sources into the electricity grid. City. Vienna is also working to protect urban biodiversity by mapping common algal bloom sites through an intelligent and participatory approach (Joss et al., 2020).

CONCLUSION

Because different cities face different climate risks and have varying levels of vulnerability, adaptation options that are effective in most of them may not be feasible in others. To manage this complexity, here, the role of technology becomes clear in the smart city model for analyzing data, improving management and setting priorities in order to reach the best practices to confront climate change. cities can focus on actions that play to their strengths (in resources, physical features and assets, and judicial oversight) and offer a high return in reducing risk. Identifying such high-impact adaptations can be daunting, given the evolving nature of the climate threat and the dizzying array of adaptation options available. Some cities around the world are pioneering this avenue, helping the development community to envision alternatives to prevailing models of urban development, and focus on creating environmentally friendly "people's cities", rather than economic growth. This Spotlight shares innovative thinking in urban planning, urban design, and urban technology to highlight some of the transformative solutions that are changing the way the world looks at cities. Cities are on the front lines of increased physical risks associated with climate change. They are home to more than half of the world's population, and by 2050, this number is expected to rise to 68 percent. 2 Urban areas are often located in places with special climate risks, such as coasts floodplains and islands. Moreover, modern urban infrastructure and operating systems are closely related.

If the smart city is the best urban solution to confront climate change, it is necessary to maintain construction incentives to transform towards a sustainable smart city and incentives for settlement, countries, communities, and individuals in order to save the earth planet.

REFERENCES

Московский стандарт реновации. (2020). *City Projects*. Moscow government. https://www.mos.ru/city/projects/ renovation/

0ESPON. (2012). *SGPTD Second Tier Cities and Territorial Development in Europe: Performance, Policies and Prospects*. ESPON. https://www.espon.eu/sites/default/les/attachments/SGPTD_

Adams, J. (2013). Collaborations: The fourth age of research. *Nature*, *497*(7451), 557–560. doi:10.1038/497557a PMID:23719446

Aleixandre-Tudo, J. L., Bolanos-Pizarro, M., & Aleixandre, J. L. (2019). Current trends in scientific research on global warming a bibliometric analysis. *International Journal of Global Warming*, *17*(2), 142–169. doi:10.1504/IJGW.2019.097858

Badyina, A., & Golubchikov, O. (2016). "Gentrication in central Moscow – a market process or a deliberate policy? Money, power and people in housing regeneration in Ostozhenka". Geograska Annaler: Series B. *Human Geographies*, *87*(2), 113–129. doi:10.1111/j.0435-3684.2005.00186.x

Blunier, T., Chappellaz, J., Schwander, J., Dällenbach, A., Stauffer, B., Stocker, T. F., Raynaud, D., Jouzel, J., Clausen, H. B., Hammer, C. U., & Johnsen, S. J. (1998). Asynchrony of Antarctic and Greenland climate change during the last glacial period. *Nature*, *394*(6695), 739–743. doi:10.1038/29447

Bonacina, L. C. W. (1947). Climatic Change and the Retreat of Glaciers. 1 The Self-Generating or Automatic Process in Glaciation. *Quarterly Journal of the Royal Meteorological Society, 73*(315-316), 85–000. doi:10.1002/qj.49707331506

Brazier, A. (2015). *climate change in Zimbabwe, facts for planners and desicion makers.* Konrad Adenauer stiftun. https://www.kas.de/c/document_ library/get_fle?uuid=6dfce726-fdd1-4f7b-72e7-e6c1ca9c9a95&group

Bronnimann, S., Appenzeller, C., Croci-Maspoli, M., Fuhrer, J., Grosjean, M., Hohmann, R., Ingold, K., Knutti, R., Liniger, M. A., Raible, C. C., Röthlisberger, R., Schär, C., Scherrer, S. C., Strassmann, K., & Thalmann, P. (2014). Climate change in Switzerland: A review of physical, institutional, and political aspects. *Wiley Interdisciplinary Reviews: Climate Change, 5*(4), 461–481. doi:10.1002/wcc.280

Bulkeley, H., Coenen, L., Frantzeskaki, N., Hartmann, C., Kronsell, A., Mai, L., Marvin, S., McCormick, K., van Steenbergen, F., & Voytenko Palgan, Y. (2016). Urban living labs: Governing urban sustainability transitions. *Current Opinion in Environmental Sustainability, 22,* 13–17. doi:10.1016/j.cosust.2017.02.003

Carbon Neutral Cities Alliance. (2019). *Carbon Neutral Cities Alliance 2019 Annual Report.* CNCA. https://carbonneutralcities.org/cities/

Charles, A. (2015). *"Can we build cities that anticipate the future?".* World Economic Forum. https://www.weforum.org/agenda/2015/10/can-we-build-cities-that-anticipate-the-future

CitiesAct. (2009). *The Copenhagen Climate Communiqué.* Clover Archive. http://www.cloverarchive.com/main/ page/3125.pdf.

City of Ottawa. (2016). Guide for Older Adults: Services and Programs oered by the City of Ottawa. City of Ottawa. https://documents.ottawa.ca/sites/documents/les/2019-.058%20Older%20Adult%20Booklet%20-%20ENG.pdf. City of Ottawa (2020).

City of Rotterdam. (2013). *Rotterdam: Climate Change Adaptation Strategy.* Rotterdam Climate Initiative.

City of Toronto. (2020). *Toronto Strong Neighbourhoods Strategy 2020.* City of Toronto. https://www.toronto.ca/city

City of Vienna. (2013). *Gender Mainstreaming in Urban Planning and Urban Development.* WIEN. https://www.wien.gv.at/stadtentwicklung/studien/pdf/b008358.pdf

City of Vienna. (2014a). *Smart City Wien.* Wien. https://smartcity.wien.gv.at/site/les/2014/09/

City of Vienna. (2014b). *Smart City Wien: Framework Strategy; Municipal Department 18.* City Development and Planning.

Clément, J.-F. (2018). *Smart City Framework.* Wien. SmartCityWien_FrameworkStrategy_english_onepage.pdf.

Clemente, M., & Salvati, L. (2017). 'Interrupted' Landscapes: Post-Earthquake Reconstruction in between Urban Renewal and Social Identity of Local Communities. *Sustainability (Basel), 9*(11), 2015. doi:10.3390u9112015

ClimateHomeNews. (2019) Demark adopts climate law to cut emissions 70% by 2030. *Climate Home News.* https://www.climatechangenews.com/2019/12/06/denmark-adopts-climate-law-cut-emissions-702030/.

Cohen, G. (2017). *What is social infrastructure?* Aberdeen Standard Investments. https://www.aberdeenstandard.com/en-us/us/investor/insights-thinking-aloud/article-page/what-is-socialinfrastructure

Connolly, C., Keil, R., & Ali, S. H.. (2021, February). Extended urbanisation and the spatialities of infectious disease: Demographic change, infrastructure and governance. *Urban Studies (Edinburgh, Scotland), 58*(2), 245–263. doi:10.1177/0042098020910873

CPH2025 Climate Plan. (2020aCity of Moscow. https://kk.sites.itera.dk/apps/kk_pub2/pdf/983_jk-P0ekKMyD.pdf.

Culwick, C., Washbourne, C.-L., Anderson, P. M. L., Cartwright, A., Patel, Z., & Smit, W. (2019). CityLab re-ections and evolutions: Nurturing knowledge and learning for urban sustainability through co-production experimentation. *Current Opinion in Environmental Sustainability, 30,* 9–16. doi:10.1016/j.cosust.2019.05.008

Dawson, A. (2017). *Extreme Cities: The Peril and Promise of Urban Life in the Age of Climate Change.* Brooklyn: Verso.

De Urbanisten. (2016). *Climate Proof ZOHO District: Work in Progress.* Rotterdam Climate Initiative.

DeVerteuil, G., & Golubchikov, O. (2016). Can resilience be redeemed? Resilience as a metaphor for change, not against change. *City, 20*(1), 143-151. doi:10.1080/13604813.2015.1125714

Drenth, P. (2001) Freedom and responsibility in science: reconcilable objectives? Joachim Jungius-Gesellschaft der Wissenschaften, Symposium "Forschungsfreiheit und ihre ethische Grenzen."

EU. (2020) 2030 climate & energy framework. EC. https://ec.europa.eu/clima/ policies/strategies/2030_en. Accessed Mar 2020

European Commission. (2015). *Closing the loop – An EU action plan for the Circular Economy. Communication from the Commission to the European Parliament, the Council, the European Economic and Social Committee and the Committee of the Regions.* EC. https://eur-lex.europa.eu/resource.html?uri=cellar:8a8ef5e8-99a0-11e5-b3b7-01aa75ed71a1.0012.02/DOC_1andformat=PDF.

European Commission. (2019). *The Digital Cities Challenge: Designing Digital Transformation Strategies for EU Cities in the 21st Century.* EC. https://www.intelligentcitieschallenge.eu/sites/default/les/2019-09/EA-04-19-483- EN-N.pdf

European Environment Agency. (2019). *Open data and e-government good practices for fostering environmental information sharing and dissemination.* UNECE. https://www.unece.org/leadmin/DAM/env/pp/a_to_i/Joint_

European Innovation Partnership on Smart Cities and Communities (EIP-SCC). (2019). *Smart City Guidance Package: A Roadmap for Integrated Planning and Implementation of Smart City Projects; Norwegian University of Science and Technology.* NTNU.

Smart Sustainable Cities to Treat Economic, Climate Repercussions

False Creek Neighborhood Energy Utility. (2020). *Property Development.* False Creek Neighborhood Energy Utility. https://vancouver.ca/homeproperty-development/southeast-false-creek-neighbourhood-energy-utility.aspx

Flohn, H. (1961). Mans activity as a factor in climatic change. *Annals of the New York Academy of Sciences, 95*(1), 271–281. doi:10.1111/j.1749-6632.1961.tb50038.x

Gastner, M. T., & Newman, M. E. J. (2004). Diffusion-based method for producing density-equalizing maps. *Proceedings of the National Academy of Sciences of the United States of America, 101*(20), 7499–7504. doi:10.1073/pnas.0400280101 PMID:15136719

German Watch. (2019). *CCPI, Climate Change Performance Index.* German Watch.

Groneberg-Kloft, B., Fischer, T. C., Quarcoo, D., & Scutaru, C. (2009). New quality and quantity indices in science (NewQIS): The study protocol of an international project. *Journal of Occupational Medicine and Toxicology (London, England), 4*(1), 16. doi:10.1186/1745-6673-4-16 PMID:19555514

IPCC. (2013). *Climate Change 2013: The physical science basis. Contribu- tion of working group i to the ffth assessment report of the intergovern- mental panel on climate change.* Cambridge University Press.

IPCC. (2019). *The Intergovernmental Panel on Climate Change.* IPCC.

Klingelhofer, D., Braun, M., Quarcoo, D., Bruggmann, D., & Groneberg, D. A. (2019). Research landscape of a global environmental challenge: Microplastics. *Water Research, 170*, 115358. doi:10.1016/j.watres.2019.115358 PMID:31816566

Kopp, R. E., DeConto, R. M., Bader, D. A., Hay, C. C., Horton, R. M., Kulp, S., Oppenheimer, M., Pollard, D., & Strauss, B. H. (2017). Evolving understanding of Antarctic ice-sheet physics and ambiguity in probabilistic sea-level projections earths. *Earth's Future, 5*(12), 1217–1233. doi:10.1002/2017EF000663

Kulp, S. A., & Strauss, B. H. (2019). New elevation data triple estimates of global vulnerability to sea-level rise and coastal fooding. *Nature Communications, 10*(1), 4844. doi:10.103841467-019-12808-z PMID:31664024

Lockyer, W. J. S. (1910). Does the Indian climate change? *Nature, 84*(2128), 178–178. doi:10.1038/084178a0

Magri, M., & Solari, A. (1996). The SCI Journal Citation Reports: A potential tool for studying journals? 1 Description of the JCR journal population based on the number of citations received, number of source items, impact factor, immediacy index and cited half-life. *Scientometrics, 35*(1), 93–117. doi:10.1007/BF02018235

March, H., & Ribera-Fumaz, R. (2016). Smart contradictions: The politics of making Barcelona a Self-sufficient city. *European Urban and Regional Studies, 23*(4), 816–830. doi:10.1177/0969776414554488

Menzel, A., Sparks, T. H., Estrella, N., Koch, E., Aasa, A., Ahas, R., Alm-Kübler, K., Bissolli, P., Braslavská, O. G., Briede, A., Chmielewski, F. M., Crepinsek, Z., Curnel, Y., Dahl, Å., Defila, C., Donnelly, A., Filella, Y., Jatczak, K., Måge, F., & Zust, A. (2006). European phenological response to climate change matches the warming pattern. *Global Change Biology, 12*(10), 1969–1976. doi:10.1111/j.1365-2486.2006.01193.x

Mesolell, K. J., Matthews, R. K., Broecker, W. S., & Thurber, D. L. (1969). Astronomi- cal Theory of Climatic Change—Barbados Data. *The Journal of Geology*, *77*(3), 250–274. doi:10.1086/627434

Mills, J. N., Gage, K. L., & Khan, A. S. (2010). Potential influence of climate change on vector-borne and zoonotic diseases: A review and proposed research plan. *Environmental Health Perspectives*, *118*(11), 1507–1514. doi:10.1289/ehp.0901389 PMID:20576580

Nature Index. (2018) *10 institutions that dominated science in 2017*. Nature Index. https://www.natureindex.com/news-blog/twenty-eighteen-annual-table s-ten-institutions-that-dominated-sciences.

Naylor, M. (2019). *How the Nordics are standing up to climate change*. STP Trans. https://www.stp-transcom/how-nordics-are-standing-up-to-clima te-change/.

ND-GAIN. (2019) *Notre Dame Global Adaptation Initiative, Country Index*. ND-Gain. https://gain.nd.edu/our-work/country-index/.

OECD. (2019). OECD Economic Surveys: China 2019. *OECD Economic Surveys: China*, *2019*. OECD. doi:10.1787/eco_surveys-chn-2019-en

One Young World. (2020). *The effects of climate change in Zimbabwe*. One Young World. https://www.oneyoungworld.com/blog/efects-climate-change-zimba

Parmesan, C., & Yohe, G. (2003). A globally coherent fngerprint of climate change impacts across natural systems. *Nature*, *421*(6918), 37–42. doi:10.1038/nature01286 PMID:12511946

Plass, G. N. (1956). The carbon dioxide theory of climatic change. *Tellus*, *8*(2), 140–154. doi:10.3402/tellusa.v8i2.8969

Pounds, J. A. (2006). Widespread amphibian extinctions from epidemic disease driven by global warming. *Nature*, *439*(7073), 161–167. doi:10.1038/nature04246 PMID:16407945

Ritchie, H., & Roser, M. (2019) Our world in data: CO2 and greenhouse gas emissions. *Our World in Data*. https://ourworldindata.org/co2-and-other-greenhouse-gasemissions.

Roblek, V. (2019). The smart city of Vienna. In L. Anthopoulos (Ed.), *Smart City Emergence: Cases from around the World* (1st ed.). Elsevier. doi:10.1016/B978-0-12-816169-2.00005-5

Root, T. L., Price, J. T., Hall, K. R., Schneider, S. H., Rosenzweig, C., & Pounds, J. A. (2003). Fingerprints of global warming on wild animals and plants. *Nature*, *421*(6918), 57–60. doi:10.1038/nature01333 PMID:12511952

Salvador, J. (2017). PPP For Cities Case Studies. Barcelona GIX: IT Network Integration (Spain). UNECPPP. https://www.uneceppp-icoe.org/people-rst-ppps-case-studies/ppps-in-it/barcelona-gix.-it-network-integrationn barcelona-spain/

Sangam, S.L. & Savitha, K.S. (2019). Climate change and global warming: a scientometric study. *Collnet J Scientomet, 13*, 199–212. doi:. doi:10.1080/09737766.2019.159800136

SIA. (2018). Les Ateliers de Renens: Construction d'un site dédié à l'emploi. *Forum Bâtir et Planier: la Ville Productive Conference*. VD. https://www.vd.sia.ch/sites/vd.sia.ch/les/20181112_Batir_Planier_Cl%C3%A9ment.pdf

Sinaee, A. (2016). *Estimating Smart City Sensors Data Generation: Current and Future Data in the City of Barcelona.* In Proceedings of the 15th IFIP Annual Mediterranean Ad Hoc Networking Workshop, Barcelona, Spain.

Stott, P.A., Stone, D.A., & Allen, M.R. (2004). Human contribution to the European heatwave of 2003. *Nature, 432*, 610–614. https://doi.org/ e03089 doi:10.1038/natur

The World Bank. (2019a) *Data, GDP (current US$).* World Bank. https://data.worldbank. org/indicator/NY.GDP.MKTP.CD.

The World Bank. (2019b) *Data, Population, total.* World Bank.

Chapter 9
Social Resilience's Significance in Managing Extreme Climatic Events in the Cities of the Eastern Area:
A Field Study

Yasmin Alaa Ali Youssef
Ain Shams University, Egypt

Ahmed Zayed Abdalla Zayed
Assuit University, Egypt

Mahmoud Zayed Abdalla Zayed
Cairo University, Egypt

ABSTRACT

This current chapter starts from a main question: What is the extent of social resilience as a function of managing extreme climatic events in the Kingdom of Saudi Arabia? The study will present three basic concepts: social resilience, crisis management, and extreme climatic phenomena. As for the approach used for the study, it is the descriptive approach, and an electronic questionnaire is relied upon as a tool for collecting data within cities in the eastern region. The study reached a number of results, the most important of which are: dehydration is one of the many manifestations of the impact of climatic phenomena on cities, including: increasing rates of economic pressures, giving priority to smart cities, and restructuring development plans for infrastructure within cities.

DOI: 10.4018/978-1-6684-8963-5.ch009

Social Resilience in Managing Extreme Climate Events in Eastern Area

FIRST: THE PROBLEM OF THE STUDY

Extreme weather events in recent years have a huge impact of greenhouse gas emissions on the global climate. Moreover, the costs of these phenomena are on the rise, and then there is a need for social resilience as one of the distinguishing features of contemporary organizations in light of climatic risks. in managing climate change. Resilience refers to the ability of a particular work or system to return to equilibrium after decline or deterioration **(Shalender,2015: 266).**

A large number of studies have been presented over the past two decades, investigating the effects of climate change risks and adaptation to them at the local, regional and global levels. This is supposed to be in response to the effects of climate change, and accordingly international agreements, treaties and conferences that dealt with the problem of climate change were held, starting with the United Nations Framework Convention on Climate Change in Rio de Janeiro in 1992, passing through the Paris Agreement in 2015, up to the twenty-sixth session. For the Conference of the Parties to the United Nations Framework Convention (COP26) on Climate Change in November 2021 in Glasgow, Scotland.

Based on the above, the problem of the study crystallizes in a main question: "What is the extent of social resilience as a function of managing extreme climatic events in the Kingdom of Saudi Arabia?" The current chapter stems from a main objective (identifying social resilience as a function of managing extreme climatic events in the Kingdom of Saudi Arabia). Several sub-objectives stem from this objective:

- Detection of extreme climatic phenomena in the Kingdom of Saudi Arabia.
- Recognizing the effectiveness of government sector institutions as an indicator of social resilience for crisis management.
- Reaching a set of proposals that would increase the effectiveness of the role of institutions in dealing with extreme climatic phenomena.

SECOND: THE IMPORTANCE OF THE STUDY

(a) **Theoretical significance:** The importance of this study was embodied in theory in support of previous research and studies.
   ○ This study helps to develop scientific knowledge on the subject of the social resilience of some institutions of extreme climatic events (such as heavy rains - dust storms - extreme heat waves).
   ○ Enriching the local and Arab libraries on this topic, as it is the first study - within the limits of the research team's knowledge - that linked the social resilience of some institutions and extreme climatic phenomena.
   b) practical significance
   ○ This study, with its findings and recommendations derived from a field study, enables the identification of the relationship between the level of social resilience of some institutions and climate changes, from which it is possible to read about the features of development and organizational resilience within those institutions in the face of extreme climatic phenomena.

- This study enables the relevant institutions and organizations to activate their role in giving clear and field indicators that help those working in them to identify some mechanisms to confront extreme climatic phenomena.
- -This study enables access to methods and strategies planned by some institutions in the government sector to deal with extreme climatic events, which will be reflected on the performance of those institutions.

THIRD: CONCEPTS

The concepts of the study are: (social resilience - crisis management - and extreme climatic phenomena).

1- **Social** resilience: Social resilience or what can be called ease of adaptation; It is defined as the ability of a person to return to his normal life after experiencing a crisis or distress, and it also means the ability of society to adapt and adapt to pressures such as social, economic or environmental change, given that humans can anticipate future conditions and prepare for them (Shalender, 2015: 266) . Social resilience can be defined procedurally as: a continuous process of change and adaptation, in conjunction with any change that may occur in the surrounding environment, and therefore its various pressures. If society has the ability to withstand, adapt and enhance resilience - as a kind of strength - then it will be able to take proactive measures to mitigate the effects of crises and quickly recover from them. However, it is not possible to define the way in which people deal with the problems they face in one framework, as the nature of each person differs from the other, as some of them spend a long time to overcome a problem, while others are able to make a new start thanks to their flexibility and their ability to withstand in the face of No position and poise in a short time compared to others.

2- **Crisis management:** A crisis is considered a transitional condition characterized by imbalance and represents a turning point in the lives of individuals or groups, and often results in a major change in the practices themselves in order to confront. The relationship, just as the crisis is considered an excellent experience for the party or parties that suffered from it, because the organization in a situation of crisis tries to make every effort to employ previous experiences for confrontation to ensure its continuity (Mark, 2016: 8-9). The procedural concept of crises is determined by the following indicators- It is an event characterized by surprise, without expectation, and a challenge to normal behavior. At its beginning, it caused shock and a high degree of tension, weakness, and the absence of organization and integration in the methods of confrontation. A succession of events that affect at the beginning of making the appropriate decision. It leads to a halt in the work movement and impedes the achievement of the required goals in the time allotted for them. Confronting the crisis and preventing its impact requires all efforts and capabilities to mitigate and contain its dangers.

3- **Extreme weather events:** They are the climatic phenomena that show sharp fluctuations in the values of their indicators, the number of days, and the periods of their occurrence. Among these phenomena are heat waves, cold waves, and rainfall in terms of abundance and scarcity, both of which cause problems, one of which is related to floods, and the other is associated with periods of drought. The weather is disturbed for three days, and these phenomena are one of the indicators of climatic changes. On the other hand, the recurrence of these phenomena exacerbates the severity of climate change, especially on environmental systems (United Nations, 2011).

FOURTH: THE THEORETICAL FRAMEWORK GUIDING THE STUDY

The theoretical framework issues that emerge from the social system theory, the risk society theory, and the ecological approach to crisis management will be addressed as theoretical approaches. Finally, the theoretical issues directed to the current study will be presented.

1-Social system theory: Parsons defines the social pattern as "a group of individuals motivated by a tendency to optimally satisfy their needs, while the prevailing relationships among the members of this group are determined according to a pattern of complex and culturally common patterns. for the antecedent analysis of social action"; Individuals are driven to better satisfy their needs that dominate defensive orientation, as Parsons defines the relationship of individuals to their social attitudes in light of their cultural patterns (Mayrhofer, 2004: 178-180).

Perhaps the term relationship refers to what is called guidance in another context; Where that part of Parsons' ideas refers to the other main components of directing the actor towards the situation, which is the complementary directive, and here the term value does not appear explicitly during the analysis, but it can be assumed that the patterns involve values, and these patterns are characterized by cultural composition and participation, and this aspect can be The social system serves as a bridge between the social and cultural systems; The social system implies something that belongs to the culture (Luhmann, 2013:25-40).

Parsons also talked about the way the social system works; He mentioned that each format must find a solution to a number of problems, and he called these problems or conditions the name of functional obligations or functional requirements, and these functions consist of

- **Adaptation:** It means that every social system has to adapt to the social and physical environment in which it exists.
- **Achieving the goal:** It refers to the methods of effective individuals in order to achieve the goal.
- **Integration:** the meaning that the system relies on a set of methods, values, and standards that link the individual to the community, so the normative integration results in the system of the general society as a whole, and integration within the sub-systems focuses on the relationships that take place within the sub-system.
- **Preserving the survival of the pattern and managing tension:** It means that the system, including the standards and values that have generality, lead to maintaining the pattern of interaction, so that it does not deviate from or deviate from the limits of the system (Lutfi, Al-Zayyat, 2009: 73).

2- Risk society theory: This theory was put forward by the German sociologist Ulrich Beck in 1986 titled Risk Society. It is a social theory that describes the production and management of risks in modern society. The concept of a risk society by itself does not mean that it is a society in which risk rates increase, as much as we mean that it is a society organized to face risks, because it is increasingly preoccupied with the future and security. The industrial society, including the wide use of machines, depending on the technology that invaded all fields of life, led to the emergence of various types of risks that were not known before, and then the modern society is a society that is exposed to a special pattern of danger that is the result of the same modernization process that changed the organization social.

The presence of risks calls for the rationalization of the process of manufacturing, managing and making industrial, technological and economic decisions in modern organizations. The need arose to develop methods for calculating risks through the development of disciplined approaches to calculating them, which are extremely important, and their issuance will depend on the accuracy of risk calculation (Ulrich Beck, 2005:6). Beck's theory holds that it is a combination of multiple risks (manufactured, environmental, and health) in a form called the global risk society. confront or adapt to it. Anthony Giddens (A. Giddens) also adds to the environmental, health and industrial risks other aspects of the overlapping variables in our contemporary social life that are the real reasons for the emergence of the risk society, including the following: (Danied, 2007: 80-82).

- Fluctuation in patterns of employment and employment within organizations.
- The growing sense of job insecurity.
- The prevalence of liberation and democracy in personal relations

Beck stresses that risks also affect people, because it is difficult to predict the nature of the skills and practical experience that will be required in the field of the ever-changing global economy

Academics have analyzed risk from a sociocultural perspective, emphasizing the role of social and cultural contexts to control and manage risk. Mary Douglas has tried to prove that cultures and societies share concepts of risk and that these concepts are not a product of individual knowledge (Windvsky Douglas, 2008: 15-20).

Risks have great challenges, as Beck assumes that the public (ordinary individuals) do not have the knowledge, and they constantly need someone to educate them about the risks through knowledgeable people, and they are the ones who direct, supervise and manage them as people of knowledge and accordingly they are permanently dependent on expert knowledge To educate them and warn them against the dangers (Ulrich Beck, 2009: 5-10).

3- The ecological approach to crisis management: This approach aims to provide indicators to identify the possibility of achieving development through the available material, political and social resources, and to initiate organized procedures for data collection to determine the available base for work mechanisms within the target entity and to provide guidelines for research and study in the field of crises. This approach also relies on Several hypotheses related to coping and managing crises are as follows (Al-Qahtani, 2003: 25).

- The target entity is the right place to study, make decisions and carry out the necessary actions to confront and manage the crisis.
- Interest in organizational, objective, planning and behavioral variables.
- Striving to achieve the ultimate goal, which is investing the available resources, developing them, improving their quality, and using them to provide the necessities and needs necessary to manage the crisis or prevent its occurrence.
- Providing information necessary for the development process on resources, their types and quantity, methods of benefiting from them and how to obtain them.
- Providing the necessary information for the development process about resources, their types and quantity, methods of obtaining them, and how to benefit from them.

Social Resilience in Managing Extreme Climate Events in Eastern Area

- Working on introducing new ideas that affect the current and future plans of individuals and organizations to confront crises so that they are commensurate with the development taking place in various fields.
- Participation of leaders and workers in identifying problems and needs. In planning and implementing preparedness programs to confront and manage crises.

The ecological framework consists of three components. Crises appear within one component as a result of differences, conflicts, natural disasters, or by the component itself, as a result of dangers or miscalculation, prediction and forecasting. Crises may also appear as a result of the overlap of the human component with other components of society or the overlap of more than one component at the same time, and these components are represented in (Al-Farazi, 2005: 60)

- **The human component:** It is the main organic component that includes human organizations and institutions and can develop, store and transfer knowledge from one team to another, and invent, build, organize and use machines to obtain, prepare, distribute and use resources. Crises are linked to the human component as a causative source and interfering with other components of society in the emergence of crises. For example, the supervisors of the work of workers and maintenance engineers in factories are part of the human cadre that affects negatively or positively in the situation of the crisis.
- **The artificial component:** It is the one that represents the physical structures and resources that are used and managed by the human being, the facility or the target entity. It is possible when studying to prepare for confronting and managing crises to identify and describe these elements quantitatively and qualitatively. As well as classifying them on the basis of their relationship to the various organizational units in the facility or the target entity in the economic, social, political and human fields. For example, computers are among the manufactured physical components that are related to crisis management, as important information is stored on them.
- **The natural component:** It represents the spatial space and the materials necessary to preserve the human component and the resources with which man manufactures the artificial component. The natural component represents one of the important components that must be precisely identified and appropriate standards should be put in place to keep it away from crises, and it is represented in the location of the crisis.

Hence, the current study proceeds from an integrative theoretical framework directed at it, and is based on a number of assumptions

- Each format must face at least four problems. In order to be able to continue and survive and achieve balance.
- The basis of society is the tendency towards balance, harmony, and the main processes that help the systems to function and interrelate with each other.
- The sources of empowerment differ from one system to another, and these systems must find ways to adapt to the requirements of society.
- The ecological framework for crisis management consists of three components. Crises appear within one component as a result of differences, conflicts, natural disasters, or by the component

itself as a result of dangers, miscalculations, predictions, and predictions. component at a time, which underscores the necessity of creating a homogeneous ecosystem.

FIFTH: LITERATURE REVIEW

1- Bulkeley and Kern's study entitled "Local Government and Climate Management in Germany and the United Kingdom" (2006). This study aims to identify how climate protection policy is formed in local administrations in Germany and the United Kingdom, and the obstacles that stand in front of local administrations in implementing policies that support Climate protection The study confirmed that financial crises and political challenges to implement climate change policy change the ability of local administrations to intervene and protect the environment from carbon emissions, so these countries seek to reduce voluntary measures while providing incentives to make energy saving attractive from a financial point of view And political support for the local authority by providing resources and incentives to work in partnership with other actors to help local governments play an important role in climate protection, and finally we find that Germany and the United Kingdom stumbled in reducing the gases responsible for global warming due to the interaction of transportation, planning and housing issues with total emissions

2- AROMAR REVI study entitled "Climate Change Risks: The Adaptation and Mitigation Agenda for Indian Cities" (2008) This study aims to highlight the importance of the participation of urban management in improving the utilization of infrastructure in light of climate crises. Although the level of exposure to risk in India is high, poor interaction contributes more to the overall risk in Indian cities. Therefore, this requires a shift in public policy, popular mobilization and support for projects that contribute to mitigation and adaptation. These shifts are based on the institutional, social, cultural and political reality of India and need to focus on the poor and the most vulnerable through a combination of policy, regulatory, scientific, financial, institutional and mobilization tools. These actions are likely to be best implemented by mainstreaming climate change risk assessment, adaptation and mitigation measures into ongoing national risk mitigation programs, and building a concrete set of linkages with urban renewal interventions being addressed in many of India's largest cities. To achieve this it is necessary to have a multi-tiered framework for climate adaptation, which operates at the national, state, city and neighborhood levels and brings together the state, private sector and civil society sectors. Robust adaptation programs in a range of pilot cities will allow important links between adaptation and mitigation to be explored.

3- Douglas' study entitled "Climate Change, Floods and Food Security in South Asia" (2009), this study aims to identify how to address the climatic challenges that affect food security in South Asia, where the poor, women and children suffer as the most vulnerable groups in South Asia. Therefore, the current measures to transfer climate adaptation funds tend to reduce the marginalization of these groups. By 2080, the situation is likely to be much worse than it is now, and therefore adaptation must encourage the management of all stages of food security from farm to consumer, whether in urban or rural areas, and that participatory measures be from the community to the international level, as many initiatives Individualism gives hope and good practices emerge, and on the other hand, institutional, economic and environmental factors all hinder the preservation and promotion of food security in South Asia. Therefore, innovative forms of food production, distribution and storage must be developed.

4- Hartmut's study entitled "Institutional Challenges for Climate Risk Management in Cities" (2010) This study confirms that cities have begun to assume many responsibilities towards identifying and analyzing local risks resulting from climate change, but there are many institutional challenges for city governments, civil society and companies The specific barriers to effective adaptation are two: understanding emerging information about climate change risks and their impacts on cities, and the impact of broader social and economic processes on urban vulnerabilities, and integrating information about climate risks and vulnerabilities into local planning processes and development agendas. Finally, the lack of adequate governance frameworks to manage climate risks in cities. These challenges reflect the limited level of city management understanding of local changes in vulnerability to climate change hazards and spatial distribution. Therefore, effective mechanisms for integrating climate change into local planning and decision-making processes are still the exception rather than the rule.

5- A study by Bord and Juhola titled "Climate Change Adaptation Governance for Urban Resilience" (2015). The study concludes that climate change poses many challenges to cities and requires new flexible forms of governance that are able to take into account climate changes. This study reviews three literary traditions. about the idea of climate transformations, and assesses to what extent these transformations include the elements of adaptive governance, by relying on the open source urban transformations project database to assess how urban experiences take into account the principles of adaptive governance, and the results showed that the responses do not give clear information about environmental knowledge, especially city leaders From local authorities, in addition to limited or absent evidence of proactive or planned partnerships and adaptation, the analysis reveals that technological, political, or environmental solutions alone are insufficient to increase our understanding of the analytical aspects of transformational thinking in urban climate management.

6- Bayulken et al's study entitled "How nature-based solutions help green cities in the context of crises such as climate change and pandemics: a comprehensive review" (2021). This study aims to identify the ways in which cities can be transformed into more resilient and sustainable areas by improving green areas within and around them. A review of 298 articles from 109 academic journals published during the period 1997-2020 was conducted. The "nature-based" changes that have been implemented in urban areas, globally to enhance their resilience and the quality of life of human beings. humankind by surrounding these areas with a rich biodiversity of locally adapted aquatic and terrestrial flora and fauna. The reviewers provided guidance for urban leaders to incorporate NBS into their policies and strategies to improve urban resilience and equity and to more effectively reduce the impacts of climate change, population growth, and pandemics on the city.

Commentary on previous studies related to the study variables.

By analyzing previous studies related to urban resilience in dealing with climate crises, the following became clear

- There are many institutional challenges facing cities in dealing with climate change, as Hurtmut's study.
- The need to develop many solutions, especially those based on nature, to facilitate the residents within the city, as a study by Bayulken.
- Epidemiological crises within cities must be dealt with and managed quickly to mitigate climate change and help the population adapt to these changes, as Bord and Juhola study.

SIXTH: THE METHODOLOGICAL FRAMEWORK OF THE STUDY

1) **Type of study:** The current study is a descriptive analytical study, which aims to describe and analyze phenomena through collecting data and information. With the aim of describing the social resilience of some institutions as a function of managing climate extremes.

2) **The method used:** the analytical descriptive method through the sample social survey method.

3) **Data collection tool:** The current study relies on the (graded questionnaire) "Triple Likert scale"; To measure "social resilience as a function of managing extreme climatic events in the Kingdom of Saudi Arabia," and it will be applied electronically in the study community.

4) The validity and reliability of the research tool

   ◦ **Apparent validity "the opinions of the arbitrators**

The questionnaire was presented to a number of arbitrators from faculty members specialized in the field of social sciences; To verify the apparent validity of it, and with the aim of judging the suitability of the questionnaire's items to the characteristics it measures, in addition to judging and expressing an opinion regarding the various elements of the questionnaire and the modification, deletion, addition or reformulation; as they see fit; In order to achieve the objectives of the current study.

- Structural validity:

   A small group was tested as a random prospective sample of "15"; In order for the study tool to be closer to accuracy and clarity, the sample members were informed; With the aim of testing the tool and applying it to them, and the need to comment on the questions and encourage them to ask questions to find out what is ambiguous or difficult to answer; To ensure the structural validity of the study tool.

- **The stability of the study tool**

   The internal consistency method was relied on, which depends on the extent to which the phrases are related to each other within the questionnaire and the correlation of the degree of each phrase with the total score of the questionnaire in general by analyzing the items. (0,360 - 0.892), and this indicates the strength of the internal coherence and consistency of the questionnaire statements.

5) **The study population:** The empirical material for the study was collected from the Municipality in the Eastern Province of the Kingdom of Saudi Arabia using the social survey method in the sample as a spatial field, and it is one of the government departments responsible for the development of the Eastern Province.

6) **Study sample:** The study sample consisted of workers in some departments in the organization, the field of study (130) individuals.

   It is clear from the results of Table No. (1) related to the characteristics of the study sample that the higher percentage of respondents are males; As their percentage reached (70%), while the percentage of females was (30%), and this may be due to the suitability of working conditions within the organization

Social Resilience in Managing Extreme Climate Events in Eastern Area

Table 1. Characteristics of the study sample

Characteristics		F	%
Gender	Male	39	30
	Female	91	70
Age	Less than 25 years old	73	56.2
	From 25-35	13	10
	From 35-45	31	23.8
	45 years old, remember	13	10
Educational Status	Intermediate education	13	10
	Above average education	13	10
	University education	91	70
	Postgraduate education	13	10
marital status	Unmarried	60	46.2
	Married	44	33.8
	Widower	13	10
	Divorced	13	10

under study for males more than for females due to the length of daily working hours in the company in the field of study, and therefore the working conditions are more suitable for males.

The results indicated that the highest percentage of ages was (56.2%) for those who reached the age of (less than 25 years), while the ages of (23.8%) of the respondents were (35-45 years), and this result can be explained In light of the fact that young fresh graduates join work with this type of organization, while the age group (35-45 years) is the mature group that has experience in the field of the company's field of study.

As for the variable (educational status), it is clear that the highest percentage was for those who received a university education and obtained a bachelor's degree. Their percentage reached (70%), while the percentage of those who received intermediate and secondary education reached (10%) of the respondents, in equal proportions, and (10%) continued post-university education and obtained postgraduate studies.

According to the variable (marital status), the results indicated that the highest percentage, which is (46.2%), are unmarried. It is also clear that (33.3%) of the respondents are married and (10%) are divorced.

Table 2. The most important climatic phenomena in the Kingdom of Saudi Arabia

climatic phenomena	F	%
temperature changes	117	90
Flood warning	26	20
Drought	87	66.9
Dust storms	104	80
Some climate elements fluctuate	72	55.4
The number of respondents	130	-

Table 3. Manifestations of the impact of climatic phenomena on cities

Manifestations	F	%
Restructuring development plans for infrastructure within cities.	102	78.5
Increasing rates of economic pressures on those affected by climate change	102	78.5
Giving priority to smart cities.	89	68.5
High rates of respiratory diseases.	101	77.7
High rates of earthquakes within cities	29	22.3
The number of respondents	130	-

THE RESULTS OF THE FIELD STUDY AND THEIR DISCUSSION

The extreme climatic phenomena in the Kingdom of Saudi Arabia.

It is clear from the data of the previous table that there are many climatic phenomena facing the eastern region of the Kingdom of Saudi Arabia. Temperature changes come at the forefront of these phenomena, accounting for 90% of the total number of respondents. This is because this area has high levels of humidity, which increases the feeling of temperature. Then the earthy emotions in the second place, with a rate of 80%, because this region constitutes a large part of the desert of the Arabian Peninsula, in which various sandy forms are spread. A group of local winds that cause disturbances and disturb the atmosphere in the region, such as the winds of Towz in Kuwait.

While the phenomenon of drought constitutes the third climatic phenomenon that the eastern region is exposed to, with a rate of 66.9%, because the long periods of drought that afflict the eastern coast of Saudi Arabia cause climatic fluctuations that affect the region, so the region is exposed to violent rainfall, which is associated with pools of water in the street that confuse traffic, and often The matter is that these rains are related to the escape of rainstorms from their natural course, but what prevails in the region in terms of periods of severe drought due to the location of the region, as it does not supervise significant bodies of water, as it is part of the desert of the Arabian Peninsula, so the phenomenon of drought is prevalent while the dear rains are the exception . The fluctuation of some climate elements in the eastern region comes fourth with 55.4% of the total number of respondents. Finally, the phenomenon of floods in climatic phenomena increased by 20%, due to the state of climatic extremism facing the eastern region. These data agree with Ulrich Beck's theory on the importance of anticipating risks within modern society, and that society has many artificial, environmental and health risks that pose a threat to the nature of social life. economics of these communities. While this result is consistent with the Douglas study, which confirms that there are many climatic phenomena that affect the nature of the human security of the population, including floods.

It is clear from the data of the previous table that there are many manifestations of the impact of climatic phenomena on the social, economic and planning life within the city. The rates of economic pressures on the population inside the cities as a result of the climatic changes that the cities are exposed to, the importance of development plans and the restructuring of the service sectors (educational, industrial and service) is increasing in a way that helps the population to adapt to the spread of extreme manifestations of the climate inside the city, for example we find interest in developing a network of public utilities related to sewage, water and electricity, so that institutions can quickly cope with climate changes from the dangers of torrential rains and fluctuations in temperature, as there is a very high humidity in cities,

Social Resilience in Managing Extreme Climate Events in Eastern Area

Table 4. Institutions that contribute to mitigating the effects of climate change

Institutions	F	%
Governmental institutions	88	67.7
Institutions of civil society.	13	10
private institutions.	13	10
international bodies.	16	12.3
Total	130	100

which affects the health of the population, especially the incidence of respiratory diseases by 77.7% of the total number of respondents. This puts pressure on the nature of health services within the city.

These challenges that cities face in the face of extreme climatic phenomena, requires a vision of his planning to build smart cities in which the technological role is activated to help residents improve their quality of life and face potential risks in the future with the acceleration of climate change worldwide, especially within cities. Therefore, the demand for building smart cities came among the priorities of the population within the cities to overcome the manifestations of the impact of climate change on the city by 68.5%. Thus, the planning trends towards building smart cities and green cities with is only solutions to improve the quality of life of residents, and overcome the negative effects of climate change on cities. Finally, we find that cities in the eastern region are exposed to house collapses as a result of the torrential rains that cities are frequently exposed to, with the climatic fluctuations that the world is exposed to. This data agrees with the AROMAR REVI study, which states that weak interaction contributes more to risks within cities, so this requires shifts in public policy and support for projects that contribute to mitigation and adaptation. While these results are consistent with Beck's risk society theory, which he argued that risks affect people, because it is difficult to predict the nature of practical skills and experience that will be required in the field of the global economy.

THE EFFECTIVENESS OF GOVERNMENT SECTOR INSTITUTIONS AS AN INDICATOR OF SOCIAL RESILIENCE FOR CRISIS MANAGEMENT

It is clear from the data of the previous table that there are many institutions that contribute to mitigating the severity of climate change on the population, and government institutions come at the forefront of these institutions with a rate of 67.7%, due to the active role of these institutions in providing services to the entire population within the Kingdom of Saudi Arabia, and then international bodies At a rate of 12.3% of the total sample size, as it contributes to international bodies in providing some distress to some of those affected by climate change.

Private institutions and civil society organizations are in the third rank, with 10% of the total sample. It is noted for this data that we find that there is an impact of government institutions on providing the nature of services, and overcoming obstacles in front of the population, with the development of many development plans that contribute to improving social and service life within the city. This result agrees with Bulkeley and Kern's study entitled "Local Government and Climate Management in Germany and the United Kingdom" (2006). Carboniferous This result is consistent with the social system theory of

Table 5. The contribution of government institutions to overcoming the negative effects of climate change

Contribution of government	F	%
Provide sample aid to affected families.	15	11.5
Structuring the service sectors.	16	12.3
Interest in afforestation spaces within cities.	73	56.2
Design awareness programs on the need to reduce carbon emissions.	13	10
Providing psychological support to students in educational institutions.	13	10
Total	130	100

Talcott Parsons, where he stated that each system must find a solution to a number of problems facing their society.

It is clear from the data of the previous table that government institutions contribute to providing many solutions to overcome the difficulties and negative effects of climate change, and interest in afforestation within cities is at the forefront of government solutions to confront climate changes by 56.2% of the total sample size, which was confirmed by the United Nations that Parks, green spaces and waterways are important public spaces in most cities, providing solutions to the health and safety impacts of rapid and unsustainable urbanization. The social and economic benefits of urban green spaces are equally important, and should be seen in the context of global issues such as climate change, and other priorities set out in the Sustainable Development Goals, including sustainable cities, public health and nature conservation.

Government institutions also contribute to the restructuring of the service sectors within the city by 12.3%. Government institutions, with their large components, are the ones that have the ability to cover services and facilities on all aspects of the Kingdom, and to reach all segments of society within the study community. While the provision of in-kind assistance to families affected by climate change comes in third place for the services provided by government institutions to the population, with a rate of 11.5%, in order to mitigate the severity of climate changes. climate in cities in the eastern region of the study population. Designing awareness programs on the need to reduce carbon emissions, and providing psychological support for students in educational institutions comes fourth in the mitigation packages offered by government institutions to help their citizens adapt to the climatic disturbances that the study community is constantly witnessing. This study agrees with the study of Bayulken et al, which argues that the implementation of nature-based solutions works in urban areas and their remote areas to expand the scope of their blue and green areas and thus reduce the effects of heat islands on their residents. While this result agrees with Ulrich Beck's risk society theory, which emphasizes the

Table 6. Measures to adapt the population to climate changes

Procedures	F	%
Giving importance to digital transformation within the vital sectors within the city.	102	78.5
Continuous infrastructure restructuring	88	67.7
Interest in designing smart cities and green cities.	117	90
Designing a network of reservoirs capable of accommodating large quantities of water.	72	55.4
The number of respondents	130	-

Social Resilience in Managing Extreme Climate Events in Eastern Area

Table 7. The effectiveness of the institutional role in confronting climate change within cities

Effectiveness	F	%
Effective	42	32.3
Medium effectiveness	73	56.2
not effective	15	11.5
Total	130	100

importance of risk management in modern society, and the rationalization of the process of making and managing industrial, technological and economic decisions in modern organizations.

Most of the responses of the study sample by 90% indicated that interest in designing smart cities and green cities is one of the main measures through which the population can adapt to climate changes, followed by giving importance to digital transformation within the vital sectors within the city, so that the percentage reached 78.5% of the total sample. Also, the continuous restructuring of the infrastructure is one of the executive measures to adapt the population of the study area to climate changes by up to 67.7%; It is clear from this that there are multiple mechanisms for the adaptation of the population in the region, the field of study, and the Eastern Region Municipality is trying to implement some of them in the face of some extreme climatic changes such as the dangers of floods and droughts, which contributes to increasing the processes of confrontation and pre-planning for any disaster, and this leads to harmony and harmonization between the organization and the field of study. and the needs of the local community, which increases the development and improvement of work strategies; Therefore, the organization seeks the field of study by the leadership role played by the managers to give attention to the initiatives related to climate change, and this result is consistent with the theoretical perception (the social format), which emphasizes the need for the survival of the institutional style with its specific values and standards and the management of the tension that arises within the institutions as a result of the obstacles internal and external, which prevent the achievement of its objectives.

It is clear from the data of the previous table that the effectiveness of the institutional role in confronting climate changes within cities is average, at a rate of 56.2% of the total sample size. Government institutions contribute a major role in adapting the population to climate changes, through the development of infrastructure and public facilities, and the restructuring of development sectors within the city. It contributes to drawing up future plans for city planning, and international bodies support the efforts made by institutions to mitigate climate change. This support extends from cash to awareness of risks and how to deal with manifestations of climate change. While 32.3% of the sample size believes that the institutional role is effective in confronting climate changes within cities. Hence, it can be said that a third of the sample size confirms that the institutions operating within the study community are effective in facilitating and supporting the processes of adaptation to the risks faced by cities as a result of supporting the social repercussions of climate change. While 11.5% of the total sample believe that the institutional role is not effective in confronting climate changes within cities. This is due to the difficulty of institutions reaching all social groups within society. These results agree with Bord and Juhola's study that climate change imposes many challenges on cities and requires new, flexible forms of governance that are able to take into account climate changes.

Table 8. Proposals to reduce climate change

Proposals	agree		sometimes		disagree		Average	Rang
	f	%	f	%	F	%		
Reliance on renewable energy sources	91	70	26	20	13	10	1.4	4
Use clean transportation.	88	67.7	29	22.3	13	10	1.423	7
Applying legal procedures to environmental violations	104	80	13	10	13	10	1.3	2
Spreading awareness among urban residents of the dangers of carbon emissions	90	69.2	27	20.8	13	10	1.408	5.5
The use of solar energy in factories and industrial complexes	90	69.2	27	20.8	13	10	1.408	5.5
Reducing food waste	73	56.2	44	33.8	13	10	1.539	8
Afforestation spaces within cities	117	90	13	10	0	0	1.1	1
Encouraging environmentally friendly investment	103	79.2	14	10.8	13	10	1.308	3

PROPOSALS THAT WOULD INCREASE THE EFFECTIVENESS OF THE ROLE OF INSTITUTIONS IN DEALING WITH EXTREME CLIMATIC PHENOMENA

The study sample showed a high response towards proposing a number of practical proposals to reduce climate change by organizing the field of study by 90%. Environment-friendly investment by 79.2%, and reliance on renewable energy sources by up to 70%, and it is clear from this that there are a number of practical proposals to limit climate changes in the eastern region because of these changes having wide-ranging and unprecedented effects Similar in size, from changing weather patterns that threaten food production, to rising sea levels that increase the risk of catastrophic floods, dust storms and droughts, and thus there is an urgent need to preserve natural resources and green spaces, develop infrastructure and flexible services in the face of Impacts of climate change and tools for disaster risk reduction, improving governance and management of action in the field of climate change, so that the Kingdom of Saudi Arabia is in the international ranking of climate change actions; To attract more investments and opportunities for climate financing, and this result is consistent with the study (Douglas, 2009), which emphasized that there should be participatory measures from the community to the international level, as many individual initiatives give hope and show good practices, and in return all hinder institutional, economic and environmental factors. To maintain and enhance food security in South Asia, innovative forms of food production, distribution and storage must be developed. This is confirmed by the theoretical conception of the ecological approach to crisis management, where the ecological framework for crisis management consists of three components. Crises appear within one component as a result of differences, conflicts, natural disasters, or by the component itself as a result of dangers, miscalculation, prediction, and forecasting. Crises may also appear as a result of the overlapping of the component. human interaction with other components of society or the overlap of more than one component at a time, which confirms the necessity of creating a homogeneous ecosystem.

REFERENCES

AntonioA. (2017): *Social Responsibility and Ethics in Organizational Management.* (IESE Business School Working Paper No. 1163-E) https://ssrn.com/abstract=2969746

Bayulken, B., Huisingh, D., & Fisher, P. M. (2021). How are nature based solutions helping in the greening of cities in the context of crises such as climate change and pandemics? A comprehensive review. *Journal of Cleaner Production, 288,* 125569. doi:10.1016/j.jclepro.2020.125569

Boyd, E., & Juhola, S. (2015). Adaptive climate change governance for urban resilience. *Urban Studies (Edinburgh, Scotland), 52*(7), 1234–1264. doi:10.1177/0042098014527483

Bulkeley, H., & Kern, K. (2006). Local government and the governing of climate change in Germany and the UK. *Urban Studies (Edinburgh, Scotland), 43*(12), 2237–2259. doi:10.1080/00420980600936491

Douglas, I. (2009). Climate change, flooding and food security in south Asia. *Food Security, 1*(2), 127–136. doi:10.100712571-009-0015-1

Fünfgeld, H. (2010). Institutional challenges to climate risk management in cities. *Current Opinion in Environmental Sustainability, 2*(3), 156–160. doi:10.1016/j.cosust.2010.07.001

Johne, D. (2009). *Risk and Manager Core Competency Model.* Professional Development Advisory Council.

Luhmann, N. (2013). *Introduction to System Theory.* Polity Press, Cambridge University.

Mahfouz Al-Fazari. (2005): *Developing crisis management in educational institutions in the Sultanate of Oman.* [unpublished master's thesis, University of Jordan, Amman].

Mayrhofer. (2004). Social System Theory as Theoretical Framework for Human Resource Management. *Management Revue, Rainer Hampp Verlag, 15*(2), 178-191.

Morsi, M. (2001). *Talcott Parsons' sociology between the theories of action and the social system.* Al-Qassim, Buraydah.

Revi, A. (2008). Climate change risk: An adaptation and mitigation agenda for Indian cities. *Environment and Urbanization, 20*(1), 207–229. doi:10.1177/0956247808089157

Samhi Al-Qahtani. (2003). *The role of public relations departments in dealing with crises and disasters.* [An unpublished master's thesis, Naif Arab University for Security Sciences, Riyadh].

Shalender, K. (2015). Organizational Flexibility for Superior Value Proposition: Implications for Service Industry. *Int J Econ Manag Sci, 4*(256), 2. doi:10.4172/2162-6359.1000256

Lutfi, T. & Al-Zayyat, K. (2009): *Contemporary Theory in Sociology, first edition, Dar Al-Gharib, Cairo.* UNESC. https://archive.unescwa.org

Ulrich, B. (2009): Living in the World Risk Society. Economy and Society Journal, 35.

Windvsky, R. S., & Douglas, P. G. (2008). Leadership Skills in Modern Organization. New York.

Chapter 10

The Relationship Between Consumer Culture and Food Security in the Context of Climate Change:
A Prospective Study in Sulaymaniya, Iraq

Nieaz Mohammed Fatah
Halabja University, Iraq

ABSTRACT

The current study is one of the future studies which attempt to evaluate what societies may go through under the influence of global and trade openness; it also evaluates the industrial development and the merger between agriculture and the food industry. The majority aspires to make greater profit through a higher level of production, unaware of the impacts of their greed on consumers' diet and food security, nor the negative impact they have on climate change. The main conclusions are the citizens do not have enough knowledge about what is offered in the markets they make their choices blindly under the urge of practicing their freedom to the fullest. At the same time, due the large number of offers presented to the people they became daily purchasers who are careless about reading the written labels on the products. Consequently, the markets became a place for the liquidation of all products which led to three major issues: the change of the culture of consumption, a high level of consumer purchases, and the misuse of nature in many aspects which has caused climate change.

INTRODUCTION

Civil and technological developments in addition to global openness are considered amongst the great achievements of humanity and the fruition of many scholars and specialists' efforts. However, we cannot deny the negative impacts they have on climate change which eventually has reflected on the life of humans around the globe. The threat should be a warning sign for us to contemplate in order to contribute

DOI: 10.4018/978-1-6684-8963-5.ch010

to securing human safety on earth. Therefore, this study emphasizes on the influential aspect of this issue, which is consumer culture and goes deeper by explaining the impact of: trade, food industry, and lack of awareness on consumer culture.

Furthermore, as the Consumer culture is one of the key terms within this study it is important to define it. One of the definitions is that the consumer culture is a form of material culture promoted by the market that eventually created a certain relationship between the consumer and the goods or even with the services he or she consumes (Milies, 2021). Also, consumer culture can be generally defined as social states, matters and activities that are emphasized on the consumption of goods and services. Meaning, consumer culture has to do with consumer's spending from an economic perspective. In addition, popular culture is viewed as a set of capitalist production processes directed by the motive of profit and selling to consumers (wang, 2020).

Food security is another key term of the study which is defined by the united nation' Committee on world food Security as a condition that all people, at all times, have social, physical, and economic access to safe and nutritious food that provides them their food preferences and dietary needs for the sake of an active and healthy life. The organization also claimed that over the years, a changing climate, growing global population and environmental stressors will have significant yet uncertain impacts on food security. But, it is also worth mentioning that food security is" important not only to hundreds of millions of hungry people, but also to the sustainable economic growth of these nations and the long-term economic prosperity of the united states" (USDA, 2011).

There is strong connection between environmental pollution and consumption. Humans by nature are consuming being that is why they have always been consumers, however, what has changed is the quality and quantity of the products consumed. Regarding the quality of the products, they are all commercially manufactured materials instead of being extracted directly from natural resources: which indicates the presence of factories and chemical uses for the sake of producing them. More importantly, the process of production releases huge amounts of greenhouse gasses which pollutes the air, also, the produced waste pollutes the water and soil. Therefore, the main question of the study is how to balance between the economic prosperity and food security while preserving quality of the products and the security of environment.

The Objective of the Study

The study achieves its aim by meeting the following objectives:

1. Knowing the consumer's level of awareness about his nutritional needs and how to choose them.
2. Securing foodstuffs in the markets, their sources, and how to control their quality.
3. How the large number of requests and attempts to meet them impacts climate change.

Study Hypothesis

1. The food industry is reflected in the reduction of food and health security.
2. Trade and advertising of food led to an increase in consumption and distorted consumer culture.
3. The whole industry and trade contributed to climate changes in various forms, including air and soil pollution.

In order to achieve these goals, the study took some scientific steps; including the analysis of the theory of Liquid-life by Zygmunt Bowman to use what is useful as a measure for the study. As for the design of the study methodology; the study is one of the future studies and it uses the Delphi method which consists of the method of a sample survey; the tools are a questionnaire form and an in-depth interview. The study analyzes the results of the collected data by comparing the most frequent answers of the survey and the data collected from the interview with the statements and theoretical studies presented previously. Then underlining the conclusions we have reached.

LITERATURE REVIEW

Previous studies recognized the negative impacts these developments have on the lives of humans, that is why each tried to explain it in their own way. For instance, (Pandey, 2021) stated that every year, a third of all food produced for human consumption is wasted, which weakens global food security and eventually to an increase in hunger around the world. Because, food waste not only reduces the availability of food, but also reduces the materials needed to produce food for future generations in a number of countries. Hence, Consumer behavior must be amended, also, the government and development partners should develop sustainable initiatives. (Vysotska and Vysotskyi, 2022) also confirms the same point as the previous study by explaining that the insufficient development of the environmental culture of society and the current consumer culture worsen the environmental situation. For this reason, increasing awareness and responsibility towards society is a valid response to this global challenge, in addition to the process of forming a green consumer. (Dwiartama et al., 2022) is a study conducted about the comparison between the poor in two different cities in Southeast Asia. The author shows how the poor can obtain the required food security and how they try to maintain their food security despite the economic shortcomings they face. This disapproves the common idea people have; which is the economic shortcoming leads to eating unhealthy foods as the people cannot afford much. (S.Colic, 2008) Offers another analysis as it analyzes the social and cultural aspects of consumption. Through his study he realized that values in the field of consumption flow into other areas of social activity, and he concluded that modern society is a consumer culture due to the key values, practices and institutions that are associated with modernity such as individualism and market relations. (Shahrin and Hussin, 2023) Shares a similar idea with the previous study but explains it from a different perspective as it states that food has entered commerce as an intangible cultural heritage due to the fact that people nowadays desire authentic food and special consumption experiences. Which eventually, led to the production of a variety of products based on the desires of people instead of high standards of quality, because the reality is traders have no knowledge about the real needs of their citizens and not to mention that they puts their personal gain as their first and foremost priority. Thus, the researcher pointed out that based on his research he came to the conclusion that in our contemporary world authenticity is given to what is a trend amongst people. Meaning, authenticity has lost its real meaning in the consumption context. This shows us a gap that needs to be filled as it raises a question of what is the real measure for the people that drives them to view a product desirable and others not. (Javeed et al., 2023) Is another study that emphasizes food security and how important it is for all humans around the world. However, the study views it from a different perspective as it argues that the temperature rise which is mainly caused by industrialization influences the food system in various ways starting from a direct impact on crops to changes in markets and food prices, in addition to the infrastructure in the supply chain. Noteworthy, the warmer and colder temperatures

Relationship Between Consumer Culture and Food Security

may lead to changes in the growing seasons, that is why the author stated that for the next 50 years and beyond, global food security will remain a global matter. He also stated that the relative importance of climate change varies from region to region. For instance, in few regions crop yield declined primarily due to poor research infrastructure and other facilities related to citing with the issue of climate change. But, in a large number of other regions rainfall shifts and temperature fluctuations are threatening agricultural development and the livelihood of the people who depend on agriculture. In addition to the agricultural threats the researcher states that food quality, access and availability may also be impacted by climate change. As for the solution the writer argues that measures from the community to the international level have to be participatory. Although he recognizes that many individual endeavors provide inspiration and beneficial methods, however, the maintenance and improvement of food security can all be hindered by institutional, environmental and economic factors. That is why it will be crucial to develop innovative strategies for food production, delivery, and storage. On the other hand (Sagi and Gokarn, 2023) depends on the multi-stakeholder theory to empirically identify the main determinants that affect one of the food losses in the mourning baskets of Indian agricultural foods and concludes the paper by linking the increasing population with the increasing interest in food and food waste. As for the solution, the author states that it is everyone's responsibility to work for the sake of achieving food security. (Sharma, 2023) disagrees with the previous study as it states that the personality type is what impacts the food security rather than the increasing numbers of population. Through his study the author examines the relationship between personality factors and customer switching for services. His main aim was how the role of personality in customer switching and building the conceptual framework on services rather than products. According to the researcher, the switching of extroverted individuals is going to be high. However, the switching is going to be less for introverted people. Similarly, consumers with high Neuroticism and agreeableness switch less, compared to the open and conscious versatility type. More importantly, this level varies with the level of a consumer's involvement and the type of service being offered based on its value. The study (S.A.Abdulmumin, 2022) calls attention to advertisements and how displaying goods to the consumer contributes to the formation of a culture around products. Meaning, social media platforms have greater purposes such as forming cultures in addition to the other services it provides as connecting people, or selling goods and products. (malik, 2014) on the other hand disagrees with the previous study as he clarifies his point through explaining the theory (the ideology of beauty) and how it contributed to affecting the consumer culture of women The researcher explains that it is true that advertisement helps with forming new cultures but not all of them are positive and he proves his point by mentioning the working women, who became victims of the theory; as they started using more beauty products, which shows us the negative impacts of advertisement. (kurylo, 2020) is another study which offers a new perspective as it revives the controversy between Adorno and Benjamin to investigate the potential of contemporary technological consumer culture to become a space for political agency and an ideal example for them. Each thinker's point of view differed at the beginning, but after the new era of technology they both agreed on a crucial point; they both confirmed that nowadays social media is used to reprogram the mind of people as it has been used by people for self-interest rather than for the benefit of the public and that is because there no filter or quality control to check the content that is being displayed on social media's platforms.

Interestingly enough, previous studies emphasized on the subject of consumption, food security, and climate change and each offered a different explanation.

For example, one focused on consumption and how it changes, and another on advertising and its impact on the culture of consumption.

On the other hand, some were concerned with the manifestations of imbalance in the distribution of products; one is large waste category and a category that has nothing to eat.

Some underlined climate change as the cause of food insecurity. But, some underlined the increasing numbers of population as the cause. However, some disagreed with both ideas as they confirmed that there are deeper causes such as the personality type. Meanwhile, other scholars have tried to find a solution through analyzing the opinions of some known scholars. Others have conducted comparative studies between countries. However, no study relied on a future line that helps with clarifying the vision for those who are responsible for making decisions, nor studying the consumer awareness to help them make conscious decisions by linking this culture to the negative impacts it has on the natural environment and climate change. The difference between the vision of this study and other studies is other studies only focused on one part of the issue and lacked the piece that connects us as humans; for the sake of solving the issue collectively. While, this study touches on the human awareness which connects people all around the world to collectively take action towards solving the issue of climate change through understanding the consumption culture. That is why the objective of this study is to fill this gap.

THE ARTICLE BACKGROUND

The study relies on the theory of liquid life by Zygmunt Bauman that explains the kind of life we all have in our contemporary world which the theorist calls the liquid-modern society. The author explains that this type of society cannot keep its shape for long periods of time due to the constant changes and uncertainties it faces. That is why the biggest worry that people have in the modern world is to fail to catch up with the fast-moving events that occur on the daily basis. More importantly, with the never-ending changes and the occurrence of series of new beginnings each day people are led toward possessing and aiming for worthless possessions. The theorist states that root cause of all of these are the tremendous technological progress, the prevalence of the logic of consumerism and lack of morals. He asks since when did the man kind turn into a commodity within the universe? While humans should operate according to their morals and ethical values, they act otherwise as they are characterized by flexibility, liquidity and disconnection from any theological and metaphysical bond. The Polish sociologist Zygmunt Bauman noted that we humans should take critical accountability for moral values in our current society as they are going to fade away. (Bauman, 2016) The reason why he is concerned with the disappearance of morals is that in his theory he highlighted the theme of "ethics" as a major importance of the contemporary human need due to existence of the world of digital rationalism and the existence of the world which is all about speediness and continuous change (Bauman, 2016). Not to mention the role of the New liberalism, the domination of individual internationalism, and the acquisition of the consumer system on humanitarian life. Moreover, the theorist attribute the legacy of the philosophy of Emmanuel Levinas which he has been using in all of his theoretical and practical speeches due to the similar ideas they share. For years he has been stating that protecting and reviving moral and ethics are the biggest challenges for post-modern era individuals. During this era the importance of a moral system appears in framing contemporary consciousness and shaping their daily lives so that they are strong enough to challenge the consumerist model that encourages selfishness and stands in the direction of morality.

It is also worth mentioning that Bowman in his work named consumerism and post-modernism talked about the transformation of society from a productive society to a consumer society that occurred in the late twentieth century. (Bauman, 2016) In it, he dive deep into the topic; according to Bowman - and in

Relationship Between Consumer Culture and Food Security

contrast to Freud's point of view he remarked that in the paradigm of evolution; protection is suspended for the sake of enjoying the highest degrees of freedom; called the freedom of abuse that provide a kind of life one enjoys to the fullest. To explain his vision he underlines some essential themes amongst them; the crisis of democracy, security and freedom, modernity and freedom.

The crisis of democracy or the shrinkage of trust: the theorist asserts that we do need politics as it gives the ability to pass on what it needs. However, a problem rises as politician turns to social media, which Bauman considers "just a trap". As it leads us to depend on social media through making us believe that we belong to the world created by it: due to the fact that it gives us the power to have some sorts of control over who we add or delete on our friend's list or the power to present ourselves in a certain way. But the reality says otherwise. Even the friendships we look highly of, is not as strong as it seems because there is no real intimacy, expression, exchange, and interaction (Bauman, 2016) . Thus, as the theorist states that technological progress led to the creation of consumers instead of active citizen who are knowledgeable enough to make choices that benefits them.

Security and freedom: the two things that have always been hard to reconcile. We are used to think that if we want security, we have to compromise some degree of freedom and if you want freedom, we have to compromise some degree of our security. Consequently, this dilemma used to exist for years. However, the current conflict revolves around the relationship of the individual to society, and it no longer concerns the issue of security, but rather the issue of freedom. Eventually, the extreme freedom opened new world for the people that in most cases they do not have enough awareness to deal with them, consequently the extreme freedom causes more insecurity more than you can imagine. In conclusion, under the influence of practicing full freedom people are putting themselves under real danger as they forgot about the safety that food security brings them.

Modernity and freedom: The mission of modernity was to unleash freedom of control and human choice, then provide extreme freedom for individuals, guarantee individuality and remove one from the iron joint of tradition. Which eventually caused people to become slaves of the unknown future and zone out from the present; as each individual aims to practice their freedom to the fullest and at the same time try to keep up with the fast moving changes of the modernity that is why they cannot make their present choice with awareness (Bauman, 2016). Thus, modernity brought a new type of culture that focuses on the trends rather than beneficial nutrition and high quality foods. This eventually, leads to the loss of food security and freedom.

In conclusion, the modern industrial society has become a threat to the values of humanity according to Zygmunt Bauman. Consequently, this led him to develop a comprehensive philosophical project in which he underlines the blind sides of the current civilization. However, what stands out for Bauman is what he calls moral blindness in the era of liquid modernity. As ethics is no longer as powerful and influential as it used to be in the past. In fact, individuals and societies that used to be morals became liquid when the consumer movement overwhelmed them with their consistent influence. Eventually, the individuals transformed from a sane moral being into a consumer animal. Thus, the theorist underlines the reviving of morals as primary goal to undermine the liquidity that drowns the current humanity. But it also worth mentioning that by reading his work we understand that the theorist sympathized with the disappearance of the church in the post-modern world, meaning that his emphasis on the morals is rooted from his religious background. That is why he aims to solve the problem through the control of morals. However, the study is aiming to show another solution which is the idea that if individuals gain a high level of awareness they can make the right choices and stop the issue from growing without the use of force or control. In another word, people live their lives to the fullest and in the healthiest way without compromising their freedom (Bauman, 2016).

Relying on the theory of Zygmund Bauman helps us gain deeper understanding of the consumer culture as it explains that all aspects of man are a consumer market according to fluid standards of life. It turns the world, with all its living and inanimate objects, into objects of consumption that lose their worth when used by the liquid consumer system; that imposes authority on culture to become a subject to its marketing standards. When in reality the cultural principles should take the leading role; production must be on cognitive foundations that transcend people's current desires. Waste is the main product, and indeed the most widespread product in modern liquid consumer society. Meaning, waste production is the 'largest industry' in consumer society and the most immune to crises, and getting rid of the waste is one of the most crucial challenges that liquid life must face and deal with. In a world full of consumers and consumption issues, life fluctuates sharply between the two spectrums.

Furthermore, use immediate consumption and assimilation into the consumption process are neither the goal of cultural products nor the criterion of their value. Culture seeks to achieve behavioral, moral, and values for humans, while the market seeks to fulfill desires and satisfy appetites. Therefore, forcing people to submit to market standards is what led to a decline in the level of cultural productions.

Bauman says, "Subjecting cultural creativity to the standards of the consumer market means forcing it to accept requirements for consumer outcomes. It will either gain its legitimacy from market value, or it will perish."

In his book, Bauman presents an analysis of the modern consumer situation, considering consumerism to be based on three pillars, so to speak:

Firstly; Consumer life, where he believes that man since ancient times has been a consumer and is by nature so, but what has differed in the current era is (the consumer syndrome), which goes beyond mere fascination with oneself, but rather creates the desire for possession that follows and is accompanied by getting rid of the old. The consumer plays a role in an endless spiral.

Secondly: the consumer body, as this machine works to rid man of the freedom to control his body and its pleasures and force him to chase after satisfying these pleasures. Then, labeling it: freedom.

Third: (consumer childhood) where modern societies seek to train children from a young age to be obedient consumers as everyone keep telling them that they need this or that product, so, that they shape their personalities into suitable ones.

Hence, the consumer culture encourages the high quantity of the products without regarding the quality; which protects the interests of the marketing interest the security of humans and the environment.

STUDY METHODOLOGY

The current study is considered one of the future studies which apply the Delphi method; depending on both the quantitative and qualitative approach to achieve the objectives of the study. The data of the sample for the two interview tools and the questionnaire sample:

As for the sample data of the questionnaire, which consisted of (100) samples, was distributed through Google Form and shared amongst a variety of people. However, the sample contains an almost balanced percentage of women and men participants which is 52% and 48% per cent. Also, the sample consists of different age stages; it starts from twenty and stops at sixty years old. We worked to ensure that the sample has different levels of educational and economical levels, as well as different types of jobs or even jobless and some with families but some single. So, that the sample is diverse and includes all groups within the community, especially the category which is called educated or uneducated. These five tables show the sample and their percentage.

Sample Data Information

In the interview, we selected two groups; the first group consisted of five people affiliated with five specialized centers for the food trade process and licensing its traders. The centers were the (Directorate of Licenses for Companies affiliated to the Ministry of Trade, Directorate of Inspection and Control, Directorate of Trade for Wheat, Directorate of Trade for Foodstuffs, and Quality Control Consultant in the Council of Ministers). However, the second group consisted of five market owners located in the same residential area which is the Raparin residential area.

On the other hand, the study relied on the use of the in-depth interview tool with specialized authorities in order to access the data that guide us to the path of securing nutrition and consumer culture.

Table 1. Gender

N	gender	duplicates	percentage
1	Female	52	52%
2	meal	48	48%
3	total	100	100%

Table 2. Age range

N	age	Duplicates	Percentage
1	20-30 year	33	33%
2	31-40 year	27	27%
3	41-50 year	26	26%
4	51-60 year	14	14%
5	total	100	100%

Table 3. Education level

N	certificate	Duplicates	Percentage
1	Junior high	8	8%
2	institute	12	12%
3	Bachelor	62	62%
4	masters	12	12%
5	PhD	6	6%
6	total	100	100%

Table 4. Occupation

N	type	Duplicates	Percentage
1	Public sector	56	56%
2	Private sector	16	16%
3	Free business	9	9%
4	Jobless	19	19%
5	total	100	100%

Table 5. Marital status

N	M.S	Duplicates	Percentage
1	Married	61	61%
2	single	39	39%
3	Total	100	100

Table 6. Interview sample data

n	workplace	position	Symbol	Years of service
1	approving the license for the companies	boss	A	3
2	a specialist in the field of quality control council of ministry	boss	B	5
3	the Directorate of Inspection and Control of Foodstuffs inside	boss	C	4
4	directorates responsible for distributing flour	boss	D	3
5	directorates responsible for distributing foodstuffs	advisor	E	3
6	Super market sork	coordinator	1	3
7	Super market hamajand	coordinator	2	4
8	Super market govand	coordinator	3	5
9	Super market lyver	coordinator	4	5
10	Super market kobany	coordinator	5	7

Discussion Chapter

Survey form that aims at understanding the relationship between consumer culture and food security. The survey data of the questionnaire is consisted of (100) samples which we distributed through Google Form and share it randomly with a variety of people. Noteworthy, in order to gain a deeper knowledge about the relationship between the two terms, we decided to balance between the questions we ask about each term; which are consists of seven questions about the consumer culture and nine questions about the food security. Through this survey we aim at underlining the level of awareness amongst the participants in order to understand how they deal with each term in their private lives. Then explain it within the frame of climate change.

Relationship Between Consumer Culture and Food Security

Table 7. Participants' purchasing system

N	time	Duplicates	Percentage
1	Daily	15	15%
2	Weekly	13	13%
3	Monthly	12	12%
4	If necessary	54	54%
5	All of them	6	6%
6	total	100	100%

Table 8. Preferred shopping destination

N	please	Duplicates	Percentage
1	large malls	17	17%
2	supermarkets	57	57%
3	Local markets	24	24%
4	Touring cars	2	2%
5	total	100	100%

The participants have different levels of educational level, as well as different jobs and different economic status including some jobless people. Also, some of them were single but some were involved in marriages. This, the sample is diverse and includes all groups within the community, as mentioned previously.

Within this survey, we asked a set of questions to clarify the consumer culture amongst them is the question of: What is your purchasing system: daily, weekly, monthly, and as we notice in Table (7) that the answers to the question about the purchasing pattern (54%) per cent of the participants chose that they buy foodstuffs when they are in need of and the rest of the daily, weekly and monthly choices did not exceed 15% per cent. These percentages are very different which shows how the consumer decides to buy.

We asked the participants: Where do you prefer to shop: large malls, supermarkets, local markets, or from touring cars. As it is displayed in the table among all the choices, (57%) per cent of the participants confirmed that they depend on the markets near their homes. However, (24%) per cent of them chose the local markets and only (17%) per cent chose the big malls. This coincides with the previous question about purchasing pattern which the highest percentage of choice was buying food when in need of. It was expected that they will choose this answer.

It was important for us to know what people care to know when buying food. So, we inquired them what are the criteria for deciding to buy food: domestic products, European products, whatever you like, to know how it is made, or they care whether it is a local product or they prefer the imported one, or they only want to know how the product was produced, or none of what mentioned before but they only care about what they crave. The tables shows that the highest percentage was to know how it was manufactured or producing which is (51%) per cent. The other (31%) percentage was the local fruits choice and the importer food choice has not been chosen by the participants. However, (18%) per cent

Relationship Between Consumer Culture and Food Security

Table 9. Chosen purchase criteria for food

N	measures	Duplicates	Percentage
1	domestic products	31	31%
2	European products	0	0%
3	whatever you like	51	51%
4	how it is made	18	18%
5	total	100	100%

Table 10. Cheap products or high quality products

N	selection	Duplicates	Percentage
1	price	42	42%
2	Quality	58	58%
3	total	100	100%

Table 11. The quality of market product

N	the market products are all of good quality	Duplicates	Percentage
1	Yes	0	0
2	No	100	100%
3	total	100	100%

of the participants chose that they buy the food as long as they crave they don't put too much thought in to other things.

We inquired to know what is the main factor they take into consideration when purchasing, is it the price they care about, or the quality of the products that is more important. So we asked: Have you ever resorted to cheaper products because of the high price? Between these two options; 58% per cent of the participants chose quality over price.

We aimed to know to what extent the participants believe that what is offered in the market is usually of good quality. So we asked them: do you think that what is put on the market is usually of good quality? interestingly enough, (100%) per cent of the participants chose (no) as the answer to the question.

We inquired another question: Do you think that the abundance of food products have changed the food culture? Yes or no. From the participant's point of view, the large number of products displayed in the markets led to a change in the consumer's purchasing culture as (94%) per cent of them chose yes as the answer, and only (6%) per cent of them believe otherwise.

We asked when buying any canned food products, do you read the labels: Yes or no. Interestingly enough, (75%) per cent of the participants answered yes they read what is written before making purchases. Only (25%) per cent chose otherwise.

We wanted to know the effects of commercial advertisements on the purchasing pattern, so, we asked: Did you ever buy a food under the advertising influence? (50.5%) per cent of the participants confirmed

Relationship Between Consumer Culture and Food Security

Table 12. The abundance of food products and its impact on the food culture

N	The abundance of food products and its impacts on the food culture	Duplicates	Percentage
1	Yes	94	94%
2	No	6	6%
3	total	100	100%

Table 13. Reading food nutrition labels

N	do you read the label	Duplicates	Percentage
1	yes	75	75%
2	No	25	25%
3	Total	100	100%

Table 14. Buying food under the influence of advertising

N	Buying food under the influence of advertising	Duplicates	Percentage
1	Yes	50.5	50.5%
2	No	49.5	49.5%
3	total	100	100%

Table 15. When purchasing fruits and vegetables

N	When buying fruits and vegetables	Duplicates	Percentage
1	Do you ask what region it belongs to	48	48%
2	just ask how it was grown	6	6%
3	you do not ask any questions	46	46%
4	Total	100	100%

that they were affected by commercial advertisements, and (49.5%) per cent demonstrated that they did not buy any type of product under their influences.

The table shows the choices related to a question about the behavior of the consumer while buying fruits and vegetables. We wanted to know whether they ask about the region from which they were produced, or they are satisfied with knowing how they are made or produced, or they do not ask any questions because they make their choice depending on their vision. We found that (48%) per cent of them ask about the place from which they were produced and (46%) per cent of them rely on their sense of vision, only (6%) per cent f them were interested in the production process.

We aimed to know the level of the people's knowledge about the symbols and signs written on the manufactured and imported food. So, we asked the participants whether they have any knowledge about

Table 16. Food symbols and their meanings

N	Do you have any knowledge	Duplicates	Percentage
1	Yes	61	61%
2	No	39	39%
3	total	100	100%

Table 17. Basic knowledge of food processing

N	What do you know about food industry	Duplicates	Percentage
1	I have deep knowledge	6	6%
2	Little knowledge	35	35%
3	I have enough information to help me make a good choice	40	40%
4	I have no knowledge	19	19%
5	total	100	100%

food symbols and their meanings: answer with yes or no. We found that (61%) per cent of them confirmed that they had knowledge about these symbols, only (39%) per cent answered otherwise.

To underline the level of people's knowledge about the food industry, we asked the participants the question of What do you know about the food industry and the steps of working in it? Interestingly enough, (40%) per cent of them confirmed that their knowledge was at a level that helps them choose the nutrition that is beneficial for them, and a percentage of (35%) chose that they have little knowledge and (19%) per cent of them chose they have no knowledge about the topic, only (6%) per cent confirmed that they have deep knowledge about the topic.

How do you view the abundance of food products in the markets regardless of the seasons? Was another question that we asked the participants to understand how they view the matter of buying whatever we desire and need despite the seasonal changes; do they view it as good, bad because it affects our lack of passion towards seasonal changes, normal, or a dangerous matter as it impacts the level of quality as they have not been naturally produced. The majority was (39%) per cent who views it as bad because we lost our desire for seasonal changes, and (36%) per cent thinks it is dangerous due to its impact on the quality of the products, only (25%) per cent off them consider it as good.

To understand the view of people on the relation between demands and productions we asked do you think that high level of demands led to high production or vise-versa. The answers were as follows:

Table 18. The abundance of food that is not in season

N	the abundance of food products	Duplicates	Percentage
1	great not worry about what is available	25	25%
2	Bad, because we lost our desire for seasonal foods	39	39%
3	Dangerous, because it impacted the level of quality	36	36%
4	Total	100	100%

Relationship Between Consumer Culture and Food Security

Table 19. The relation between demand and production

N	Your view on the relation between demands and productions	Duplicates	Percentage
1	High demand has led to high production	49	49%
2	High production has led to high demand	51	51%
3	Total	100	100%

Table 20. High production and its effect on the quality of products

N	high production and its impact on the quality of products	Duplicates	Percentage
1	Yes	96	96%
2	No	4	4%
3	total	100	100%

Table 21. High production and its impact on climate change

N	High production's impact on climate change	Duplicates	Percentage
1	Yes	84	84%
2	No	16	16%
3	Total	100	100%

Table 22. Sources of information about food

N	What do you use as source	Duplicates	Percentage
1	Participated in a course	3	3%
2	Media channels	11	11%
3	Internet	75	75%
4	Consulate the specialists	11	11%
5	total	100	100%

(51%) per cent confirmed that excessive consumption led to increased production and (49%) per cent assumes otherwise.

We asked the people: Does high production has an impact on the quality of products: Yes, no. The answers were as following: (96%) per cent of the participants confirmed that it has an impact on the quality of products and only (4%) per cent chose no as an answer.

The table shows the results of the question of: In your opinion, do you think that the excessive production impacted climate changes or not? Interestingly enough, (84%) per cent of the participants affirmed that it has impacted climate change and only (16%) per cent chose no as the answer.

We asked the people: What do you use as a source to gain information about food? To understand what is the source that they rely on to raise their awareness. We found that (75%) per cent uses the Internet as a reliable source, (11%) per cent rely on satellite channels, and (11%) per cent consult specialists, only (3%) per cent have entered special courses.

Interview Interpretation

In the interview, we selected two groups. The first group consisted of five people affiliated with five specialized administration for the food trade process and licensing the traders (Directorate of Licenses for Companies affiliated to the Ministry of Trade, Directorate of Inspection and Control, Directorate of Trade for Wheat, Directorate of Trade for Foodstuffs, and Quality Control Consultant in the Council of Ministers). The second group is the owners of the five supermarkets (sork, hamajand, govand, lyver, kobany) affiliated to the residential area of the Raparin.

The first group:

The first interview was with (A), who is responsible for approving the license for the companies. We asked him about the companies that import and export food; what are the red lines limit their freedom, how they are allowed to take certain actions, and what are the conditions for granting them license. The answers were as follows: first and foremost, the only condition required of them is to implement all the instructions issued by the quality control and the Ministry of Health. In addition to that information, the official confirmed that they are following up the progress of companies' work through the inspection and oversight committees periodically. Other than that the committee does not take any kind of action to limit their freedom and they do not have crucial laws as license ban to implement in the time violation of the rules.

We conducted another interview, this time with (B), who was a specialist in the field of quality control. We asked him what conditions you set for the import-exported food process and how companies comply. He informed us that as a member of the quality control committee so far they have not been able to reach the bar they set for themselves due to the trade openness caused by globalization; which has opened a gateway for importing and introducing a variety of nutrition across the globe and the committee has not been able to put nutrition label on the food cans in the Kurdish language so that the consumer will be aware of the components of the foods they buy. As for the level of validity of use and its duration he stated that they have reached a good level of governing. However, they have not been able to conduct a special examination of pesticides inside agricultural fruits whether local or imported due to the lack of required equipment and specialized cadres in this field. To this day, those working in the field of quality control are not specialists and do not have sufficient awareness of the existing risks.

On the other hand, we interviewed (C) the official of the Directorate of Inspection and Control of Foodstuffs inside the markets. We asked him a set of questions about their work patterns, such as how to select committees and workers, the form of inspection, and the stages of procedures they take in the event of unusable food. His answers were as follows: They have committees Competent and work with the committees of other parties such as the Department of Health and the Governorate to work periodically in keeping up with the shops and markets in order to know what they offer to the consumers. So, when there is any violation and any invalid product we close the place and find the owner, then announce it so that it becomes a lesson for others. In the answer to another question he stated that they do not have mediators because they do believe in the equality of everyone. He even mentioned that he worked there for many years but no one has ever visited him with the intention of bribing him for sake of forgiving a violator. In addition to that he stated that for him as the director, controlling the markets requires continuous work from them and they need more committees and more accurate follow-up, however, with the help of the citizens their work have been easier as the people consistently contact them for state an

Relationship Between Consumer Culture and Food Security

official complaint about a violators. In another word the citizens contributed to covering the shortcomings as they became a third eye for them everywhere.

We also interviewed both the (D) directorates responsible for distributing flour and (E) directorates responsible for distributing foodstuffs to citizens with the so-called insurance materials. They confirmed that the materials are under their control and they check them continuously from the warehouses until they reach the hands of the distributor. Not only that but also they try hard to ensure that the level of distributed materials is of high-level quality. More importantly, their job also includes controlling prices in the markets. However, concerning the decision of imported materials or relying on local production, they confirmed that it is not included in the list of their responsibilities rather the Special Ministry in Baghdad takes it as their responsibility.

From all the interviews, we concluded that control and discipline are the only modes in their work and this is what contradicts the open and democratic state in which we live in. In the large geographical areas which consists of large population control is a difficult process and this what all of them confirmed. Since, their level of capabilities is also not at the required level, why not think about making a balance in their work using other means, such as, raising the level of awareness and culture of their consumers.

The second group:

Due to the answer of the participants in the questionnaire form that confirmed their dependence on the markets near their homes we decided to interview five markets within a specific area in the center of the city of Sulaymaniyah and we asked a set of questions and presented them as follows: We asked them what is the basic criterion for determining what they offer in their shop, they all stated that the criterion is consumers' desire. The next question was about whether they would offer food that was not approved by quality control or not, their answer was that they stick to what is approved by them. Amongst all the information they shared with us, they pointed out something noteworthy and it was the fact that they noticed the most important thing that concerns the majority of consumer is the price only few of them care about the quality; their evidence was despite the fact that they had excellent quality foodstuffs, yet they have not been sell them due their high prices. We also asked whether the customers read the labels written on the products, all the them demonstrated that they do not have consumers who read the labels, but rather they rely on their words and the trust they have in them, and all of them confirmed that they aim not to break the consumer's trust at any cost. We asked some other question concerning their clients and their purchasing patterns; some of them confirmed that they have customers outside their geographical area and the rest confirmed that their customers live in same neighborhood. Concerning their purchasing patterns they stated that most of them make the purchase on a daily basis. More importantly, they shared another useful information as they all pointed out that most of their products are imported due to the high cost of the local products and their consumer's dislike. Equally important, the import companies bring them the products they desire and there are committees that visit them from time to time for checking their products.

CONCLUSION

General Results and Achievement of Research Objectives

A. After examining all the data collected from the sample of consumers, we find that there is a big gap between what they should do and what they actually do. As we saw in the results that 56% per cent of the participants purchase food when they are in need of and this contradicts with the abundance of production we witness on the daily basis. Equally important, 58% per cent of them chose quality over price, while 42% per cent chose otherwise. It is worth mentioning that 57% per cent of the participants answered they rely on the market near their homes and we find that all of these answers contradict with each other because at the time of need and in a market in your neighborhood, can we find good quality products?

Another interesting point is while 94% per cent of the participants chose that the abundance of products affects the quality but in a following question 50.5% per cent of them admitted that they are affected by advertisements when they buy products. More importantly, concerning food safety, we found that although the majority confirmed that they read the written labels on the products, but in another question, 55% per cent answered they have little knowledge about the symbols printed on the food products and only 6% percent confirmed that they have deep knowledge. On the other hand, the quality control consultant confirmed that all the imported products are not attached with the required data in the Kurdish language so that the consumers know what they are buying. Interestingly enough, the data from all the market owners we met in the market near the houses confirmed that no one reads written labels but rather they rely on the price and on the trust they have in the market owners. Everything that is presented previously shows the consumer's lack of awareness as they buy what is offered in the markets, they purchase the foods that are trending, more importantly, they buy what they can afford not the ones of high quality. Interestingly enough, this conclusion agrees with the presented study that was conducted in 2008 by (S.colic) as the study explained how the field of consumption flows into other areas of social action, because despite the existence of different cultural, practical and economical levels, we find that the nutritional culture is the same due to the reason that people act according to the trends rather than awareness. However, it is also significant to mention that the previous study disagrees with the presented study that was conducted in 2003 by (Sharma) which explained that each individual act according to their type of personality and how each detail derives them to act in a certain way. For instance, the introverted type of individuals switches less, meaning less waste. This contradicts with the other study that generalizes the whole society under the name of modernity. Thus, we have achieved the first objective of the study.

B. If the consumers do not have sufficient awareness, then what did the concerned authorities do to secure their lives?

It is worth mentioning that in an answer to the question of to what extent the participants trusts the quality of the products that are offered in the markets, 100% per cent confirmed their lack of trust and chose that not everything that is offered is of good quality. On the other hand, from the data we collected from the concerned authorities, we find that the majority worked on enhancing laws, defining guidelines and instructions, but under the influence of trade openness, freedom of importing and exporting and the influence of advertisements, controlling the markets may be a difficult process. Equally important, the main goal of the responsible authorities is examination and follow-up in the time of doubt or when receives report from the citizen. However, we did not find any link between the authorities and the citizens to raise awareness, also, we have not found publish instructions that provide the consumer with necessary information. This point was also confirmed by 75% per cent of the participants as they admitted that they depend on the Internet as their source of information in order to help them to gain information. Interestingly, this result agrees with the result of the presented study, which was conducted in the year 2022 by Vysotska and Vysotskyi that emphasizes the importance of consumer awareness. By explaining the insufficient development of the environmental culture of society and how the current consumer culture worsens the environmental situation. For this reason, the authors underlined the increasing awareness and responsibility towards society is a valid response to this global challenge.

C. What is the connection between of all this with climatic changes? 84% per cent of the participants confirmed that the abundance of production and consumption has an impact on climatic change. Also, 96% per cent of them agreed that the abundance of production affected the quality of nutrition. Equally important, 49% per cent of the participants confirmed that the abundance of products and offers led to an increase in consumption. Moreover, the market owners whom we interviewed stated that there are products that are expired due to lack of sale. In addition, those we interviewed from the competent authorities did not mention any step they take to reduce the large number of products that is produced every day and even how to destroy them when they are harmful for the sake of protecting the environment. This appears to agree with the study which was conducted in the year 2022 about the waste of food products that was done by Vysotska and Vysotskyi (2022) with a group of researchers about the impact of food waste and its reflection on the existence of great human poverty and hunger and its reflection on climate changes directly. However, this study contradicts with the study conducted by Javeed et al. (2023) which underlines the climate change as the threat to food security, the author states that In addition to the agricultural threats food quality, access and availability may also be impacted by climate change. Hence, we achieved the third goal of the study.

Theoretical Implications

A. The loss of citizens between what is of low cost, what is available and what is popular has led to a Confucian amongst the people, consequently, the instability of the principles of choice. This is also confirmed by the theory of liquid life and that the citizen has become a mere consumer on the globe and this is what the opinions of the participants, as well as the owners of the markets, and the large number of markets and products proves to us.

B. On the other hand, the theory of liquid life affirmed that the moral blindness of people in the contemporary world has led them to lose control. So, in order to regain control, people need to stick to morals and ethics. However, this study came to the conclusion that with the presence of laws, administrative institutions, laboratories specialized in examinations, and instructions we may be somewhat control the situation. As the study argues that in the contemporary world consumers cannot protect themselves from everything only through sticking to morals but rather they need the modern solution that was mentioned above.

C. The theory's main emphasis is on what humans suffers in the unstable world we live in. It is so focused on the sufferings of humans, that it forgot what nature and climate suffer under the pressures of human desires and ambitions in the name of securing their needs: on one hand his nutritional needs and on the other hand material and economic needs. This statement is confirmed by data that has been presented previously; as the consumers admitted they need a good commodity with a lower price, on the other hand the producers and the seller want more sales to gain a higher income, and the authority aims at balancing between them. But no one pays attention to the environment and the negative impacts they have on it.

The Practical Implications

1. The citizens must be fully aware of their body and their nutritional requirement, which helps them determine their nutritional purchases, and thus blocks the way for those who try to exploit them or label them as consumers.
2. If the income of the citizens is limited, then he must have a prior planning of what they need and what is their cost. To help them secure for themselves the required quantity and a high level of quality, without resorting to undesirable choices just because they cannot afford the price.
3. The citizens in the globalized world are considered just a number at all levels and will continue to be if they continued to not work to understand their requirements and are not aware of all the laws that help them to protect their health and security.
4. The contradictions that we find within our Eastern societies stem from our lack of understanding of our identity in the globalized society. More importantly, the only way to dissolve the contradictions is through raising awareness and the enhancement of the necessary laws from competent authorities.
5. The diets we follow and everything we eat daily has an impact on many aspects of our lives; such as our physical, mental, and even our mood health, so paying attention to it and working to understand its requirements must be one of our main priorities.
6. All of this has adverse effect on the environment and its natural cycle of wasting energy, its sources and the increase of the waste percentage; which they all threaten the security of humans. However, the consumer awareness may help in the reduction of wastes at all stages and the increase of quality of the products; which leads to environmental balance. In other words, if the humans were aware of the actions they take and the impacts their actions have on the environment they will be more likely to act more responsible and reduce their negative impacts in the environment. For instance, if a person knew the products and the companies they support, have negative impacts on the environment they stop supporting them and giving them feedbacks for producing products that serve both humans and the environment.

REFERENCES

Abdulmumin, S. A. (2022). Hausa in Abrertising and consumer clutture. *Zaria university*.

Bauman, Z. (2016). *Liquid Life*. Egyot: hajaj abu jbr.

Colic, S. (2008). *Sociocltural aspects of cohsumption consumer cultuer and society*.

Cuzick, J. o. (2008). Overview of Human Papillomavirus-Based and Other Novel Options for Cervical. *Vaccine*. www.elsevier.com/locate/vaccine

Dwiartama, A., Kelly, M., & Dixon, J. (2023). Linking food security, food sovereignty and foodways in urban Southeast Asia: Cases from Indonesia and Thailand. *Food Security*, *15*(2), 505–517. doi:10.100712571-022-01340-6

Galvin, C. J. (2020). COVID-19 preventive measures showing an unintended decline in taiwan. *International Journal of Infectious Diseases*. www.elsevier.com/locate/ijid

Wang, y. (2020). *Multidisciplinary perspectives on media fandom.*GlobalIGI.

Javeed, H. M. R., Ali, M., Qamar, R., Sarwar, M. A., Jabeen, R., Ihsan, M. Z., & Sabagh, A. E. (2023). Food Security Issues in Changing Climate. In *Climate Change Impacts on Agriculture: Concepts, Issues and Policies for Developing Countries* (pp. 89–104). Springer International Publishing. doi:10.1007/978-3-031-26692-8_6

Kurylo, B. (2020). Technologised Consumer Culture; the Adorno-Benjamin Debate and the Reverse side of Politicisation. *Journal Of Consumer Culture*.

Li, H. (2021). Follow or not follow?: The relationship between psychological entitlement. *Personality and Individual Differences*. www.elsevier.com/locate/paid

Malik, S. (2014). Women's Objectification by Consumer Culture. *International Journal of Gender&Womenes*.

Milies, S. (2021). *Consumer Culture*. Oxford.

Pandey, A. (2021). Food wastage: Causes, Impacts and Solutions. *Science Henritage Journal*.

Sagi, V., & Gokarn, S. (2023). Determinants of reduction of food loss and waste in Indian agri-food supply chains for ensuring food security: A multi-stakeholder perspective. *Waste Management & Research*, *41*(3), 575–584. doi:10.1177/0734242X221126421 PMID:36218223

Schmid, H. o. (2021). Preventive measures for accompanying caregivers of children in. *Clinical Microbiology and Infection*: www.clinicalmicrobiologyandinfection.com

Shahrin, N., & Hussin, H. (2023). Negotiating food heritage authenticity in consumer culture. *Tourism and Hospitality Management*, *29*(2), 183–193. doi:10.20867/thm.29.2.3

Sharma, A. (2023). *Moderating role of Involvement and Value of Service*. Indian Instiute of Technology Kanpur.

USDA. (2011). Food Security. *U.S.department of Agriculture*, 38.

Vysotska, O. Y., & Vysotskyi, O. Y. (2022). Green consumer culture as a factor of sustainable development of society. Journal of Geology. *Geography and Geoecology*, *31*(1), 171–185. doi:10.15421/112217

Chapter 11

The Social Responsibility of Third Sector Institutions in Planning to Face Climate Change in the Kingdom of Saudi Arabia

Yasmin Alaa Ali
Ain Shams University, Egypt

Asmaa Hassan Omran Hassan
Helwan University, Egypt

Yasmin Alaa Ali
Ain Shams University, Egypt

ABSTRACT

This current chapter stems from a main question: "What is the reality of the social responsibility of third-sector institutions in planning to face climate change in the Kingdom of Saudi Arabia?" The study will present three basic concepts: social responsibility, third-sector institutions, and climate change. It has a theoretical approach consisting of four theoretical approaches: the theory of social system, the theory of social action, the theory of risk society, and finally the theory of social capital. The current study is one of the descriptive analytical studies, and an electronic questionnaire is relied upon as a tool for collecting data. The study reached a number of results, the most important of which are: that there is a social responsibility for third sector institutions in planning to face climate changes in the eastern region.

DOI: 10.4018/978-1-6684-8963-5.ch011

Copyright © 2024, IGI Global. Copying or distributing in print or electronic forms without written permission of IGI Global is prohibited.

INTRODUCTION

The phenomenon of climate change is a global phenomenon, but it has local effects, that is, it varies from place to place on the surface of the globe due to the nature of environmental systems in light of the dangerous changes that may result from them that threaten the future of man on earth.

Where climate changes constitute one of the most important threats to sustainable development, and many economies of the world still depend mainly on sectors dependent on climatic conditions, such as agriculture, fishing, forest exploitation and other natural resources, and tourism (Fawaz, Suleiman, 2015:1).

Human-induced climate change causes serious and widespread disturbances, affecting the lives of billions of people around the world, and scientists emphasized in the latest report of the Intergovernmental Panel on Climate Change (IPCC) in February 2022 AD that people's health, lives and livelihoods are increasingly affected by the dangers caused by waves Heat, storms, droughts, and floods as well as slow-onset changes, including sea level rise (UNEP report, 2022).

Development literature confirms that human resources are the most valuable wealth owned by the state, and therefore their protection and investment requires a measure of social responsibility (Perrson and Thomas, 2008).

A sense of social responsibility is extremely important for all societies, and without it chaos spreads, there is no cooperation, and individual interests prevail over the public interest. Therefore, it is something refined by awareness and a sense of duty, and leads to adherence to social and human standards that lead to the unity and development of society (Al-Harthy, 2001: 3).

Social responsibility includes all of the institution's economic, legal, moral and charitable obligations towards society (Shugair, Khalil, 1436 AH, 8).

The third sector in the Kingdom of Saudi Arabia is one of the main arms, along with the government and private sectors, to meet the needs of society, contribute to solving its problems, and confront changes affecting human living conditions. Therefore, the current study attempts to determine the reality of the social responsibility of the third sector in planning to confront climate changes. The problematic of the study is the main question: "**What is the reality of the social responsibility of third sector institutions in planning to face climate change in the Kingdom of Saudi Arabia?**".

Second: Study objectives

The current study is based on a main objective (**determining the reality of social responsibility for third sector institutions in planning to face climate change in the Kingdom of Saudi Arabia**), and several sub-objectives branch out from this objective:

i. Determining the role of charities in planning to confront climate change in the eastern region.
ii. Monitoring the difficulties facing charities in planning to deal with climate change
iii. Develop a set of proposals necessary for the effective participation of charities in planning to face climate change.

The importance of this chapter was embodied in theory in support of previous research and studies. The importance of this study is evident in that it sheds light on the reflection of the social responsibility of third sector institutions in planning to confront climate changes in the eastern region through:

(a) Theoretical significance

-This study helps in developing scientific knowledge on the subject of the reality of social responsibility of third-sector institutions, as well as climate changes.

- Enriching the local and Arab libraries on this subject, as it is the first study - within the limits of the knowledge of the two researchers - that linked the social responsibility of some non-profit institutions and climate change.
- Draw attention to the experience of the role of the third sector and its responsibility in planning to confront climate changes in the eastern region of the Kingdom of Saudi Arabia through programs, projects and initiatives that achieve human security and stability.
  (b)   Applied Significance
- This chapter, with its findings and recommendations derived from a field study, enables the identification of the relationship between the level of social responsibility of third-sector institutions and climate changes, from which it is possible to read about the features of development and organizational flexibility within those institutions in the face of climatic phenomena. This study is based on giving a broad view of the most important Patterns of social responsibility of NGOs in planning to face climate change in field reality.
  -This chapter enables the relevant institutions and organizations to activate their role in giving clear and field indicators that help those working in them to identify some mechanisms to confront climate changes.
- This chapter enables access to methods and strategies planned by third-sector institutions to deal with extreme climatic events, which will be reflected on the performance of those institutions.

BACKGROUND

The Study's Concepts

The concepts of the study are: (Social Responsibility - Third Sector Institutions - and Climate Change).

i.   Social Responsibility (Concept, Theoretical Implications)

Today, the concept of social responsibility occupies a leading position in economic, legal, social and political analyses. Many definitions put forward by researchers and authors in this field appeared, and they express different affiliations and ideologies that resulted in a strong scientific heritage in this concept, so that each concept includes a criterion (or new criteria) associated with it, and it has moved towards a general consensus at the turn of the millennium. Third. Business organizations have also begun to show their social responsibilities more seriously in managing their strategies and social reports to stakeholders. In addition to highlighting it through different names, as they all refer to social responsibility, including corporate accountability Corporate Ethics, Corporate Citizenship, and Corporate Obligations (Najat, Hayat, 2012: 3).

Khamidullina, F. I., & Ilnurovich, G. R. (2022) define social responsibility as an ideological construct that originated in the evolution of industrial society and evolved more from economic to legal to social content, a social commitment to promote the common good in every possible way.

And (Gaski, J. F. (2022) believes that responsibility needs to dismantle its vocabulary in order to reach a scientific concept that is entrenched in the minds and collective conscience.

Bikeeva, M.V. (2017) that it is institutional altruism, i.e. recognition of the duty of institutions to make a significant voluntary contribution to improve the quality of life of individuals. Since institutions are not socially isolated but are essentially open structures, because they produce goods and services for all, they must, therefore, be active participants in all social projects that contribute to solving pressing social problems.

Maslova, T. (2019)) showed that it is the duty of businessmen to make decisions and pursue desirable activities from the point of view of general goals and values in society.

Nurtdinova (2015) realized that the concept of social responsibility must "take into account the interests of the entire society (its sustainable development) in addition to businessmen, as social responsibility helps maintain social peace and prevent social unrest.

Dickson, Marsha A. & Eckman, Molly 2006, p 181) has explained that it is "a comprehensive and integrated set of policies, practices, and programs that are integrated into business operations and decision-making processes".

Lim, Alwyn, & Tsutsui, Kiyoteru (2012)) believe that it is "the individual's awareness of commitment to a cognitive basis of the need to direct his voluntary actions towards the group."

Common elements between the previous definitions:

- Increasing social support among the segments of society and creating a sense of belonging by individuals and different groups.
- Achieving social stability as a result of providing a level of social justice.
- Improving the quality of life in the community.
- Taking rational decisions to ensure the performance of the roles and general objectives of the institution with integrity and transparency.
- Enhancing community participation and forming positive attitudes towards it.
- Facing the difficulties that prevent achieving the desired social responsibility goals.

Procedurally, social responsibility means:

The responsibility of third-sector institutions to play multiple roles and confront the problems that prevent them from achieving them and effective planning for them, through the design and implementation of a number of community initiatives in the face of climate change, and the role of institutions in assessing the impact, in managing the risks resulting from climate changes, and in taking adaptation measures with climate change.

ii- Climate Change

The term refers to climate changes that are attributed directly or indirectly to human activity that leads to the observed change in the composition of the global atmosphere, in addition to the natural variability of climate, over similar periods of time (Khaled, 2021:15).

Climate change also occurs in response to the interaction of three basic variables: human actions or activities, the atmosphere, and natural processes (Issa, 2019: 228).

It also means any significant and long-term change in the average weather condition that occurs in a specific region, which includes average temperatures, precipitation rate, and winds, as the frequency and magnitude of comprehensive climate changes in the long term lead to enormous impacts on natural vital systems (Hawraa, 2019: 3).

There are many reasons that led to the development of the phenomenon of climatic changes and the emergence of what is known as global warming, and these reasons are generally divided into human and natural, and human activity is the main reason behind this sudden change due to the emission of greenhouse gases in the atmosphere and one of the features of climate changes at the present time Drought and severe desertification ravaging some regions of the world, and heavy rains causing floods and devastating torrents in other places (Abdullah, 2016:136).

Procedural definition of climatic changes: It is a change in the elements of the climate for a long period of time, and is associated with the occurrence of major changes in the natural environment in many of its vegetative and vital components such as: changes in temperature, dangers of torrential rains, droughts, and dust storms.

iii- Third Sector:

The concept of the third sector appeared at the hands of the American meeting "Etzione" as a sector that is distinguished from the public and private sectors, and is also labeled as the civil society sector, the voluntary sector, the charitable sector, the non-profit sector, the sector that serves a specific social group, the civil sector, and the social and solidarity economic sector. This diversity of concepts expresses a diversity of perceptions, as well as a divergence in its components and parts. Some analysts focus on broader definitions of non-governmental or charitable organizations, individual actions and initiatives, and value systems adopted by societies (Abdel Razek, 2017: 4).

The Cambridge Dictionary defines it as the part of the economy that consists of charities.

And the Oxford dictionary defines it as the part of the economy that consists of non-governmental, not-for-profit associations, including charities and voluntary associations.

We refer to institutions of the third sector procedurally in this study are:

These well-known charitable institutions affiliated to the Ministry of Human Resources and Social Development, and their main objective and activities are caring for the environment and preserving it through the social responsibility of these associations in confronting the phenomenon of climate change in the eastern region of the Kingdom of Saudi Arabia.

Fifth: The theoretical starting point of the study

The theoretical framework issues that emerge from the theory of social system, the theory of social action, the theory of risk society and the entrance of social capital will be dealt with as theoretical approaches. In the end, the theoretical issues directed to the current study will be presented.

i- Social System Theory

Parsons defines the social pattern as "a group of individuals motivated by a tendency to optimally satisfy their needs, while the prevailing relationships among the members of this group are determined according to a pattern of complex and culturally common patterns. for the antecedent analysis of social action"; Individuals are driven to better satisfy their needs that dominate defensive orientation, as Parsons defines the relationship of individuals to their social attitudes in light of their cultural patterns (Wolfgang, 2004: 178-180).

Responsibility of Third Sector Institutions in Climate Change

Perhaps the term relationship refers to what is called guidance in another context; Where that part of Parsons' ideas refers to the other main components of directing the actor towards the situation, which is the complementary directive, and here the term value does not appear explicitly during the analysis, but it can be assumed that the patterns involve values, and these patterns are characterized by cultural composition and participation, and this aspect can be The social system serves as a bridge between the social and cultural systems; The social system implies something that belongs to the culture (Luhmann, 2013:25-40).

Parsons also talked about the way the social system works; He mentioned that each format must find a solution to a number of problems, and he called these problems or conditions the functional obligations or functional requirements, and these functions consist of:

Adaptation: It means that every social system has to adapt to the social and physical environment in which it exists.

Achieving the goal: It refers to the methods of effective individuals in order to achieve the goal.

Integration: the meaning that the system relies on a set of methods, values, and standards that link the individual to the community, so the normative integration results in the system of the general society as a whole, and integration within the sub-systems focuses on the relationships that take place within the sub-system.

Preserving the survival of the pattern and managing tension: It means that the system, including the standards and values that have generality, lead to maintaining the pattern of interaction, so that it does not deviate from or deviate from the limits of the system (Talaat Lutfi, Kamel Al-Zayyat, 2009: 73).

ii- Social Action Theory

Sociologist Weber defines the social act as he sees that the action is the one that carries a meaning and takes into account the actor whether the actor is an individual or a group, and Weber takes the other interviewers into consideration whether they are an individual or a group, and in addition to that he believes that individual actions are the basis of what is formed from Buildings, systems and groups, and based on the above, it includes the reality of social reality, and it is evident in the meanings carried by the actors, and that discovering that reality is linked to trying to understand it through interpretation and deduction. It can be said that Weber classified social action into ideal patterns, as indicated by (Omar Maan, 2006: 386).

The rational social act associated with an end: that is, it takes place for a certain purpose, and he has given it other names such as: the final or instrumental act. And in addition to the rational social action associated with a value: that is, the agent, when choosing the course of action, resorts to a directed social value. The same applies to traditional social action: it is the one in which the experiences of the actor are directed by the traditions, norms, and values of the group to which he belongs.

Finally, the social and affective action: it is the action that is directed by feelings and emotions. As for Parsons, he believes that the social act represents the basic unit of social life, and in addition to that it represents the forms of social interaction between people, and based on the above, the social act at Parsons represents the link between individuals and groups. For Parsons, verb is the unit through which the researcher can monitor social phenomena and explain the problems that individuals and institutions suffer from according to their development. It can be said that Parsons' special position on the issue of social action defines it as the "voluntary model" and this model is based on the efforts of individual actors and groups in shaping social life, and in achieving the principle of complementarity and differentiation or differentiation in performance.

In addition, it includes measures of behavior and different patterns of social and cultural orientation of the actor. It should be noted that the first type of ideas indicates that they are: a composite of realistic phenomena and experimental research methods, how to apply them and the criteria for judging their validity and stability, in addition to non-empirical ideas that are studied in the light of approaches that are appropriate to their nature. As for the second type of normative, It refers to the levels, principles and rules that define behavior in the light of standards and values as approved by the social heritage and forms of systems. We should not fail to point out that imaginary ideas include perceptions related to future, exemplary and ideal existence, and are not related to a specific program of action, but express emotions and trends that go beyond the limits of positive and effective performance, and are of less importance (Muhammad Morsi, 2001: 55-78).

iii- Risk Society Theory

This theory was put forward by the German sociologist Ulrich Beck in 1986 titled Risk Society. It is a social theory that describes the production and management of risks in modern society. The concept of a risk society by itself does not mean that it is a society in which risk rates increase, as much as we mean that it is a society organized to face risks, because it is increasingly preoccupied with the future and security. The industrial society, including the wide use of machines based on technology that invaded all fields of life, led to the emergence of various types of risks that were not known before, Hence, modern society is a society exposed to a special type of danger that is the result of the process of modernization itself that changed the social organization. The presence of risks calls for rationalizing the process of manufacturing, managing and making industrial, technological and economic decisions in modern organizations. The need arose to develop methods for calculating risks through the development of disciplined approaches to calculating them, which are extremely important, and their issuance will depend on the accuracy of risk calculation (Ulrich Beck, 2005:6). It is called the World Risk Society, as the emergence of this society is due to the acceleration of technological developments, which led to the emergence of new types of risks to humans, and they must face or adapt to them. Anthony Giddens (A. Giddens) also adds to the environmental, health and industrial risks other aspects of the overlapping variables in our contemporary social life that are the real reasons for the emergence of the risk society, including the following: (Johne Danied, 2007: 80-82)

- Fluctuation in patterns of employment and employment within organizations.
- The growing sense of job insecurity.
- The prevalence of liberation and democracy in personal relations

Beck stresses that risks also affect people, because it is difficult to predict the nature of the skills and practical experience that will be required in the field of the ever changing global economy.

Academics have analyzed risk from a sociocultural perspective, emphasizing the role of social and cultural contexts to control and manage risk. Mary Douglas has tried to prove that cultures and societies share concepts of risk and that these concepts are not a product of individual knowledge (Windvsky Douglas, 2008: 15-20).).

Responsibility of Third Sector Institutions in Climate Change

Risks have great challenges, as Beck assumes that the public (ordinary individuals) do not have the knowledge, and they constantly need someone to educate them about the risks through knowledgeable people, and they are the ones who direct, supervise and manage them as people of knowledge and accordingly they are permanently dependent on expert knowledge To educate them and warn them against the dangers (Ulrich Beck, 2009: 5-10).

iv- Social Capital Theory

Social capital refers to the ability of individuals to work together within networks of common relationships, in a way that facilitates social action to face the problems that these individuals may encounter in the context of their movement within society, and that collective action requires balances of social cohesion, the ability to integrate, and trust in others, Tolerance and acceptance of the other, and all of this is one of the manifestations of social capital, which is evident through the behavior and attitudes of individuals, meaning that the balances of social capital are closely related to social construction, but in order to fulfill their purpose they must be linked to social action, and these sources are represented in (Zayed,2006:9):

- Networks and relationships established by individuals to achieve specific goals such as: public benefit associations, factional associations, and other relationships that establish a civil life.
- A system of values, on top of which are the values of trust, transparency, tolerance of the other, cooperation, rationality, flexibility and other values of modernity.

The network of relationships represents an essential axis of social capital, and the main idea in the approach of the network of relationships is that interaction helps to build societies and each individual feels committed to the other, and thus it supports the social fabric, a sense of belonging, and the prevalence of trust and tolerance relationships that can be of great benefit to individuals, and become As a common value, virtues, and expectations within society in general, networks of relationships constitute a resource available to achieve social interests and provide its members with collective capital and qualifications that enable them to obtain credit, through multiple channels: (Khalaf Al-Shazly, 2008: 119):

- Information flow (exchanging information and ideas in different fields)
- Criteria for reciprocity or mutual assistance that depend on the network of horizontal and vertical relationships and include: networks related to customs and traditions, networks related to ties within groups, and networks related to bridging ties between groups.
- Collective work that depends on social networks.
- Trust that leads to mutual benefit.
- Solidarity and cooperation that social networks support and help transform from "me" to "we."

And about forms of social capital: there is formal social capital, which includes social ties and relationships that are formed within the framework of formal social structures such as government institutions and NGOs, and there is informal social capital, which is formed within the framework of traditional, informal social structures (such as family relationships, kinship networks, and social relations). Neighborhood and friends) and contribute to the formation of trust that stimulates community participation, and there are three ideal types of social capital networks, which are (Humnath Bhandri, 2010: 4-10):

- A pattern of an economic nature, in which the transactions of individuals are linked to economic factors.
- A status-based pattern, in which individuals are motivated by the variables of reputation, status, and centers of power.
- A pattern based on promiscuity, i.e. it is motivated by the tendency to intermingle and relate to others.

Hence, social scientists tend to consider social capital as an analytical tool both at the individual level and at the level of small groups.

Hence, the current study proceeds from an integrative theoretical framework directed at it, and is based on a number of assumptions:

- Each format must face at least four problems. In order to be able to continue and survive and achieve balance.
- The foundation of society's institutions is the tendency towards balance, harmony, and the main processes that help the systems to function and interrelate with each other.
- The sources of empowerment differ from one system to another, and these systems must find ways to adapt to the requirements of society.
- Social capital constitutes the raw material on which leaders rely within the third sector institutions for empowerment, and this is produced through daily interactions between members and third sector institutions in order to plan to confront climate changes, and monitor the reality of the practice of social responsibility among the leaders and workers in those institutions.
- The institutions of the third sector, as official organizations, carry out social action and voluntary behavior directed towards social responsibility, through community initiatives that they provide to the local community with regard to climate changes. Social action can be classified into four ideal patterns, which are as follows: Third sector institutions, as non-profit institutions, carry out community initiatives for a purpose such as improving their reputation or improving their efficiency in them. In addition, the traditional social action indicates that these institutions carry out community initiatives based on the prevailing customs and traditions in society. As for emotional social action, indicates that third-sector institutions carry out community initiatives in order to achieve balance between the layers of society, by reducing feelings of deprivation and meeting the needs of individuals in the local community, and finally the rational social action associated with value, as the organizations in the field of study carry out community initiatives through representation The prevailing social values in society such as cooperation and social solidarity.
- Ulrich Beck confirmed in the dominance of collective perception that the more the individual is characterized by a rational awareness of the huge risks, the greater his ability to overcome crises and risks.

Sixth: Literature Review

- Abraham's study 2008

Entitled: Climate Change and the Business Sector: Opportunities and Challenges The study aimed to shed light on the opportunities and challenges that the climate change issue represents for the business sector and reviews the most important global initiatives of major companies (Toyota - Shell - General Electric - Ford .... etc.) to face these challenges and seize These opportunities to transform the problem into more investment opportunities, technology development and job creation for millions of young people worldwide.

The results of the study proved that the opportunities available to the business sector were represented in new opportunities for investment and the development of a wide range of technologies, in addition to the establishment of a promising market that did not exist before (the global carbon market), which included the Kyoto mechanisms represented by the Clean Development Mechanism.

- Dari study 2008

Entitled: Climate Change and its Impact on the Environment, the study aimed to shed light on the factors that lead to climate change and its negative impacts, and how to confront them.

The results of the study confirmed the impact of climate changes on environmental accidents, as they cause severe waves of heat, frost, floods and severe storms, which are undoubtedly extreme phenomena that affect human life.

- A study by Wajdi and Mona 2009

Titled: Climate Change and its Effects on the Egyptian Economy. The study aimed to know the impact of climate change on the Egyptian economy in several areas of agriculture, industry, tourism, health and human resources.

This study relied on a number of future studies that were concerned with setting a vision and foreseeing the future and used the Delphi methodology for a number of personal interviews with experts in various fields, in addition to analyzing scientific reports and attending conferences on climate.

The results of the study also demonstrated the impact of climate change on human health, which varies in dealing with it between rich and poor countries, which requires the need to develop an integrated strategy among countries to enable the population to improve the situation to deal with potential disasters such as floods and weather events.

The study reached a set of recommendations, the most important of which are: the need to prepare a national program for the agricultural system, crops and agricultural seasons exposed to the risks of climate change, and the preparation of a national program to confront the dangers of flooding lands, coasts and population migration.

- Saif Study 2010

Entitled: The Social Responsibility of Private Sector Companies and their Role in Supporting Non-Governmental Associations in the Sultanate of Oman, an applied field study of private sector institutions and non-governmental charitable societies in the Governorate of Muscat.

The study relied on the descriptive method. The results of a study confirmed the recent interest of companies in social responsibility and its application.

The study recommended the need to draw landmarks for how to employ the social responsibility program to achieve the objectives of social care and the provision of community services.

- Study by Mahmoud and Sarhan 2015

Entitled: An economic study of climate change and its effects on sustainable development in Egypt, the study aimed to identify the phenomenon of climate change and its global and local dimensions and to stand on the effects of climate change on the productivity of the most important agricultural crops and natural resources and the expected scenarios for the impact of climate change on the cultivated areas of agricultural land.

The study relied on the descriptive analytical method to find out the current situation and the future vision.

The results of the study showed that climatic changes and the resulting increase in temperature negatively affect the productivity of agricultural crops, and also proved that there is a continuous deficit in both the Egyptian agricultural and food trade balance during the study period.

The study also found ways to confront climate change, which are: mitigation by reducing greenhouse emissions, fuel replacement, and the use of renewable energies.

Threat: through the exposure of a particular ecosystem to the risks of climate change.

Adaptation: by responding to the repercussions of climatic changes, coexisting with the conditions resulting from them, and making the best use of them through codified policies.

- A study by Amani, Tarek and Naha 2017

Entitled: Climate Change and Public Insurance in Egypt. The study aimed to shed light on the issue of climate change and its relationship to general insurance in Egypt and to measure the effects of climate through the use of the value-at-risk method.

The results of the study confirmed the existence of a strong relationship between the rise in temperatures on the surface of the globe and the increase in the recurrence rates of natural accidents and the spread of epidemic and infectious diseases.

- Kazem, Ali and Bushra study 2019

Entitled: The Impact of Climate Changes on the Sustainable Development of Environmental Resources: An Applied Study on Basra Governorate.

The results of the study proved that climate change had a negative impact on the characteristics of water resources through the rise in temperatures, the occurrence of drought and the lack of rainfall, which negatively affected and led to a decrease in revenues.

-Sumaya Study 2020

Entitled: Climate Change and its Impact on Human Security, the study aimed to know the serious effects of environmental imbalances on the social side, the study relied on the analytical method.

The study concluded that governments have a very important and effective role by cooperating to find appropriate solutions and focusing on a global agreement for safe migration and protecting the rights of people displaced by disasters and climate change.

The study recommended asking the international community to find urgent solutions to this issue for the sake of stability and balance in various fields and the need for cooperation between the various institutions to adapt to this issue.

- Mona Study 2020

Entitled: Climate Change and the stakes of international environmental policy, the study aimed to identify the concept of climate change and how the international community deals with this phenomenon in light of the challenges it poses, and to search for mechanisms that would confront this phenomenon and ensure human development in its current reality and expected future according to a sustainable security perspective.

The results of the study confirmed the repercussions and risks resulting from climate change and the associated threat to the existence of humanity, present and future, which requires a radical change in policies and the need to adapt to this situation.

The study recommended the need for international cooperation to find solutions to this issue, with the intensification of research and studies, the holding of conferences, the signing of agreements that would unite international efforts and confront this problem in order to achieve sustainable environmental security for all populations, and the need to take the necessary legal procedures and measures to protect the environment locally and internationally.

- Hanan Study 2023

Entitled: The Climate Change Crisis and the Future of the Egyptian State, the study aimed to answer the following questions: What is the impact of climate change in Egypt, what are the foundations of the Egyptian position in climate negotiations, how does the Egyptian state deal with the climate change crisis, and what are the proposed scenarios about Egypt's ability to confront the crisis ?

The results of the study confirmed the existence of three scenarios for dealing with climate change.

The study recommended the need to intensify the state's efforts to address climate changes, whether in adaptation, mitigation, or adaptation to development, and the need for countries to integrate the climate dimension into development plans.

By analyzing previous studies related to the study variables, it becomes clear that:

The interest of many of the study, such as (Hanan, Mona, Kazem) in the need for a future forward-looking vision to address the dangers of climate change.

Somaya's study also focused on the need for cooperation between institutions to address and confront the phenomenon of climate change.

The current study agrees with previous studies in its interest in the planning variable to confront climate changes as part of the social responsibility of associations that provide services to the surrounding community.

Seventh: The Methodological Framework of the Study

- **Type of study:** The current study is one of the descriptive analytical studies, which aims to describe and analyze phenomena by collecting data and information. With the aim of describing the social responsibility of third-sector institutions in facing climate changes in the eastern region.

- **The method used:** the analytical descriptive method through the comprehensive social survey method.
- **Data collection tool:** The current study relies on (stepwise questionnaire) to "measure" the reality of the practice of social responsibility of third-sector institutions in the face of climate change, and it was applied electronically in the institutions of the study community.
- **Study population:** Data was collected from the leaders and employees of the third-sector institutions in the eastern region of the Kingdom of Saudi Arabia through a comprehensive survey, which was represented in (Environmental Friends Association - Environmental Protection Association - Green Suman Association).

THE VALIDITY AND RELIABILITY OF THE RESEARCH TOOL

a- Apparent validity "the opinions of the arbitrators

The questionnaire was presented to a number of arbitrators from faculty members specialized in the field of social sciences; To verify the apparent validity of it, and with the aim of judging the suitability of the questionnaire's items to the characteristics it measures, in addition to judging and expressing an opinion regarding the various elements of the questionnaire and the modification, deletion, addition or reformulation; as they see fit; In order to achieve the objectives of the current study.

b- Structural validity:

A small group was tested as a random prospective sample of "25"; In order for the study tool to be closer to accuracy and clarity, the sample members were informed; With the aim of testing the tool and applying it to them, and the need to comment on the questions and encourage them to ask questions to find out what is ambiguous or difficult to answer; To ensure the structural validity of the study tool.

The Stability of The Study Tool

The internal consistency method was relied on, which depends on the extent to which the phrases are related to each other within the questionnaire and the correlation of the degree of each phrase with the total score of the questionnaire in general by analyzing the items. (0,340 - 0.872), and this indicates the strength of the internal coherence and consistency of the questionnaire statements.

- **Study sample:** The study sample was taken by snowball method, and it is a deliberate sample consisting of (120) From the administrative leaders of the associations and employees, the field of study, and based on the demographic characteristics of the study sample, it was the highest percentage of respondents is male; Their percentage was (65%), while the percentage of females was (35%). This may be due to the suitability of working conditions within the associations under study for males more than for females due to the long daily working hours in the organizations in the field of study, and therefore the working conditions are more suitable for males.

Responsibility of Third Sector Institutions in Climate Change

The results indicated that the highest percentage of ages was (41.7%) for those who reached the age of (more than 50 years), while the ages of (23.3%) of the respondents were in the two categories (from 30-40 years, 40-50 years).

As for the variable (educational status), it is clear that the highest percentage was for those who received higher education and obtained a master's degree. Their percentage was (47.5%), while the percentage of those who received a university education and obtained a bachelor's degree was (23.3%) of the respondents.

According to the variable (marital status), the results indicated that the highest percentage, which is (46.2%), are unmarried. It is also clear that (33.3%) of the respondents are married and (10%) are divorced.

While the variable of scientific specialization for workers in associations, the field of study, came first in psychological sciences with a rate of (35%), followed by social sciences with a rate of (24.2%), then administrative sciences with a rate of (23.3%) of the total sample size. While we find years of work experience, the workers who spent 15 years or more at work ranked first with a rate of (41.7%), followed by those who spent (less than 5 years, and from 5-10 years) with a rate of (23.3%). %)

As for the profession variable, we find that the workers in the societies under study occupy a rate of (47.5%), followed by secretaries of the societies under study with a rate of (23.3%), then the category of executive managers to reach (11.7%) of the total sample size.

THE RESULTS OF THE FIELD STUDY AND THEIR DISCUSSION

Determining The Role of Charities in Planning to Confront Climate Change in The Eastern Region

It is clear from the data of the previous table that the highest percentage of achieving the areas of social responsibility for associations is the field of study for the issue of climate change, (presenting initiatives aimed at protecting human health and safety by 100%), which was revealed through planning community initiatives in a clear and appropriate manner for all development partners, in addition to youth empowerment The workers of these associations design these community initiatives, which works to develop their spirit of loyalty, belonging and citizenship towards society and directs them to distance themselves from extremist ideology, and this is confirmed by the (Social Responsibility Analysis Report 2019) that social responsibility has specific standards within institutions related to rights and duties.

The second ranking of the role of associations came in the field of study in impact assessment (interest in raising the level of environmental performance to improve the quality of life by 76.7%). This is considered an important aspect of the quality of life program in the Kingdom of Saudi Arabia and according to the Kingdom's Vision 2030, which emphasized two dimensions: Lifestyles by activating the participation of individuals in environmental, sports, recreational and cultural activities, and the other dimension is improving the infrastructure through upgrading services, infrastructure, urban design and the environmental system, which helps to achieve an environmental balance between the natural and vital elements within the community.

Then came the third ranking of the role of associations in the field of study in assessing the impact (contributing to the development of climate information for the surrounding community by 65%), given the size of climate changes, we find that they affect many areas of life, and adaptation must also be done on a larger scale . Society as a whole has to become more resilient to climate impacts, and this will

Responsibility of Third Sector Institutions in Climate Change

Table 1. The role of field-of-study associations in impact assessment

The role of associations in impact assessment	Yes		To some extent		No		Mean	Arrangement
	F	%	F	%	F	%		
Carrying out studies and research to find out the polluting emissions to the environment	43	35,8	28	23,3	49	40,8	2,05	16
Presenting initiatives aimed at protecting human health and safety	120	100	0	0	0	0	1	1
Cooperate with the authorities concerned with controlling violations and imposing penalties for the environment system	43	35.8	49	40.8	28	23.3	1.88	15
Seeking to provide the necessary measures to protect the environment	78	65	28	23.3	14	11.7	1.47	6
The Foundation is trying to take the necessary proactive steps to address climate change	78	65	28	23.3	14	11.7	1.47	6
Interest in studying renewable energy resources to measure their impact in light of climate changes in the future	50	41.7	56	46.7	14	11.7	1.7	12
Enhancing human capacity building to respond to environmental emergencies	57	47.5	49	40.8	14	11.7	1.64	9.5
Using different media platforms to disseminate individual and societal initiatives related to the environment	78	65	28	23.3	14	11.7	1.47	6
Providing programs for young people to qualify them to deal with relief work in order to avoid natural disasters and emergency crises	50	41.7	56	46.7	14	11.7	1.7	12
Interest in raising the level of environmental performance to improve the quality of life	92	76.7	28	23.3	0	0	1.23	2
Promote a culture of social innovation to solve environmental problems.	75	60	28	23.3	14	11.7	1.47	4
Contribute to the development of climatic information for the surrounding community	78	65	42	35	0	0	1.35	3
Disseminate warnings of the dangers of climate change among all segments of society	70	58	28	23.3	14	11.7	1.47	5
Contribute to the localization of environmentally friendly industries	57	47.5	28	23.3	35	29.2	1.82	14
Activate the role of the watchdog against wrong behaviors in the environment	57	47.5	49	40.8	14	11.7	1.64	9.5
Contribute to supporting the infrastructure of the local community in the event of climate changes	64	53.3	28	23.3	28	23.3	1.7	12

require large-scale efforts, many of which will need to be coordinated by governments. We may need to build roads and bridges that are adapted to withstand higher temperatures and stronger storms within the Eastern Province of Saudi Arabia.

Some coastal cities may have to install flood prevention systems on the streets and in underground transportation facilities. Mountainous regions may require ways to reduce landslides and floods caused by melting glaciers.

And in the fourth place (promoting a culture of social innovation to solve environmental problems by 60%), through strategies, ideas, and systems, to develop the response of third-sector institutions, to find sustainable solutions to the challenges of climate change and the problems that may result from it, and to confront societies constantly with this change, and on Third sector institutions planning multi-dimensional community projects and initiatives between economic, social and environmental, with a

Responsibility of Third Sector Institutions in Climate Change

Table 2. The role of associations field of study in risk management

The role of associations in risk management	Yes		To some extent		No		Mean	Arrangement
	F	%	F	F	%	F		
Spreading the culture of environmental risk management among the company's employees	106	88.3	14	11.7	0	0	1.1167	1
The association strives to build plans to address potential risks among workers	42	35	29	24.2	49	40.8	2.0583	4
Determine the tasks and responsibilities for managing risks in climate change among workers	14	11.7	78	65	28	23.3	2.1167	2
Interest in mitigating the risks resulting from climate change in society	29	24.2	63	52.5	28	23.3	1.9917	3
Monitoring the risks arising from climate change in terms of their magnitude and severity.	0	0	92	76.7	28	23.3	2.2333	5

clear plan to ensure the continuity of the results of community initiatives. This is in agreement with the study (Saif, 2010), which emphasized the need to draw landmarks for how to employ the social responsibility program within the institutions of the third sector to achieve the objectives of social care and the provision of community services. Balance, harmony, and the main processes that help the systems to function and relate to each other.

It is clear from the data of the previous table the role of associations in the field of study in risk management within societies, where the vast majority of the study sample indicated, (spreading the culture of environmental risk management among the employees of the organization, by 88.3%), by developing strategies to spread the culture of environmental risk management These strategies are implemented through initiatives, workshops, and panel discussions that emphasize the concepts of risks, the ecosystem, and environmentally friendly products. This results in the implementation of tasks in a more streamlined and effective manner, which makes workers feel their professional selves as a result of fusion in the required tasks and in accordance with the rules and regulations of the associations in the field of study and the dissemination of a culture of risk management in them. The second ranking of the role of associations came in the field of study in risk management (identifying tasks and responsibilities for risk management in climate change among workers by 65%), through outlining methods for risk management in climate change (temperature changes, dust storms, and drought in some areas) that It must be used in the stages of the risk management process, which consists of four steps, which are identifying risks, analyzing risks, responding to risks, tracking risks and reporting on them, and thus the performance of good management is measured in containing the risks and challenges facing the employees of the institutions.

Then came the third ranking of the role of associations in the field of study in risk management (attention to mitigating the risks resulting from climate changes in society by 52.5%), as climate change actually affects health in many ways, including causing death and disease as a result of extreme weather events that are becoming more frequent Such as heat waves, storms, floods, disruption of food systems, the increase in zoonotic diseases, foodborne diseases, water and vectors, and mental health problems. This is in agreement with the study (Mona, 2020), which emphasized the repercussions and risks resulting from climate change and the associated threat to the existence of humanity, present and future, which requires a radical change in policies and the need to adapt to this situation, and also this result is

Responsibility of Third Sector Institutions in Climate Change

Table 3. The role of the field of study associations in taking action to adapt to climate changes

The role of associations in taking adaptation measures	Yes		To some extent		No		Mean	Arrangement
	F	%	F	F	%	F		
The association seeks to support environmental initiatives to combat climate change.	75	62	28	23.3	14	11.7	1.4667	3.5
Availability of the necessary capabilities to support programs offered to the community.	78	65	42	35	0	0	1.35	1.5
Attention to encourage environmentally friendly technologies.	57	47.5	49	40.8	14	11.7	1.6417	5
Ensure waste recycling within the association.	75	62	28	23.3	14	11.7	1.4667	3.5
Urging the external community to rationalize energy consumption.	78	65	42	35	0	0	1.35	1.5
Contribute to the implementation of the thermal insulation program in buildings.	50	41.7	28	23.3	42	35	1.9333	6
Interest in restoring homes for different climate changes.	14	11.7	78	65	28	23.3	2.1167	7

consistent with one of the arguments of the theoretical framework, Introduction The risk society, which believes that there is a dominance of the collective perception that the more the individual is characterized by a rational awareness of the huge risks, the greater his ability to overcome crises and risks.

It is clear from the data of the previous table the role of associations in the field of study in taking adaptation measures, as the vast majority of the study sample indicated, (the availability of the necessary capabilities to support programs provided to the community and urging the external community to rationalize energy consumption by 65%), and it is clear from this that there is Positive adaptation to climate changes through a number of mechanisms that enhance the ability of members of society to adapt by taking actions at the national, regional and global levels, and cooperation across various sectors and initiatives (humanitarian assistance, economic development, disaster risk reduction, food security, rationalization of energy consumption and others).

And in the second order came the role of associations in the field of study in taking adaptation measures (the association seeks to support environmental initiatives to confront climate changes and to ensure waste recycling within the association by 62%), where institutional building is one of the important axes for dealing with the issue of climate change to preserve natural resources and ecosystems from the impacts of climate change, by improving their adaptive capacity, and promoting a linkage approach between efforts to address biodiversity loss, climate change, land degradation and desertification, and conservation of protected areas.

Then, in the third place, the role of associations came in the field of study in taking adaptation measures (interest in encouraging environmentally friendly technologies by 47.5%), as these technologies give great attention to making the future environmentally sustainable, as third-sector institutions seek to make a positive impact on the environment, whether at the local or global level. This can be achieved through many practices and strategies, from recycling to sourcing local products to boosting energy efficiency. This is in agreement with the study (Hanan, 2023), which recommended the need to intensify the state's efforts to address climate changes, whether in adaptation, mitigation, or adaptation to development, and the need for countries to integrate the climate dimension into development plans, Also, this

Responsibility of Third Sector Institutions in Climate Change

Table 4. Difficulties faced by the associations in the field of study in performing their role towards the issue of climate change

Difficulties facing associations	Yes		To some extent		No		Mean	Arrangement
	F	%	F	F	%	F		
The lack of a clear strategy to deal with climate change crises.	78	65	28	23.3	14	11.7	1.4667	2
Shortcomings in the necessary resources and capabilities.	78	65	28	23.3	14	11.7	1.4667	2
A shortage of technical personnel necessary to implement activities and initiatives.	49	40.8	57	47.5	14	11.7	1.7083	6.5
The concept of social responsibility of the association is not clear to some.	28	23.3	50	41.7	42	35	2.1167	9
Duplication of powers and responsibilities.	64	53.3	28	23.3	28	23.3	1.7	4.5
Weak turnout of young people to volunteer work	21	17.5	57	47.5	42	35	2.175	11
Weak financial resources to face climate change.	78	65	28	23.3	14	11.7	1.4667	2
Confusion between charitable work and social responsibility.	64	53.3	28	23.3	28	23.3	1.7	4.5
Some members are not convinced of the importance of social responsibility	14	11.7	78	65	28	23.3	2.1167	9
Complicated in procedures and regulations.	0	0	78	65	42	35	2.35	12
There are no clear criteria for evaluating social responsibility in charities	14	11.7	78	65	28	23.3	2.1167	9
Lack of cadres specialized in social responsibility.	49	40.8	57	47.5	14	11.7	1.7083	6.5

result is consistent with one of the arguments of the theoretical framework, which considers that social capital constitutes the raw material upon which the leaderships within the third-sector institutions rely for empowerment, and this is produced through daily interactions between members and third-sector institutions in order to plan to confront climate changes, and monitor the reality of Practicing social responsibility among the leaders and employees of these institutions.

Monitoring The Difficulties Faced by Charities in Planning to Deal with Climate Change

Most of the responses of the study sample by 65% indicated that there are difficulties and challenges that prevent the associations in the field of study from performing their role towards the issue of climate change, including, respectively, the lack of a clear strategy to deal with the crises of climate change, the lack of resources and the necessary capabilities, and the weakness of financial resources to confront these changes. Climate, duplication of powers and responsibilities within the assigned departments, confusion between charitable work and social responsibility, lack of necessary technical personnel To implement activities and initiatives, there is a lack of specialized cadres in the areas of social responsibility within the institutions of the third sector, and these results are consistent with the report (Strategic Center in the United States of America 2018), which stated that there are many obstacles that prevent achieving optimal performance in the areas of social responsibility in institutions such as weak participation

Responsibility of Third Sector Institutions in Climate Change

Table 5. Proposals to enhance the social responsibility of the associations, the field of study to confront the issue of climate change

Proposals	Yes		To some extent		No		Mean	Arrangement
	F	%	F	F	%	F		
Promoting the dissemination of a culture of volunteer work and social responsibility towards society.	77	64.2	14	11.7	29	24.2	1.6	6.5
Partnerships with stakeholders to deliver community programs.	77	64.2	29	24.2	14	11.7	1.475	3
Existence of a clear strategy to deal with the crisis of climate change	78	65	14	11.7	28	23.3	1.5833	4.5
Encouraging research and innovation in monitoring and evaluation of environmentally friendly technologies.	77	64.2	14	11.7	29	24.2	1.6	6.5
Raise the level of environmental awareness of the phenomenon of climate change and adaptation and mitigation measures	78	65	28	23.3	14	11.7	1.4667	2
The need for cooperation between the association and other institutions in dealing with climate issues.	106	88.3	14	11.7	0	0	1.1167	1
Building youth capacities to deal with climate change effectively.	78	65	14	11.7	28	23.3	1.5833	4.5

within the environment organizational and outside motivational decline among employees; The report stressed the need to remove these obstacles and search for the best ways to achieve the goals of social responsibility. In this context, Parsons emphasized in his theoretical conception (the social system) the preservation of the institutional pattern, including specific values and standards, and the management of the tension that arises within institutions as a result of internal and external obstacles that prevent the achievement of its objectives.

Developing A Set of Proposals Necessary for The Effective Participation of Charities in Planning to Confront Climate Change

The study sample showed a high response towards the proposals to enhance and activate the social responsibility of the associations, the field of study to confront the issue of climate change, by 88.3%. Including, establishing partnerships with stakeholders to provide community programs, having a clear strategy to deal with the crisis of climate change, building youth capacities to deal with climate change effectively, Promoting the dissemination of a culture of voluntary work and social responsibility towards society, encouraging research and innovation in monitoring and evaluation of environmentally friendly technologies, and this is confirmed by Weber's theoretical vision of social action, as he saw that third sector institutions as official organizations carry out social action and voluntary behavior directed towards social responsibility. Through community initiatives that you provide to the local community regarding climate change, Social action can be classified into four ideal types, which are as follows: Rational social action associated with a purpose, that is, third-sector institutions, as non-profit institutions, take community initiatives for a purpose, such as improving their reputation or improving their efficiency in them. In addition, traditional social action indicates that these institutions perform community initiatives based on the prevailing customs and traditions in society, As for the emotional social action, it indicates that the third sector institutions take community initiatives in order to achieve a balance between the layers of society, by reducing feelings of deprivation and meeting the needs of individuals in the local community, and finally the rational social action associated with value, as the organizations in the field of study take initiatives community by embodying the prevailing social values in society such as cooperation and social solidarity.

References

Abbas, B., & Saeed, A. R. (2017). *The concept of the third sector and the Anglo-Saxon and European cognitive problems, a workshop in the economics of charitable work.*

Abdel Jalil, I. (2008): Climate Change and the Business Sector: Opportunities and Challenges. *World of Thought, 37,* 125-255. https://search.mandumah.com/Record/ 138847

Abno, A. S. (2016). The impact of climate changes on the rural area. *Journal of Historical and Social Studies.* http:// search.mandumah.com/Record/782604

Abu Sakin, H. K. (2023): The crisis of climate change and the future of the Egyptian state. *Western Journal of Political Science, 2*(7), 109-131. https://search.mandumah.com/Record/

Ajwa, A. M., Abdel-Hamid, N., & Al-Qanawati, T. (2017). Climate Change and General Insurance in Egypt. *The Egyptian Journal of Commercial Studies, 41*(2). search.mandumah.com/Record/ 847327

Al-Ajami, D. N. (2008). Climate changes and their impact on the environment. *World of Thought, National Council for Culture, Arts and Literature, 37,* 183-157. https://search.mandumah.com/Record/ 138852

Al-Harthy, Z. (2001), The reality of personal social responsibility among Saudi youth and ways to develop it. Naif Arab Academy for Security Sciences, Riyadh.

Al-Saadi, S. N. (2010). *The social responsibility of private sector companies and their role in supporting civil societies in the Sultanate of Oman: an applied field study of private sector institutions and charitable non-governmental organizations in Muscat Governorate,* [unpublished master's thesis, Sultan Qaboos University]. https://search.mandumah.com/Record/ 961026

AntonioA. (2017). *Social Responsibility and Ethics in Organizational Management.* (IESE Business School Working Paper No. 1163-E). https://ssrn.com/abstract=2969746

Bhandri, H. (2010). What is Social Capital? A Comprehensive Review of the Concept. *Asian Journal of Social Science.* https://www.researchgate.net/publication/233546004

Choucair, J. & Khalil, E. (1436). The Impact of Social Responsibility Activities on the Organization's Reputation: An Empirical Study on Riyadh Banks. *Journal of Human and Social Sciences.*

-Tawahriya, M. (2020). Climate Change and the Stakes of International Environmental Policy. *Journal of North African Economics, 16.* https://search.mandumah.com/Record/ 1234598

El-Shazly, K. K. (2005). Modern Theoretical and Methodological Attitudes in the Study of Social Capital. In *Modern Attitudes in Sociology, Dar Al-Tayseer for Printing and Publishing*. Minya.

Fawaz, M. M., & Suleiman, S. A. (2015). An economic study of climate change and its effects on sustainable development in Egypt. *Egyptian Journal of Agricultural Economics*. http://www.research gate. net//publication/283516139

Hassan, K. A.-S. (2021). *Climate Change and the Global Goals for Sustainable Development* (1st ed.). Ward Island Library.

Jairi, I. (2019): The effectiveness of international efforts in confronting climate change. Ramah Journal for Research and Studies, 34, 225-242.

Johne, D. (2009). *Risk and Manager Core Competency Model*. Professional Development Advisory Council.

Layla, A. (2013). *Arab civil society issues of citizenship and human rights*. Anglo Egyptian Bookshop.

Luhmann, N. (2013). *Introduction to System Theory*. Polity Press, Cambridge University.

Luhmann, N. Persson, H. & Thomas, R. (2008). Social capital and Social Responsibility in Denmark. *International Review For The Sociology of Sport, 43*, 135–51.

Lutfi, T. I., & Al-Zayyat, K. A.-H. (2009). *Contemporary Theory in Sociology* (1st ed.). Dar Al-Gharib.

Mayrhofer, W. (2004). Social System Theory as Theoretical Framework for Human Resource Management, Management Revue. *Rainer Hampp Verlag, 15*(2), 178–191.

http: //en.oxforddicitionary.com/definition/third-sector

Morsi, M. (2001). *Talcott Parsons' sociology between the theories of action and the social system*. Al-Qassim, Buraydah.

Ocean, S. (2020). Climate Change and its Impact on Human Security: Migration and Displacement as a Model. *Al-Nadwa Journal for Legal Studies*. https://search.mandumah.com/Record1282941/

Riad, W., & Mourad, M. (2009). Climate changes and their impact on the Egyptian economy. *Arab Center for Education and Development, 15*. https://search.mandumah.com/Record/ 43363

Syed, H. A. (2019). *Climate change, its causes and consequences* (5th ed.). Academic Journal for Scientific Research and Publication.

Ulrich, B. (2009). Living in the World Risk Society. Economy and Society Journal, 35(3).

Windvsky, R. S., & Douglas, P. G. (2008). Leadership Skills in Modern Organization, New York.

Zayed, A. (2006) *Social capital among the professional segments of the middle class, Center for Research and Social Studies, Faculty of Arts* (1st ed.). Cairo University.

Compilation of References

0ESPON. (2012). *SGPTD Second Tier Cities and Territorial Development in Europe: Performance, Policies and Prospects*. ESPON. https://www.espon.eu/sites/default/les/attachments/SGPTD_

Abbas, B., & Saeed, A. R. (2017). *The concept of the third sector and the Anglo-Saxon and European cognitive problems, a workshop in the economics of charitable work.*

Abbas, F. (2004). *Self-destruction among the perpetrators of traffic accidents.* [Master's thesis, College of Arts, University of Baghdad].

Abd al-Salam Ali, A. (2005). *Social support and its practical applications in our daily lives.* Egyptian Renaissance Bookshop.

Abdel Jalil, I. (2008): Climate Change and the Business Sector: Opportunities and Challenges. *World of Thought, 37*, 125-255. https://search.mandumah.com/Record/ 138847

Abdullah, A (2012). *Eltaghayorat Elmonakheyah.* Arab Press Agency.

Abdullah, A. (2016). *Ozone.* Arab Press Agency.

Abno, A. S. (2016). The impact of climate changes on the rural area. *Journal of Historical and Social Studies.* http://search.mandumah.com/Record/782604

Aboudouh, K, K.(2023). Civil society and its role Climate change issues. مقال منشور في مجلة آفاق مستقبلية، العدد الثالث يناير ٢٠٢٣، مركز دعم واتخاذ القرار . ج . م . ع.

Abu Dayyah, A. (2010). *Elbey'aa Fi 200 Soal.* Dar Alfarabi Publishing and Distributing.

Abu Halawa, M. S. (2013). The factorial structure and discriminatory analysis of psychological defeat in the light of some psychological variables among university students. A proposed model. Arab studies in education and psychology. *Association of Arab Educators., 37*(3), 128–171.

Abu Sakin, H. K. (2023): The crisis of climate change and the future of the Egyptian state. *Western Journal of Political Science, 2*(7), 109-131. https://search.mandumah.com/Record/

Adams, J. (2013). Collaborations: The fourth age of research. *Nature, 497*(7451), 557–560. doi:10.1038/497557a PMID:23719446

Ajwa, A. M., Abdel-Hamid, N., & Al-Qanawati, T. (2017). Climate Change and General Insurance in Egypt. *The Egyptian Journal of Commercial Studies, 41*(2). search.mandumah.com/Record/ 847327

Al-Ajami, D. N. (2008). Climate changes and their impact on the environment. *World of Thought, National Council for Culture, Arts and Literature, 37,* 183-157. https://search.mandumah.com/Record/ 138852

Al-Amir, N. Kamal. (November, 2022). The Egyptian Vision towards Climate Change Political and Environmental Issues and Development Goals. *Journal of International Politics.*

Al-Arabawi, R. (2022) *Cinema Al-Monakh Towage Kawareth Eltabeya'a Behazehe Eltareeqa.* Akhbar Al-Youm Portal.

Al-Assimi, R. (2012). Contradictions of self-perception and its relationship to social anxiety and depression among Damascus University students. *Damascus University Journal.*, 28(3), 1–69.

Al-Attar, M. M. (2019). Positive self-talk and its relationship to psychological flow and psychological defeat among students of the Faculty of Education. The Psychological Journal of Psychological Studies. *The Egyptian Association for Psychological Studies.*, 29(102), 388–432.

Al-Attiyah, A. (2010) *Asbab Eltaghayor Elmonakhy,* [Thesis, Aleppo, Faculty of Arts and Humanities, Geography Department].

Alcock, F. (2008). Conflicts and coalitions within and across the ENGO community. *Global Environmental Politics*, 8(4), 66–91. doi:10.1162/glep.2008.8.4.66

Al-Dahry, S. (2010). *Asaseyat Elm Elegtema' Elnafsy Eltarbawy Wa Nazareyatoh.* Amman: Dar Alhamed Publishing and Distributing.

Al-Deken, R. (2022). *Elcinema Elmasreya. Hal Ghabat Kadaya Elbeya'a.* Elseyasa Eldawleya https://www.siyassa.org.eg/News/18395.aspx

Al-Droubi, A., Janad, I., & Al-Seb, M. (2008). Climate Change and its Impact on Water Resources in the Arab Region. *Arab Center for the Studies of Arid Zones and Dry Lands, (ACSAD), Arab Ministerial Conference on Water.* Research Gate.

Aleixandre-Tudo, J. L., Bolanos-Pizarro, M., & Aleixandre, J. L. (2019). Current trends in scientific research on global warming a bibliometric analysis. *International Journal of Global Warming*, 17(2), 142–169. doi:10.1504/IJGW.2019.097858

Alexandri, G. (2014). Reading between the lines: Gentrification tendencies and issues of urban fear in the midst of Athens' crisis. *Urban Studies (Edinburgh, Scotland)*, 52(9), 1631–1646. doi:10.1177/0042098014538680

al-Haloul, I. E. (2015). Security anxiety and its relationship to psychosomatic symptoms among Palestinian police personnel in the Gaza Strip, The International Specialized Educational Journal, Al-Aqsa University. *Palestine*, 7(4), 34–58.

Al-Harthy, Z. (2001), The reality of personal social responsibility among Saudi youth and ways to develop it. Naif Arab Academy for Security Sciences, Riyadh.

Al-Hawari, A (2019). Monakh Alard 2150. Bibliomania Publishing.

Ali, L. H. (2013). *The Self-Defeated Personality and its Relationship to Psychological and Social Status.* [Master Thesis, College of Arts, University of Baghdad].

Al-Jamous, N. A.-H. (2009). *Psychosomatic disorders.* Al-Bazuri.

al-Kashef, T. M.-F. (2014). *Modeling Climate Change in Egypt, A Study in Applied Climate Geography, Using Geographic Information Systems and Remote Sensing.* College of Arts - South Valley University.

Al-Kufi, H. (2015). Zaherat Al-Enterar Al-Kawni w Elaqhatha Benashatat Alensan wal Kawareth Altabey'ea. Academic Book Center

Al-Masry Al-Youm, (July, 2023a). *The full truth behind the subsidence of the Alexandria Corniche, the collapse has nothing to do with any earthquake.* 2/7/2023.

Al-Saadi, H. (2020). Elm Elbey'aa, Jordan: Dar AL-YAZORI for Publishing and Distribution

Compilation of References

Al-Saadi, S. N. (2010). *The social responsibility of private sector companies and their role in supporting civil societies in the Sultanate of Oman: an applied field study of private sector institutions and charitable non-governmental organizations in Muscat Governorate*, [unpublished master's thesis, Sultan Qaboos University]. https://search.mandumah.com/Record/ 961026

Al-Sabahi, N. (2022). Eltaghayor Elmonakhy wa Atharoh Ala Elsera'at Fe Sharq Africa, Egypt: Al Arabi Publishing and Distributing

Al-Saee, S.-D. F., & Al-Qahtan, M. S. (2016). Studying some of the Environmental, Economic and Social Effects of Climate Change on the Fisheries Sector from the Perspective of Specialists. *Journal of Agricultural Economics and Social Science, 7*(2).

Al-Ta'I, T. (2021). Tahadeyatoh gheir Altakleedeyah w Afaqoh Almostaqbaleyah. Egypt: Dar Academics for Publishing & Distributing Co.

Al-Taher, F. A.-H. (2008). Climate Changes and their Impact on Food, Water and Energy Shortages and the Role of Standards in Mitigating this impact. *National Conference on the role of standards in facing climate changes and food, water and energy shortages, 25.* Cairo.

Al-Zrad, F. K. (2007). *Mental-physical illnesses. I (2).* Dar Al-Nafees.

Andres, S. (2011). Communication, The Essence of Management of A Nonprofit Organization. *Annals of Eftimie Murgu University Resita, Fascicle II. Economic Studies, 1*(1), 121–130.

Angotti, T., & Irazábal, C. (2017). Planning Latin American cities: Dependencies and "Best Practices.". *Latin American Perspectives, 44*(2), 4–17. https://www.jstor.org/stable/26178807. doi:10.1177/0094582X16689556

AntonioA. (2017). *Social Responsibility and Ethics in Organizational Management.* (IESE Business School Working Paper No. 1163-E). https://ssrn.com/abstract=2969746

AntonioA. (2017): *Social Responsibility and Ethics in Organizational Management.* (IESE Business School Working Paper No. 1163-E) https://ssrn.com/abstract=2969746

Antonova, N. L., & Grunt, E. V. (2019, December). Citizens' role in formation of urban environment design. In *IOP Conference Series: Materials Science and Engineering.* IOP Publishing. 10.1088/1757-899X/687/5/055053

Arnstein, S. R. (1969). A ladder of citizen participation. *Journal of the American Institute of Planners, 35*(4), 216–224. doi:10.1080/01944366908977225

Arsel, M. (2023). Climate change and class conflict in the Anthropocene: Sink or swim together? *The Journal of Peasant Studies, 50*(1), 67–95. doi:10.1080/03066150.2022.2113390

Asher, S., & Novosad, P. (2020). Rural roads and local economic development. *The American Economic Review, 110*(3), 797–823. doi:10.1257/aer.20180268

Asian Disaster Preparedness Center. (2008). *Building on Local Knowledge for Safer Homes.* ADPC. http://www.adpc.net/v2007/IKM/ONLINE%20DOCUMENTS/downloads/2008/3_CaseStudyShelterl.pdf

Attia, A. M. (2011). Academic resilience and its relationship to self-esteem among a sample of open education students. *Journal of Psychological Studies., 21*(4), 571–621.

Austin, G. (2016). *New uses of Bourdieu in film and media studies.* New York: berghahn

Avelino, F., Wittmayer, J. M., Pel, B., Weaver, P., Dumitru, A., Haxeltine, A., Kemp, R., Jørgensen, M. S., Bauler, T., Ruijsink, S., & O'Riordan, T. (2019). Transformative Social Innovation and (Dis)empowerment. *Technological Forecasting and Social Change, 145*(August), 195–206. doi:10.1016/j.techfore.2017.05.002

Badyina, A., & Golubchikov, O. (2016). "Gentrication in central Moscow – a market process or a deliberate policy? Money, power and people in housing regeneration in Ostozhenka". Geograska Annaler: Series B. *Human Geographies, 87*(2), 113–129. doi:10.1111/j.0435-3684.2005.00186.x

Bailey, D., & Solomon, G. (2004). Pollution prevention at ports: Clearing the air. *Environmental Impact Assessment Review, 24*(7-8), 749–774. doi:10.1016/j.eiar.2004.06.005

Balunde, A., Perlaviciute, G., & Steg, L. (2019). The relationship between people's environmental considerations and pro-environmental behavior in Lithuania. *Frontiers in Psychology, 10*, 2319. doi:10.3389/fpsyg.2019.02319 PMID:31681111

Baraka, A. I. (2019). *The impact of climate change on the natural, economic and social environment (the Republic of Chad as a model).*

Barca, F. (2008). *An agenda for a reformed cohesion policy: A place-based approach to meeting European Union challenges and expectations.* (No. EERI_RP_2008_06). Economics and Econometrics Research Institute (EERI), Brussels. https://ec.europa.eu/regional_policy/archive/policy/future/barca_en.htm

Barker, D. (2012). Caribbean agriculture in a period of global change: Vulnerabilities and opportunities. *Caribbean Studies (Rio Piedras, San Juan, P.R.), 40*(2), 41–61. https://www.jstor.org/stable/41917603. doi:10.1353/crb.2012.0027

Bauman, Z. (2016). *Liquid Life.* Egyot: hajaj abu jbr.

Bayulken, B., Huisingh, D., & Fisher, P. M. (2021). How are nature based solutions helping in the greening of cities in the context of crises such as climate change and pandemics? A comprehensive review. *Journal of Cleaner Production, 288*, 125569. doi:10.1016/j.jclepro.2020.125569

Becker, A. H., Acciaro, M., Asariotis, R., Cabrera, E., Cretegny, L., Crist, P., & Velegrakis, A. F. (2013). A note on climate change adaptation for seaports: A challenge for global ports, a challenge for global society. *Climatic Change, 120*(4), 683–695. doi:10.100710584-013-0843-z

Beer, S. (1984). The viable system model: Its provenance, development, methodology and pathology. *The Journal of the Operational Research Society, 35*(1), 7–25. doi:10.1057/jors.1984.2

Belous, O., & Gulbinskas, S. (2008). Klaipėda deep-water seaport development. In Conflict resolution in coastal zone management (Environmental Education, Communication and Sustainability). Peter Lang Pub Inc.

Bennett, K. (2021, November 9). *NHT defends sky-high Ruthven Towers prices.* Jamaica WI-The Gleaner.com. https://jamaica-gleaner.com/article/lead-stories/20211109/nht-defends-sky-high-ruthven-towers-prices

Berger A., & Loutre, M. F. (2004). Astronomical theory of climate change. *Journal de Physique IV.*

Berney, R. (2011). Pedagogical urbanism: Creating citizen space in Bogotá, Colombia. *Planning Theory, 10*(1), 16–34. https://www.jstor.org/stable/26165894. doi:10.1177/1473095210386069

Bhandri, H. (2010). What is Social Capital? A Comprehensive Review of the Concept. *Asian Journal of Social Science.* https://www.researchgate.net/publication/233546004

Bhatt, R., Kukal, S. S., Busari, M. A., Arora, S., & Yadav, M. (2016). Sustainability issues on rice–wheat cropping system. *International Soil and Water Conservation Research, 4*(1), 64–74. doi:10.1016/j.iswcr.2015.12.001

Compilation of References

Bleyen, P., Klimovský, D., Bouckaert, G., & Reichard, C. (2017). Linking budgeting to results? Evidence about performance budgets in European municipalities based on a comparative analytical model. *Public Management Review, 19*(7), 932–953. doi:10.1080/14719037.2016.1243837

Bloodgood, E. A., Tremblay-Boire, J., & Prakash, A. (2014). National styles of NGO regulation. *Nonprofit and Voluntary Sector Quarterly, 43*(4), 716–736. doi:10.1177/0899764013481111

Blunier, T., Chappellaz, J., Schwander, J., Dällenbach, A., Stauffer, B., Stocker, T. F., Raynaud, D., Jouzel, J., Clausen, H. B., Hammer, C. U., & Johnsen, S. J. (1998). Asynchrony of Antarctic and Greenland climate change during the last glacial period. *Nature, 394*(6695), 739–743. doi:10.1038/29447

Bogotá.gov.co. (2019). *Bogotá's Master Plan Targets Major Global Development Agendas.* Bogotá.gov.co. https://Bogotá.gov.co/internacional/Bogotás-master-plan-targets-major-global-development-agendas

Bonacina, L. C. W. (1947). Climatic Change and the Retreat of Glaciers. 1 The Self-Generating or Automatic Process in Glaciation. *Quarterly Journal of the Royal Meteorological Society, 73*(315-316), 85–000. doi:10.1002/qj.49707331506

Boyd, E., & Juhola, S. (2015). Adaptive climate change governance for urban resilience. *Urban Studies (Edinburgh, Scotland), 52*(7), 1234–1264. doi:10.1177/0042098014527483

Brazier, A. (2015). *climate change in Zimbabwe, facts for planners and desicion makers.* Konrad Adenauer stiftun. https://www.kas.de/c/document_library/get_fle?uuid=6dfce726-fdd1-4f7b-72e7-e6c1ca9c9a95&group

Brereton, P. (2012). *Smart cinema, DVD Add- Ons and New Audience pleasures.* UK: Palgrave Macmillan.

Bronnimann, S., Appenzeller, C., Croci-Maspoli, M., Fuhrer, J., Grosjean, M., Hohmann, R., Ingold, K., Knutti, R., Liniger, M. A., Raible, C. C., Röthlisberger, R., Schär, C., Scherrer, S. C., Strassmann, K., & Thalmann, P. (2014). Climate change in Switzerland: A review of physical, institutional, and political aspects. *Wiley Interdisciplinary Reviews: Climate Change, 5*(4), 461–481. doi:10.1002/wcc.280

Bulkeley, H. B., & Betsill, M. M. (2003). *Cities and Climate Change. Urban sustainability and global environmental governance.* Routledge.

Bulkeley, H., Coenen, L., Frantzeskaki, N., Hartmann, C., Kronsell, A., Mai, L., Marvin, S., McCormick, K., van Steenbergen, F., & Voytenko Palgan, Y. (2016). Urban living labs: Governing urban sustainability transitions. *Current Opinion in Environmental Sustainability, 22*, 13–17. doi:10.1016/j.cosust.2017.02.003

Bulkeley, H., & Kern, K. (2006). Local government and the governing of climate change in Germany and the UK. *Urban Studies (Edinburgh, Scotland), 43*(12), 2237–2259. doi:10.1080/00420980600936491

Burkšienė, V., & Dvorak, J. (2022). Local NGO e-communication on environmental issues. In *The Routledge Handbook of Nonprofit Communication* (pp. 269–278). Routledge. doi:10.4324/9781003170563-32

Burksiene, V., Dvorak, J., & Burbulytė-Tsiskarishvili, G. (2020). City Diplomacy in Young Democracies: The Case of the Baltics. In A. Sohaela & E. Sevin (Eds.), *City Diplomacy* (pp. 305–330). Palgrave Macmillan Series in Global Public Diplomacy. Palgrave Macmillan. doi:10.1007/978-3-030-45615-3_14

Burkšiene, V., Dvorak, J., & Duda, M. (2019). Upstream Social Marketing for Implementing Mobile Government. *Societies (Basel, Switzerland), 9*(3), 54. doi:10.3390oc9030054

Burskyte, V., Belous, O., & Stasiskiene, Z. (2011). Sustainable development of deep-water seaport: The case of Lithuania. *Environmental Science and Pollution Research International, 18*(5), 716–726. doi:10.100711356-010-0415-y PMID:21104330

Cagle, C. (2017). *Sociology on film; postwar Hollywood's prestige commodity, New Jersey*. Rutgers university press.

Campbell, H. (2005). Reflections on the post-colonial Caribbean state in the 21st century. *Social and Economic Studies*, *54*(1), 161–187. https://www.jstor.org/stable/27866408

Campbell, Y. (2020). *Citizenship on the margins*. Springer International Publishing. doi:10.1007/978-3-030-27621-8

Campellone, T. R., Sanchez, A. H., & Kring, A. M. (2016). Defeatist performance beliefs, negative symptoms, and functional outcome in schizophrenia: A meta-analytic review. *Schizophrenia Bulletin*, *42*(6), 1343–1352. doi:10.1093chbulbw026 PMID:26980144

Carbon Neutral Cities Alliance. (2019). *Carbon Neutral Cities Alliance 2019 Annual Report*. CNCA. https://carbon-neutralcities.org/cities/

Cárdenas, M., Bonilla, J. P., & Brusa, F. (2021). *Climate policies in Latin America and the Caribbean: Success stories and challenges in the fight against climate change*. Publications.iadb.org. https://publications.iadb.org/publications/english/viewer/Climate-policies-in-latin-america-and-the-caribbean.pdf

Carmen, E., Fazey, I., Ross, H., Bedinger, M., Smith, F. M., Prager, K., McClymont, K., & Morrison, D. (2022). Building Community resilience in a context of climate change. The Role of Social Capital. *Ambio*, *51*(6), 1371–1387. doi:10.100713280-021-01678-9 PMID:35015248

Causes and effects of climate change. (n.d.). The United Nations. https://www.un.org/ar/climatechange/science/causes-effects-climate-change

Chandra, M., Karkun, A., & Matthew, S. (2021). Hot and Flooded: What the IPCC Report Forecasts for India's Development Future. *The Wire*. https://science.thewire.in/politics/ government/what-the-ipcc-report-forecasts-for-india-development-future/

Charles, A. (2015). *"Can we build cities that anticipate the future?"*. World Economic Forum. https://www.weforum.org/agenda/2015/10/can-we-build-cities-that-anticipate-the-future

Chilton, L. (2021). *10 best movies about climate change, from Avatar to The Day After Tomorrow*. Independent website Arabic Edition

Choucair, J. & Khalil, E. (1436). The Impact of Social Responsibility Activities on the Organization's Reputation: An Empirical Study on Riyadh Banks. *Journal of Human and Social Sciences*.

Christophers, B. (2016). For real: Land as capital and commodity. *Transactions of the Institute of British Geographers*, *41*(2), 134–148. https://www.jstor.org/stable/45147008. doi:10.1111/tran.12111

CitiesAct. (2009). *The Copenhagen Climate Communiqué*. Clover Archive. http://www.cloverarchive.com/main/page/3125.pdf.

City of Ottawa. (2016). Guide for Older Adults: Services and Programs oered by the City of Ottawa. City of Ottawa. https://documents.ottawa.ca/sites/documents/les/2019-.058%20Older%20Adult%20Booklet%20-%20ENG.pdf. City of Ottawa (2020).

City of Rotterdam. (2013). *Rotterdam. Climate Change Adaptation Strategy*. Rotterdam Climate Initiative.

City of Toronto. (2020). *Toronto Strong Neighbourhoods Strategy 2020*. City of Toronto. https://www.toronto.ca/city

City of Vienna. (2013). *Gender Mainstreaming in Urban Planning and Urban Development*. WIEN. https://www.wien.gv.at/stadtentwicklung/studien/pdf/b008358.pdf

Compilation of References

City of Vienna. (2014a). *Smart City Wien.* Wien. https://smartcity.wien.gv.at/site/les/2014/09/

City of Vienna. (2014b). *Smart City Wien: Framework Strategy; Municipal Department 18.* City Development and Planning.

Clarke, C. (2006). Politics, violence and drugs in Kingston, Jamaica. *Bulletin of Latin American Research, 25*(3), 420–440. https://www.jstor.org/stable/27733873. doi:10.1111/j.0261-3050.2006.00205.x

Clément, J.-F. (2018). *Smart City Framework.* Wien. SmartCityWien_FrameworkStrategy_english_onepage.pdf.

Clemente, M., & Salvati, L. (2017). 'Interrupted' Landscapes: Post-Earthquake Reconstruction in between Urban Renewal and Social Identity of Local Communities. *Sustainability (Basel), 9*(11), 2015. doi:10.3390u9112015

Climate Change. (n.d.). The United Nations. https://www.un.org/ar/climatechange/science/causes-effects-climate-change

ClimateHomeNews. (2019) Demark adopts climate law to cut emissions 70% by 2030. *Climate Home News.* https://www.climatechangene ws.com/2019/12/06/denmark-adopts-climate-law-cut-emissions-702030/.

Cohen, G. (2017). *What is social infrastructure?* Aberdeen Standard Investments. https://www.aberdeenstandard.com/en-us/us/investor/insights-thinking-aloud/article-page/what-is-socialinfrastructure

Colic, S. (2008). *Sociocltural aspects of cohsumption consumer cultuer and society.*

Columbia Events. 2021. (2021, December 10). *World Report: Colombia.* Human Rights Watch. https://www.hrw.org/world-report/2022/country-chapters/colombia

Comfort, S. E., & Hester, J. B. (2019). Three Dimensions of Social Media Messaging Success by Environmental NGOs. *Environmental Communication, 13*(3), 281–286. doi:10.1080/17524032.2019.1579746

Connolly, C., Keil, R., & Ali, S. H.. (2021, February). Extended urbanisation and the spatialities of infectious disease: Demographic change, infrastructure and governance. *Urban Studies (Edinburgh, Scotland), 58*(2), 245–263. doi:10.1177/0042098020910873

Consult, I. (Ed.). (2022, May). The National Strategy for Climate Change in Egypt (2050).

Corell, E., & Betsill, M. M. (2001). A comparative look at NGO influence in international environmental negotiations: Desertification and climate change. *Global Environmental Politics, 1*(4), 86–107. doi:10.1162/152638001317146381

CPH2025 Climate Plan. (2020aCity of Moscow. https://kk.sites.itera.dk/apps/kk_pub2/pdf/983_jkP0ekKMyD.pdf.

Cross, J. (2017, October 6). *Dons want in: Downtown Kingston area leaders call meeting to discuss role in redevelopment.* Jamaica WI-The Gleaner.com. https://jamaica-gleaner.com/article/lead-stories/20171010/dons-want-downtown-kingston-area-leaders-call-meeting-discuss-role

Cuevas, M. C. (2021). *Meet the Puerto Rican sisterhood reinventing the island's future after Maria.* CNN. https://www.cnn.com/2018/09/19/us/iyw-puerto-rico-women-rebuild-trnd/index.html

Culwick, C., Washbourne, C.-L., Anderson, P. M. L., Cartwright, A., Patel, Z., & Smit, W. (2019). CityLab re-ections and evolutions: Nurturing knowledge and learning for urban sustainability through co-production experimentation. *Current Opinion in Environmental Sustainability, 30*, 9–16. doi:10.1016/j.cosust.2019.05.008

Curtin, P. D. (1955). *Two Jamaicas: The role of ideas in a tropical colony, 1830–1865.* Harvard University Press.

Cuzick, J. o. (2008). Overview of Human Papillomavirus-Based and Other Novel Options for Cervical. *Vaccine.* www.elsevier.com/locate/vaccine

D'amato, G., Vitale, C., De Martino, A., Viegi, G., Lanza, M., Molino, A., & D'amato, M. (2015). Effects on asthma and respiratory allergy of Climate change and air pollution. *Multidisciplinary Respiratory Medicine, 10*(1), 1–8. PMID:26697186

Dapilah, F. (2020). The role of social networks in building adaptive capacity and resilience to Climate change: A case study from North Ghana. *Climate and Development, 12* (1), 42-56.

Dawson, A. (2017). *Extreme Cities: The Peril and Promise of Urban Life in the Age of Climate Change.* Brooklyn: Verso.

de la Casa, J. M. H., Álvarez-Villa, À., & Mercado-Sáez, M. T. (2018). Communication and effectiveness of the protest: Anti-fracking movements in Spain. *Zer: Revista de estudios de comunicación= Komunikazio ikasketen aldizkaria, 23*(45).

De Urbanisten. (2016). *Climate Proof ZOHO District: Work in Progress.* Rotterdam Climate Initiative.

Deng, D., Zhao, Y., & Zhou, X. (2017). Smart city planning under the climate change condition. *IOP Conference Series. Earth and Environmental Science, 81*(1), 012091. doi:10.1088/1755-1315/81/1/012091

DeVerteuil, G., & Golubchikov, O. (2016). Can resilience be redeemed? Resilience as a metaphor for change, not against change. *City, 20*(1), 143-151. doi:10.1080/13604813.2015.1125714

Diken, B. (2007). *Sociology through the projector.* London and New York: Routledge Taylor & Francis group.

Dirlik, A. (2003). Globalization, indigenism, and the politics of place. *ARIEL: A Review of International English Literature, 34*(1).

Dolšak, N. (2013). Climate change policies in the transitional economies of Europe and Eurasia: The role of NGOs. *Voluntas, 24*(2), 382–402. doi:10.100711266-012-9260-6

Douglas, I. (2009). Climate change, flooding and food security in south Asia. *Food Security, 1*(2), 127–136. doi:10.100712571-009-0015-1

Drenth, P. (2001) Freedom and responsibility in science: reconcilable objectives? Joachim Jungius-Gesellschaft der Wissenschaften, Symposium "Forschungsfreiheit und ihre ethische Grenzen."

Duer, M., & Vegliò, S. (2019). Modern-colonial geographies in Latin America. *Journal of Latin American Geography, 18*(3), 11–29. https://www.jstor.org/stable/48618849. doi:10.1353/lag.2019.0058

Dutch, S. I. (2010). *Encyclopedia of global warming* (Vol. 245). Salem Press.

Duverger, M. (2020). *Introduction to the social sciences.* Routledge.

Dvorak, J. (2015). The Lithuanian Government's Policy of Regulatory of Regulatory Impact Assessment. *Management and Business Administration. Central Europe, 23*(2), 129–146. doi:10.7206/mba.ce.2084-3356.145

Dwair, M. (2022). "Elbattaria Da'eifa", Egyptian short movie addresses pollution and climate change. *Sky News Arabia.*

Dwiartama, A., Kelly, M., & Dixon, J. (2023). Linking food security, food sovereignty and foodways in urban Southeast Asia: Cases from Indonesia and Thailand. *Food Security, 15*(2), 505–517. doi:10.100712571-022-01340-6

Eissa, M. M. (2007, December). New statistical study for Global temperature. *Meteorological Research Bulletin, 22,* 17.

Ejdys, J. (2020). Trust-Based Determinants of Future Intention to Use Technology. *Foresight and STI Governance, 14*(1), 60–68. doi:10.17323/2500-2597.2020.1.60.68

Elliott, R. (2018). The sociology of climate change as a sociology of Loss. European journal of sociology. *Archives Europeees de Socioligie, 59*(3) 301-337.

El-Shazly, K. K. (2005). Modern Theoretical and Methodological Attitudes in the Study of Social Capital. In *Modern Attitudes in Sociology, Dar Al-Tayseer for Printing and Publishing*. Minya.

EU. (2020) 2030 climate & energy framework. EC. https://ec.europa.eu/clima/ policies/strategies/2030_en. Accessed Mar 2020

European Commission. (2015). *Closing the loop – An EU action plan for the Circular Economy. Communication from the Commission to the European Parliament, the Council, the European Economic and Social Committee and the Committee of the Regions*. EC. https://eur-lex.europa.eu/resource. html?uri=cellar:8a8ef5e8-99a0-11e5-b3b7-01aa75ed71a1.0012.02/ DOC_1andformat=PDF.

European Commission. (2019). *The Digital Cities Challenge: Designing Digital Transformation Strategies for EU Cities in the 21st Century*. EC. https://www.intelligentcitieschallenge.eu/sites/default/les/2019-09/EA-04-19-483- EN-N.pdf

European Environment Agency. (2019). *Open data and e-government good practices for fostering environmental information sharing and dissemination*. UNECE. https://www.unece.org/leadmin/DAM/env/pp/a_to_i/Joint_

European Innovation Partnership on Smart Cities and Communities (EIP-SCC). (2019). *Smart City Guidance Package: A Roadmap for Integrated Planning and Implementation of Smart City Projects; Norwegian University of Science and Technology*. NTNU.

Facing a Changing World: Women, Population and Climate. (2009). UNFPA. https://www.unfpa.org/publications/ state-world-population-2009

Fahim, M. A. H. M. (2013). Climate Change Adaptation Needs for Food Security in Egypt. *Science Pub, 11*(12). https:// www.sciencepub.net/nature

False Creek Neighborhood Energy Utility. (2020). *Property Development*. False Creek Neighborhood Energy Utility. https://vancouver.ca/homeproperty-development/southeast-false-creek-neighbourhood-energy-utility.aspx

Fanø, J. J. (2019). Enforcement of the 2020 sulphur limit for marine fuels: Restrictions and possibilities for port States to impose fines under UNCLOS. *Review of European, Comparative & International Environmental Law, 28*(3), 278–288. doi:10.1111/reel.12306

FAO. IFAD, UNICEF, WFP, & WHO. (2019). The State of Food Security and Nutrition in the World 2019- Safeguarding against economic slowdowns and downturns. FAO.

Fawaz, M. M., & Suleiman, S. A. (2015). An economic study of climate change and its effects on sustainable development in Egypt. *Egyptian Journal of Agricultural Economics*. http://www.research gate.net//publication/283516139

Fawaz, M. M., & Suleiman, S. A. (2015, September). An Economic Study of Climate Change and Its Effects on Sustainable Development in Egypt. *The Egyptian Journal of Agricultural Economics, 25*(3), 3.

Fay, M. (2005). *The Urban Poor in Latin America*. World Bank Publications. doi:10.1596/0-8213-6069-8

Flohn, H. (1961). Mans activity as a factor in climatic change. *Annals of the New York Academy of Sciences, 95*(1), 271–281. doi:10.1111/j.1749-6632.1961.tb50038.x

Forbath, W. E. (1999). Caste, class, and equal citizenship. *Michigan Law Review, 98*(1), 1–91. doi:10.2307/1290195

Fünfgeld, H. (2010). Institutional challenges to climate risk management in cities. *Current Opinion in Environmental Sustainability, 2*(3), 156–160. doi:10.1016/j.cosust.2010.07.001

Fung, A. (2006). Varieties of participation in complex governance. *Public Administration Review, 66*(s1), 66–75. https:// www.jstor.org/stable/4096571. doi:10.1111/j.1540-6210.2006.00667.x

Galvin, C. J. (2020). COVID-19 preventive measures showing an unintended decline in taiwan. *International Journal of Infectious Diseases*. www.elsevier.com/locate/ijid

Gangaiah, B. (2019). Agronomy-Kharif Crops. NISCAIR. http://nsdl.niscair.res.in/jspui/bitstream/123456789/527/1/Millets%20(Sorghum%2c%20Pearl%20Millet%2c%20Finger%20 Millet)%20-%20%20Formatted.pdf

García Fernández, C., & Peek, D. (2020). Smart and sustainable? Positioning adaptation to climate change in the European smart city. *Smart Cities*, *3*(2), 511–526. doi:10.3390martcities3020027

Gastner, M. T., & Newman, M. E. J. (2004). Diffusion-based method for producing density-equalizing maps. *Proceedings of the National Academy of Sciences of the United States of America*, *101*(20), 7499–7504. doi:10.1073/pnas.0400280101 PMID:15136719

Gender Climate Tracker. (2010). *Gender and the Climate Change Agenda. The impacts of climate change on women and public policy*. Gender Climate Tracker. https://genderclimatetracker.org/sites/default/files/Resources/Gender-and-the-climate-change-agenda-212.pdf

German Watch. (2019). *CCPI, Climate Change Performance Index*. German Watch.

Geyer, R. F., & Van der Zouwen, J. (Eds.). (2014). *Sociocybernetics: An actor-oriented social systems approach* (Vol. 1). Springer.

Gido, J., Clements, J., & Baker, R. (2018). *Successful Project Management*. Cengage Learning.

Gomez- Muñoz, P. (2023). Science fiction cinema in the twenty- first century; Transnational futures, cosmopolitan concerns. New York: Routledge

Grainge, P. (2007). *Film Histories an introduction and reader*. Edinburgh: Edinburgh university press.

Greenpeace. (2009). *Meet Ulamila: Climate Activist in the Pacific*. Green Peace. https://www.greenpeace.org.au/blog/meet-ulamila-climate-activist-in-the-pacific/

Groneberg-Kloft, B., Fischer, T. C., Quarcoo, D., & Scutaru, C. (2009). New quality and quantity indices in science (NewQIS): The study protocol of an international project. *Journal of Occupational Medicine and Toxicology (London, England)*, *4*(1), 16. doi:10.1186/1745-6673-4-16 PMID:19555514

Gunnarsson-Östling, U., & Svenfelt, Å. (2017). Towards social-ecological justice. In *The Routledge Handbook of Environmental Justice* (p. 160). Routledge.

Guzman, L., & Bocarejo, J. P. (2017). Urban form and spatial urban equity in Bogotá, Colombia. *Transportation Research Procedia*, *25*, 4491–4506. doi:10.1016/j.trpro.2017.05.345

Hall, N. L., & Taplin, R. (2007). Solar festivals and climate bills: Comparing NGO climate change campaigns in the UK and Australia. *Voluntas: International journal of voluntary and nonprofit organizations, 18*(4), 317–338. https://doi:org/ doi:10.1007/s11266-007-9050-8

Hamid, A. (2012). *Salah*. Algafaf w Altasahor, Almakhater w Aleyat Almokafha, Arabian Heba Nile for Publishing & Distribution.

Hamza, Y. G., Ameta, S. K., Tukur, A., & Usman, A. (2020). Overview on Evidence and Reality of Climate Change. *IOSR Journal of Environmental Science, Toxicology and Food Technology (IOSR-JESTFT)*, *14*(7).

Harrison, F. V. (1988). The politics of social outlawry in urban Jamaica. *Urban Anthropology and Studies of Cultural Systems and World Economic Development*, *17*(2/3), 259–277. https://www.jstor.org/stable/40553119

Compilation of References

Hashim, S. M., Issa, M. M., & Maghribi, N. (2021, June). Climate and its Impact on Human Comfort in the Nile River Delta in Egypt for the Period (1986-2005), a Study in Applied Climate. *Journal Research,* (6).

Hassan, K. A.-W., Yassin, B. R., & Kazem, A. A.-Z. (2019). The Impact of Climate Changes on the Sustainable Development of Water Resources: an applied study in Basra Governorate. *Peer Journal, 44*(4).

Hassan, K. E. (2021). *Climate Change and the Global Goals for Sustainable Development.* Al Jazeera Library.

Healey, P. (2003). Collaborative planning in perspective. *Planning Theory, 2*(2), 101–123. doi:10.1177/14730952030022002

Hegazy, M. (2005). Social backwardness: an introduction to the psychology of the oppressed person. 9th Edition. Morocco: Casablanca.

Ho, E. (2017). Smart subjects for a Smart Nation? Governing (smart) mentalities in Singapore. *Urban Studies (Edinburgh, Scotland), 54*(13), 3101–3118. doi:10.1177/0042098016664305

Horvath, R. J. (1972). A definition of colonialism. *Current Anthropology, 13*(1), 45–57. https://www.jstor.org/stable/2741072. doi:10.1086/201248

Hossain, A. T., & Masum, A. A. (2022). Does Corporate Social responsibility help mitigate firm Level Climate change risk? *Finance Research Letters, 47,* 102791. doi:10.1016/j.frl.2022.102791

Hricko, A. (2012). Progress & Pollution Port Cities Prepare for The Panama Canal Expansion. *Environmental Health Perspectives, 120*(12), 470–473. doi:10.1289/ehp.120-a470 PMID:23211315

http: //en.oxforddicitionary.com/definition/third-sector

Hue, D. T. (2017). Fourth Generation NGOs: Communication Strategies in Social Campaigning and Resource Mobilization. *Journal of Nonprofit & Public Sector Marketing, 29*(2), 119–147. doi:10.1080/10495142.2017.1293583

Hulme, M. (2009). *Why We Disagree About Climate Change; Understanding controversy, inaction and opportunity.* Cambridge, New York: Cambridge university press.

Ibrahim, F. I. (2022). Climate Change and Food Security in Egypt. *Scientific Journal of Economics and Trade, 52*(1), 221–262.

Ibtisam, M. A. A.-A. (2010). *Blood groups and some psychological disorders in a sample of male and female students with learning difficulties and normal in the primary stage in Makkah Al-Mukarramah.* [PhD thesis, Umm Al-Qura University, Saudi Arabia].

Inglis, D & Almila, A. (2016). *The SAGE handbook of cultural sociology.* Los Angeles/London/ New Delhi: SAGE reference

Innes, J. E., & Booher, D. E. (2018). *Planning with complexity.* Routledge. doi:10.4324/9781315147949

Inter-American Development Bank. (2015). *Panamá metropolitana: Sostenible, humana y global.* IDB. https://www.iadb.org/en/urban-development-and-housing/emerging-and-sustainable-cities-program

IPCC. (2013). *Climate Change 2013: The physical science basis. Contribu- tion of working group i to the ffth assessment report of the intergovern- mental panel on climate change.* Cambridge University Press.

IPCC. (2019). *The Intergovernmental Panel on Climate Change.* IPCC.

IRGC. (2018, September). *IRGC Guidelines for the Governance of Systemic Risks.* doi:10.5075/epfl-irgc-257279

Islam, S. N., & Winkel, J. (2017). *Climate change and social inequality.* UN Department of Economic and Social Affairs. https://digitallibrary.un.org/record/3859027?ln=en

Jadallah, A. M., & Abdel-Meguid, E. M. (2021). Rural women's awareness of the effects of climate change on health security and how to confront them: A study in the village of Sanhour Al-Madina, Desouk Center, Kafr El-Sheikh Governorate. *Agriculture Economics and Rural Development, Al-Jam'iya Scientific. Agriculture Sciences*, 7(1).

Jairi, I. (2019): The effectiveness of international efforts in confronting climate change. Ramah Journal for Research and Studies, 34, 225-242.

Jarvier, I. (2013). *Towards a sociology of the cinema (ILS 92)*. New York: Routledge.

Javeed, H. M. R., Ali, M., Qamar, R., Sarwar, M. A., Jabeen, R., Ihsan, M. Z., & Sabagh, A. E. (2023). Food Security Issues in Changing Climate. In *Climate Change Impacts on Agriculture: Concepts, Issues and Policies for Developing Countries* (pp. 89–104). Springer International Publishing. doi:10.1007/978-3-031-26692-8_6

Jerslev, A. (2002). *Realism and 'Reality' in film and media*. Museum Tusculanum press, University of Copenhagen

Johne, D. (2009). *Risk and Manager Core Competency Model*. Professional Development Advisory Council.

Jones, E. (2010). Contending with local governance in Jamaica: Bold Programme, Cautionary Tales. *Social and Economic Studies*, 59(4), 67–95. https://www.jstor.org/stable/41803728

Jones, M., Hunt, H., Kneale, C., Lightfoot, E., Lister, D., Liu, X., & Motuzaite-Matuzeviciute, G. (2016). Food Globalisation in prehistory: The agrarian foundations of an interconnected continent. *Journal of the British Academy, 4*, 73–87. doi:10.5871/jba/004.073

Kabeer, M., Peterson, N., & Waldron, D. (2021). Resilient Farmers: Investing to Overcome the Climate Crisis. Acumen and Busara Center for Behavioral Science, 16.

Kamarck, E. (2019, September). *The challenging politics of climate change*. Brookings. https://www.brookings.edu/research/the-challenging-politics-of-climate-change/

Kamel, S. (2018). *Sorat Elsahafy Fil Cinema; Mashahed Sahafeya Fil Aflam Elarabeya Khelal Elfatra Men 1952 Hata 2009*. Egypt: Al Arabi Publishing and Distributing.

Kaseb, M. (2020). *Almasoleya Aldawleya Lehemayet Altanawoa Alehyaey w Beat Alfada' Alkharegy men Altalawoth Fe Etar Almoaahadat Aldawleya, Cairo*. Egyptian publishing house and distribution.

Kelman, I. (2020). *Disaster by choice*. Oxford University Press.

Keulemans, S. (2021). Rule-following identity at the frontline: Exploring the roles of general self-efficacy, gender, and attitude towards clients. *Public Administration*. doi:10.1111/padm.12721

Khan, M. (2022, March 27). Examining gentrification: A new internal colonialism. *Inverse Journal*. https://www.inverse-journal.com/2022/03/27/examining-gentrification-a-new-internal-colonialism-an-academic-essay-by-m-moosa-khan/

Klaipeda City Municipality. (2021). *Klaipėdos miesto savivaldybės 2021 - 2030 metų strateginis plėtros planas*. Klaipėda [Strategy of Klaipeda city municipality 2021-2030]. https://www.klaipeda.lt/lt/planavimo-dokumentai/klaipedos-miesto-savivaldybes-2021-2030-metu-strateginis-pletros-planas/8827

Klingelhofer, D., Braun, M., Quarcoo, D., Bruggmann, D., & Groneberg, D. A. (2019). Research landscape of a global environmental challenge: Microplastics. *Water Research, 170*, 115358. doi:10.1016/j.watres.2019.115358 PMID:31816566

Konapur, A., Gavaravarapu, M. S., Gupta, S. D., & Nair, K. M. (2014). Millets in Meeting Nutrition Security:Issues and Way Forward for India. *The Indian Journal of Nutrition and Dietetics, 51*, 306–321.

Compilation of References

Konstantinaviciute, I. (2003). Climate change mitigation policies in Lithuania. *Energy & Environment, 14*(5), 725–736. https://www.jstor.org/stable/43734595. doi:10.1260/095830503322663429

Kopp, R. E., DeConto, R. M., Bader, D. A., Hay, C. C., Horton, R. M., Kulp, S., Oppenheimer, M., Pollard, D., & Strauss, B. H. (2017). Evolving understanding of Antarctic ice-sheet physics and ambiguity in probabilistic sea-level projections earths. *Earth's Future, 5*(12), 1217–1233. doi:10.1002/2017EF000663

Korten, D. C. (1990). *Getting to the 21st Century: Voluntary Action and the Global Agenda.* Kumanian Press.

Kotseva-Tikova, M., & Dvorak, J. (2022). Climate Policy and Plans for Recovery in Bulgaria and Lithuania. *SSRN, 22*(2), 79–99. doi:10.2139srn.4294988

Krippendorff, K. (2013). *Content Analysis. An Introduction to Its Methodology* (3rd ed.). Sage Publications.

Kulp, S. A., & Strauss, B. H. (2019). New elevation data triple estimates of global vulnerability to sea-level rise and coastal fooding. *Nature Communications, 10*(1), 4844. doi:10.103841467-019-12808-z PMID:31664024

Kumar, A., Tomer, V., Kaur, A., Kumar, V., & Gupta, K. (2018). Millets: A solution to agrarian and nutritional challenges. *Agriculture & Food Security, 7*(1), 31. https://agricultureandfoodsecurity.biomedcentral.com/articles/10.1186/s40066-018-0183-3. doi:10.118640066-018-0183-3

Kurylo, B. (2020). Technologised Consumer Culture; the Adorno-Benjamin Debate and the Reverse side of Politicisation. *Journal Of Consumer Culture.*

Lahangir, S. (2021). Odiya tribes discover the wonders of millets. *Villagesquare.in.* https://www.villagesquare.in/odiya-tribes-discover-the-wonders-of-millets/

Layla, A. (2013). *Arab civil society issues of citizenship and human rights.* Anglo Egyptian Bookshop.

Lee, T., Johnson, E., & Prakash, A. (2012). Media independence and trust in NGOs: The case of postcommunist countries. *Nonprofit and Voluntary Sector Quarterly, 41*(1), 8–35. doi:10.1177/0899764010384444

Leisher, C., Temsah, G., Booker, F., Day, M., Samberg, L., Prosnitz, D., Agarwal, B., Matthews, E., Roe, D., Russell, D., Sunderland, T., & Wilkie, D. (2016). Does the gender composition of forest and fishery management groups affect resource governance and conservation outcomes? A systematic map. *Environmental Evidence, 5*(1), 6. doi:10.118613750-016-0057-8

Li, H. (2021). Follow or not follow?: The relationship between psychological entitlement. *Personality and Individual Differences.* www.elsevier.com/locate/paid

Liu, B. F. (2012). Toward a better understanding of nonprofit communication management. *Journal of Communication Management (London), 16*(4), 388–404. doi:10.1108/13632541211279012

Lockyer, W. J. S. (1910). Does the Indian climate change? *Nature, 84*(2128), 178–178. doi:10.1038/084178a0

Luhmann, N. Persson, H. & Thomas, R. (2008). Social capital and Social Responsibility in Denmark. *International Review For The Sociology of Sport, 43,* 135–51.

Luhmann, N. (2012). *Theory of society* (Vol. 1; R. Barrett, Trans.). Stanford University Press.

Luhmann, N. (2013). *Introduction to System Theory.* Polity Press, Cambridge University.

Lutfi, T. & Al-Zayyat, K. (2009): *Contemporary Theory in Sociology, first edition, Dar Al-Gharib, Cairo.* UNESC. https://archive.unescwa.org

Lutfi, T. I., & Al-Zayyat, K. A.-H. (2009). *Contemporary Theory in Sociology* (1st ed.). Dar Al-Gharib.

M. A. D. R. E. (n.d.). *Resources and Results for Women Worldwide*. Nicaragua: Harvesting Hope. https://www.madre.org/page/nicaraguaharvesting-hope-34.html

M. A. D. R. E. (n.d.). *Resources and Results for Women Worldwide*. Sudan: Women Farmers Unite. https://www.madre.org/page/sudan-women-farmers-unite-41.html

Madani, M., Abdel-Gayed, S., & Murad, M. (2011, January). *The Future Effects of Climate Change on the Agricultural Sector in Egypt: Cost Estimation*. Research Gate.

Magd, A. (2015). Altanzim Aldostoury Lelhoqoq w Alhorreyat Alektesadeya: Derasa Tatbeykeya Ala Alnezam Aldostoury (Altaadelat Alakhera w Afaq Altanmeya). Cairo, The National Center for Legal Publications.

Magdy, S. (2022, November 4). Climate Cahnge and Sea Level Rise Threaten Food Security in Egypt. *Independent Arabia*. https://www.independentarabia.com: https://www.independentarabia.com/node/388946/

Magri, M., & Solari, A. (1996). The SCI Journal Citation Reports: A potential tool for studying journals? 1 Description of the JCR journal population based on the number of citations received, number of source items, impact factor, immediacy index and cited half-life. *Scientometrics*, *35*(1), 93–117. doi:10.1007/BF02018235

Mahfouz Al-Fazari. (2005): *Developing crisis management in educational institutions in the Sultanate of Oman*. [unpublished master's thesis, University of Jordan, Amman].

Mahmoud, L. A. (2021). *Psychological fragility and its relationship to defense mechanisms among unmarried women*. [Master's thesis, College of Education, Jordan].

Mahmoud, M. H. (2022). *An indicative program based on clinical indications for the T.A.T Subject Comprehension Test to reduce psychosomatic disorders among mothers of children with cerebral palsy (a clinical-psychometric study),* [PhD thesis, Faculty of Education, Alexandria University].

Mahmoud, H. A.-M. (2014). An Analytical Economic Study of the Current Situation and the Future of Wheat Self-Sufficiency in Egypt. *The Egyptian Journal of Agricultural Research*, *92*(2), 781–801. doi:10.21608/ejar.2014.156315

Malik, S. (2014). Women's Objectification by Consumer Culture. *International Journal of Gender&Womenes*.

Mansour, E. (2016). *Elmadkhal Ela Elm Elegtema*. Amman: Dar Arabian Gulf Publishing House

March, H., & Ribera-Fumaz, R. (2016). Smart contradictions: The politics of making Barcelona a Self-sufficient city. *European Urban and Regional Studies*, *23*(4), 816–830. doi:10.1177/0969776414554488

Marín, J. C., Raga, G. B., Arévalo, J., Baumgardner, D., Córdova, A. M., Pozo, D., Calvo, A., Castro, A., Fraile, R., & Sorribas, M. (2017). Properties of particulate pollution in the port city of Valparaiso, Chile. *Atmospheric Environment*, *171*, 301–316. doi:10.1016/j.atmosenv.2017.09.044

Marquand. J, F, A. & Elsasser, J, p . (2023) Institutionalizing climate change mitigation in the Global South: trends and future research Earth system Governance, 15, 100163.

Maurer, S., & Rauch, F. (2019). Economic geography aspects of the Panama Canal. *Centre for Economic Performance, 1633*. https://cep.lse.ac.uk/pubs/download/dp1633.pdf

Mayrhofer. (2004). Social System Theory as Theoretical Framework for Human Resource Management. *Management Revue, Rainer Hampp Verlag, 15*(2), 178-191.

Mayrhofer, W. (2004). Social System Theory as Theoretical Framework for Human Resource Management, Management Revue. *Rainer Hampp Verlag, 15*(2), 178–191.

Mayring, P. (2014). *Qualitative content analysis: theoretical foundation, basic procedures and software solution*. Austria: Klagenfurt. https://nbn-resolving.org/urn:nbn:de:0168-ssoar-395173

Mazahrh, A. (2016). *Elbey'aa Wal Mogtama'*. Amman: Dar Alshorooq Publishing and Distributing.

Mc Carthy, J. (2020). *Understanding Why Climate Change Impacts Women more than Men?* Global Citizen. https://www.globalcitizen.org/en/content/how-climate-change-affects-women/

McSweeney, K., & Jokisch, B. (2007). Beyond rainforests: Urbanisation and emigration among lowland Indigenous societies in Latin America. *Bulletin of Latin American Research*, *26*(2), 159–180. doi:10.1111/j.1470-9856.2007.00218.x

Megahed, N. (n.d.). *Eltarbeyah Ala Qeyam Elmowatanah Alalameya Lemowagahat Mogtama' Elmakhater*. Alexandria: University Education Publishing and Distributing.

Menzel, A., Sparks, T. H., Estrella, N., Koch, E., Aasa, A., Ahas, R., Alm-Kübler, K., Bissolli, P., Braslavská, O. G., Briede, A., Chmielewski, F. M., Crepinsek, Z., Curnel, Y., Dahl, Å., Defila, C., Donnelly, A., Filella, Y., Jatczak, K., Måge, F., & Zust, A. (2006). European phenological response to climate change matches the warming pattern. *Global Change Biology*, *12*(10), 1969–1976. doi:10.1111/j.1365-2486.2006.01193.x

Merriam, S. B. (1998). *Qualitative research and case study applications in education*. Jossey-Bass.

Mesolell, K. J., Matthews, R. K., Broecker, W. S., & Thurber, D. L. (1969). Astronomi- cal Theory of Climatic Change— Barbados Data. *The Journal of Geology*, *77*(3), 250–274. doi:10.1086/627434

Michie, J. & Cooper, C.L. (2015). *Why the Social Sciences Matter*. NY: Palgrave Macmillan

Mihai, R. L. (2017). Corporate Communication Management. A Management Approach. *Valahian Journal of Economic Studies*, *8*(22), 103–110. doi:10.1515/vjes-2017-0023

Milies, S. (2021). *Consumer Culture*. Oxford.

Mills, G. (1997). *Westminster style democracy: The Jamaican experience*. Grace Kennedy Foundation Lecture. http://gracekennedy.com/lecture/GKF1997Lecture.pdf

Mills, J. N., Gage, K. L., & Khan, A. S. (2010). Potential influence of climate change on vector-borne and zoonotic diseases: A review and proposed research plan. *Environmental Health Perspectives*, *118*(11), 1507–1514. doi:10.1289/ehp.0901389 PMID:20576580

Ministry of Works and Transport (MOWT)- Trinidad and Tobago. (2020). *The Port of Spain Flood Alleviation Project*. MOWT. https://www.mowt.gov.tt/Divisions/Programme-For-Upgrading-Roads-Efficiency-(PURE)-Un/Projects/The-Port-of-Spain-Flood-Alleviation-Project

Moore, F. C. (2009). Climate Change and Air Pollution: Exploring the Synergies and Potential for Mitigation in Industrializing Countries. *Sustainability (Basel)*, *1*(1), 43–54. doi:10.3390u1010043

Mora, C., Frazier, A. G., Longman, R. J., Dacks, R. S., Walton, M. M., Tong, E. J., Sanchez, J. J., Kaiser, L. R., Stender, Y. O., Anderson, J. M., Ambrosino, C. M., Fernandez-Silva, I., Giuseffi, L. M., & Giambelluca, T. W. (2013). The projected timing of climate departure from recent variability. *Nature*, *502*(7470), 183–187. doi:10.1038/nature12540 PMID:24108050

Moreno, S. H. (1993). Impact of development on the Panama Canal environment. *Journal of Interamerican Studies and World Affairs*, *35*(3), 129–150. doi:10.2307/165971

Morsi, M. (2001). *Talcott Parsons' sociology between the theories of action and the social system*. Al-Qassim, Buraydah.

Muhammad, M. S. (2023). *The Impact of Climate Change on Sustainable Development and the Labor Market in the Arab World.*

Muhammed, L. (2022). *8 Aflam Alameya Tosalet Eldoa Ala Tahdeedat Eltaghayorat Elmonakheya Legawaneb Elhayah.* Youm 7 Portal.

Muhammed, M. (2014) Eldabt Eledary Wa Dawroh Fi Hemayet Elbey'aa, Comparative Study, Riyadh: Law and Economics Library

Muhammed, N. (2023). *Khetat Eltanmeya Elmostadama: Derasa Fil Elaqat Elmasreya Elafrikeya.* Al Arabi Publishing and Distributing.

Mullings, J., Dunn, L., Sue Ho, M., Wilks, R., & Archer, C. (2018). Urban renewal and sustainable development in Jamaica: Progress, challenges and new directions. *An Overview of Urban and Regional Planning.* doi:10.5772/intechopen.79075

Muszynska, K., Dermol, K., Trunk, V., Đakovic, A., & Smrkolj, G. (2015, May). Communication management in project teams–practices and patterns. In *Joint International Conference* (pp. 1359-1366).

Nabhan, Y. (2012). *Elehtebas Elharary Wa Ta'theeroh Ala Elbey'aa.* Amman: Dar Konooz for Publishing and Distribution

National Housing Policy (Draft). (2019). Ministry of Economic Growth and Job Creation.

Nature Index. (2018) *10 institutions that dominated science in 2017.* Nature Index. https://www.natureindex.com/news-blog/twenty-eighteen-annual-table s-ten-institutions-that-dominated-sciences.

Naylor, M. (2019). *How the Nordics are standing up to climate change.* STP Trans. https://www.stptranscom/how-nordics-are-standing-up-to-clima te-change/.

ND-GAIN. (2019) *Notre Dame Global Adaptation Initiative, Country Index.* ND-Gain. https://gain.nd.edu/our-work/country-index/.

Neumayer, E., & Plümper, T. (2007). The Gendered Nature of Natural Disasters: The Impact of Catastrophic Events on the Gender Gap in Life Expectancy, 1981–2002. *Annals of the Association of American Geographers, 97*(3), 551–566. doi:10.1111/j.1467-8306.2007.00563.x

Nulman, E., & Özkula, S. M. (2016). Environmental nongovernmental organizations' digital media practices toward environmental sustainability and implications for informational governance. *Current Opinion in Environmental Sustainability, 18*, 10–16. doi:10.1016/j.cosust.2015.04.004

Ocean, S. (2020). Climate Change and its Impact on Human Security: Migration and Displacement as a Model. *Al-Nadwa Journal for Legal Studies.* https://search.mandumah.com/Record1282941/

OECD. (2019). OECD Economic Surveys: China 2019. *OECD Economic Surveys: China, 2019.* OECD. doi:10.1787/eco_surveys-chn-2019-en

OneYoungWorld. (2020). *The effects of climate change in Zimbabwe.* One Young World. https://www.oneyoungworld.com/blog/efects-climate-change-zimba

Orru, H., Ebi, K. L., & Forsberg, B. (2017). The Interplay of Climate Change and Air Pollution on Health. *Current Environmental Health Reports, 4*(4), 504–513. doi:10.100740572-017-0168-6 PMID:29080073

Orwell, G. (1945). *Animal Farm: A fairy story and essays collection.* (75th anniversary). Penguin Publishing Group.

Compilation of References

Ota, A. B. (2020). Tribal Atlas of Odisha. Academy of Tribal Languages and Culture & Scheduled Castes & Scheduled Tribes Research and Training Institute ST & SC Development Department, Government of Odisha. Commissioner-cum-Director, SCSTRTI & Member Secretary, ATLC, Bhubaneswar.

Padulosi, S., Mal, B. C., King, O. I., & Gotor, E. (2015). Minor Millets as a Central Element for Sustainably Enhanced Incomes, Empowerment, and Nutrition in Rural India. *Sustainability (Basel), 7*(7), 1–30. doi:10.3390u7078904

Pahl-Wostl, C. (2009). A conceptual framework for analysing adaptive capacity and multi-level learning processes in resource governance regimes. *Global Environmental Change, 19*(3), 354–365. doi:10.1016/j.gloenvcha.2009.06.001

Palttala, P., Boano, C., Lund, R., & Vos, M. (2012). Communication gaps in disaster management: Perceptions by experts from governmental and non-governmental organizations. *Journal of Contingencies and Crisis Management, 20*(1), 2–12. doi:10.1111/j.1468-5973.2011.00656.x

Pandey, A. (2021). Food wastage: Causes, Impacts and Solutions. *Science Henritage Journal.*

Panigrahi, J. K. (2016). Coastal Ecosystems of Odisha–Health and Nutritional Challenges Consequent to Climate Change. Directorate of Economics and Statistics, Odisha.

Parmesan, C., & Yohe, G. (2003). A globally coherent fngerprint of climate change impacts across natural systems. *Nature, 421*(6918), 37–42. doi:10.1038/nature01286 PMID:12511946

Parsons, T. (2005). *The social system* (B. Turner, Ed.; 2nd ed.). Routledge., https://voidnetwork.gr/wp-content/uploads/2016/10/The-Social-System-by-Talcott-Parsons.pdf

Patel, A. M., & Jha, M. K. (2007). *Weapons of the Weak -Field Studies on Claims to Social Justice in Bihar & Orissa.* Mahanirban Calcutta Research Group, Kolkata, India http://www.mcrg.ac.in/pp13.pdf

Patnaik, B. K. (2016). *Impact of Global Warming on Agriculture with Reference to Odisha.* Special Issue on Agriculture and Farmer's Welfare, Directorate of Economics and Statistics.

Pavlovic, J., Lalic, D., & Djuraskovic, D. (2014). Communication of Non – Governmental Organizations via Facebook Social Network. *Inzinerine Ekonomika-Engineering Economics, 25*(2), 186–193. doi:10.5755/j01.ee.25.2.3594

Planning Institute of Jamaica (PIOJ). (2009). *Urban planning and regional development: sector plan 2009–2030.* PIOJ.

Plass, G. N. (1956). The carbon dioxide theory of climatic change. *Tellus, 8*(2), 140–154. doi:10.3402/tellusa.v8i2.8969

Pörtner, H. O., Roberts, D. C., Tignor, M. M. B., Poloczanska, E., Mintenbeck, K., Alegría, A., Craig, M., Langsdorf, S., Löschke, S., Möller, V., Okem, A., & Rama, B. (Eds.). (2022). *Climate Change 2022: Impacts, Adaptation and Vulnerability.* IPCC Assessment Report. https://www.ipcc.ch/report/ar6/wg2/

Pounds, J. A. (2006). Widespread amphibian extinctions from epidemic disease driven by global warming. *Nature, 439*(7073), 161–167. doi:10.1038/nature04246 PMID:16407945

Powell, S. (2017). *Sustainability in the Public Sector: An Essential Briefing for Stakeholders.* Routledge. doi:10.4324/9781351275729

Rabie, S. (2022). *Climate Change and Population Distribution in Egypt.* Central Agency for Public Mobilization and Statistics, Egypt.

Rajdali, M. (2020). The Mediterranean Basin, Advantages and Challenges of Sustainable Development. 5.

Rajhans, K. (2018). Effective Communication Management: A Key to Stakeholder Relationship Management in Project-Based Organizations. *The IUP Journal of Soft Skills, XII*(4), 47–66. https://ssrn.com/abstract=3398050

Raupp, J., & Hoffjann, O. (2012). Understanding strategy in communication management. *Journal of Communication Management (London)*, *16*(2), 146–161. doi:10.1108/13632541211217579

Refaat, A. (2023, January). Evaluation of the Effectiveness of Financing Programs to Confront Climate Change. *College of Politics and Economics,* (17).

Regdali, M. (2020). *The Mediterranean Basin.* Advantages and Challenges of Sustainable Development.

Revi, A. (2008). Climate change risk: An adaptation and mitigation agenda for Indian cities. *Environment and Urbanization*, *20*(1), 207–229. doi:10.1177/0956247808089157

Riad, W., & Mourad, M. (2009). Climate changes and their impact on the Egyptian economy. *Arab Center for Education and Development, 15*. https://search.mandumah.com/Record/ 43363

Ripple, w, J, M, (etal). (2022) Six steps to integrate climate mitigation with adaptation for social Justice. *Environmental science & Policy, 128*, 41-44.

Risks Associated with Cimate and Environmental Changes in the Mediterranean Region. (2019). *Network of Experts on Climate and Environmental Changes in the Mediterranean Region.*

Ritchie, H., & Roser, M. (2019) Our world in data: CO2 and greenhouse gas emissions. *Our World in Data.* https://ourworldindata.org/co2-and-other-greenhouse-gasemissions.

Roblek, V. (2019). The smart city of Vienna. In L. Anthopoulos (Ed.), *Smart City Emergence: Cases from around the World* (1st ed.). Elsevier. doi:10.1016/B978-0-12-816169-2.00005-5

Rondinelli, D. A. (1990). Housing the urban poor in developing countries: The magnitude of housing deficiencies and the failure of conventional strategies are worldwide problems. *American Journal of Economics and Sociology*, *49*(2), 153–166. https://www.jstor.org/stable/3487429. doi:10.1111/j.1536-7150.1990.tb02269.x

Root, T. L., Price, J. T., Hall, K. R., Schneider, S. H., Rosenzweig, C., & Pounds, J. A. (2003). Fingerprints of global warming on wild animals and plants. *Nature*, *421*(6918), 57–60. doi:10.1038/nature01333 PMID:12511952

Runk, J. V. (2012). Indigenous land and environmental conflicts in Panama: Neoliberal multiculturalism, changing legislation, and human rights. *Journal of Latin American Geography*, *11*(2), 21–47. doi:10.1353/lag.2012.0036

Ryan, D. (1999). Colonialism and hegemony in Latin America: An introduction. *The International History Review*, *21*(2), 287–296. https://www.jstor.org/stable/40109004. doi:10.1080/07075332.1999.9640860

Saadet, B. (2015). *The Effects of Climate Change on Sustainable Development in Algeria (a Prospective Study)* [PhD Thesis, Boumerdes, Algeria: Faculty of Economic, Commercial and Facilitation Sciences, University of M'hamed Bougherra].

Sagi, V., & Gokarn, S. (2023). Determinants of reduction of food loss and waste in Indian agri-food supply chains for ensuring food security: A multi-stakeholder perspective. *Waste Management & Research*, *41*(3), 575–584. doi:10.1177/0734242X221126421 PMID:36218223

Salmon, D. (2020, June. *A Jamaican tale of two cities.* Jamaica WI-The Gleaner. https://jamaica-gleaner.com/article/focus/20200621/david-salmon-jamaican-tale-two-cities

Salvador, J. (2017). PPP For Cities Case Studies. Barcelona GIX: IT Network Integration (Spain). UNECPPP. https://www.uneceppp-icoe.org/people-rst-ppps-case-studies/ppps-in-it/barcelona-gix.-it-network-integrationn barcelona-spain/

Samhi Al-Qahtani. (2003). *The role of public relations departments in dealing with crises and disasters.* [An unpublished master's thesis, Naif Arab University for Security Sciences, Riyadh].

Compilation of References

Samir, F. (2013). *Hemayet Albeaa w Mokafhet Altalawoth w Nashr Althaqafah Albe'eya*. Dar Al-Hamed Publishing.

Sangam, S.L. & Savitha, K.S. (2019). Climate change and global warming: a scientometric study. *Collnet J Scientomet, 13*, 199–212. doi:. doi:10.1080/09737766.2019.159800136

Satpathi, S., Saha, A., & Basu, S. (2020). *Millets as a Policy Response to the Food and Nutrition Crisis—Special Reference to the Odisha Millets Mission*. BRLF. https://www.brlf.in/brlf2/wp-content/uploads/ 2020/01/Odisha-Millet-Mission.pdf

Satyavathi, C. T., Ambawat, S., Khandelwal, V., & Srivastava, R. K. (2021). Pearl Millet: A Climate-Resilient Nutricereal for Mitigating Hidden Hunger and Provide Nutritional Security. *Frontiers in Plant Science, 12*, 659938. doi:10.3389/fpls.2021.659938 PMID:34589092

Schmid, H. o. (2021). Preventive measures for accompanying caregivers of children in. *Clinical Microbiology and Infection*: www.clinicalmicrobiologyandinfection.com

Schoburgh, E. D., & Gatchair, S. (2016). Managing development in local government: frameworks and strategies in Jamaica. Developmental Local Governance: A Critical Discourse in 'Alternative Development. doi:10.1057/9781137558367_8

Schoburgh, E. D. (2017). Is a self-help orientation sufficient basis for local [economic] development? *Journal of Human Values, 23*(3), 151–166. doi:10.1177/0971685817713287

Shahrin, N., & Hussin, H. (2023). Negotiating food heritage authenticity in consumer culture. *Tourism and Hospitality Management, 29*(2), 183–193. doi:10.20867/thm.29.2.3

Shalender, K. (2015). Organizational Flexibility for Superior Value Proposition: Implications for Service Industry. *Int J Econ Manag Sci, 4*(256), 2. doi:10.4172/2162-6359.1000256

Sharma, A. (2023). *Moderating role of Involvement and Value of Service*. Indian Instiute of Technology Kanpur.

Sharma, D. C. (2006). Ports in a Storm. *Environmental Health Perspectives, 114*(4), 222–231. doi:10.1289/ehp.114-a222 PMID:16581529

Sherry, S. B., Stoeber, J., & Ramasubbu, C. (2016). Perfectionism explains variance in self-defeating behaviors beyond self-criticism: Evidence from a cross-national sample. *Personality and Individual Differences, 95*, 196–199. doi:10.1016/j.paid.2016.02.059

SIA. (2018). Les Ateliers de Renens: Construction d'un site dédié à l'emploi. *Forum Bâtir et Planier: la Ville Productive Conference*. VD. https://www.vd.sia.ch/sites/vd.sia.ch/les/20181112_Batir_Planier_ Cl%C3%A9ment.pdf

Siggelkow, N. (2007). Persuasion with case studies. *Academy of Management Journal, 50*(1), 20–24. Retrieved July 14, 2023, from https://journals.aom.org/doi/10.5465/amj.2007.24160882#:~:text=If%20one's%20conceptual%20argument%20is,is%20usually%20much%20more%20appealing. doi:10.5465/amj.2007.24160882

Sigler, T., & Wachsmuth, D. (2016). Transnational gentrification: Globalisation and neighbourhood change in Panama's Casco Antiguo. *Urban Studies (Edinburgh, Scotland), 53*(4), 705–722. https://www.jstor.org/stable/26151056. doi:10.1177/0042098014568070

Sinaee, A. (2016). *Estimating Smart City Sensors Data Generation: Current and Future Data in the City of Barcelona*. In Proceedings of the 15th IFIP Annual Mediterranean Ad Hoc Networking Workshop, Barcelona, Spain.

Singh, S. & Marwah, R. (2023). *Politics of climate change: Crises, conventions and cooperation*. USA: World Scientific publishing

Soafer, S. (1999). Qualitative methods: What are they and why use them? *Health Services Research, 34*(5 Pt 2), 1101–1118. PMID:10591275

Speer, I., Pazsegaran, M., & Heidegger, M. M. (n.d.). *Adapting to Climate Change - The New Challenge for Development in the Developing World* (K. Essin & R. Asi, Eds.).

Stiglitz, J. E. (2012). *The price of inequality: How today's divided society endangers our future.* WW Norton & Company.

Stone, C. (1973). *Class, race and political behaviour in urban Jamaica.* Institute of Social and Economic Research.

Stott, P.A., Stone, D.A., & Allen, M.R. (2004). Human contribution to the European heatwave of 2003. *Nature, 432,* 610–614. https://doi.org/ e03089 doi:10.1038/natur

Syed, H. A. (2019). *Climate change, its causes and consequences* (5th ed.). Academic Journal for Scientific Research and Publication.

Tahamineh, L. (etal), (2022). *Application of Machine Learning deep Learning Methods for Climate change Mitigation and Adaptation.*

Taleb, H. (2014). *Introduction to the psychosomatic approach.* University of Ouargla.

Tanimoto, K. (2012). The emergent process of social innovation: Multi-stakeholders perspective. *International Journal of Innovative Research and Development, 4*(June), 267. doi:10.1504/IJIRD.2012.047561

-Tawahriya, M. (2020). Climate Change and the Stakes of International Environmental Policy. *Journal of North African Economics, 16.* https://search.mandumah.com/Record/ 1234598

Telešienė, A., & Kriaučiūnaitė, N. (2008). Trends of Nongovernmental Organizations' Environmental Activism in Lithuania. *Public Policy and Administration, 1*(25), 94–103.

The Arab Strategic Report. (2010). Center for Political and Strategic Studies. Cairo: Al-Ahram.

The UNCCD files . (2018, June). UNCCD. https://www.unccd.int/: https://www.unccd.int/sites/default/files/2018-06/GLO%20Arabic_Full_Report_rev1.pdf

The World Bank. (2019a) *Data, GDP (current US$).* World Bank. https://data.worldbank. org/indicator/NY.GDP.MKTP.CD.

The World Bank. (2019b) *Data, Population, total.* World Bank.

Thomas, G. (2011). A Typology for the Case Study in Social Science Following a Review of definition, Disclosure, and Structure. *Qualitative Inquiry, 17*(6), 511–521. doi:10.1177/1077800411409884

Tovar-Restrepo, M., & Irazábal, C. (2014). Indigenous women and violence in Colombia. *Latin American Perspectives, 41*(1), 39–58. doi:10.1177/0094582X13492134

Tsaaior, J. T. (2011). History, (re) memory and cultural self-presencing: The politics of postcolonial becoming in the Caribbean novel. *Journal of Caribbean Literatures, 7*(1), 123–137. https://www.jstor.org/stable/41939271

Tumulytė., I. (2012). Darnaus vystymosi komunikacija. Pilietinės iniciatyvos aplinkosaugos komunikacijoje. [Communication of sustainable development. Citizen initiatives in environmental communication]. *Informacijos mokslai, 62,* 7–17.

U. N. (2016). *The new urban agenda.* In *The United Nations Conference on housing and sustainable urban development (Habitat III) held in Quito, Ecuador.* UN. https://habitat3.org/the-new-urban-agenda/

U. N. Women. (2022). Explainer: How gender inequality and climate change are interconnected. UN. https://www.un-women.org/en/news-stories/explainer/2022/02/explainer-how-gender-inequality-and-climate-change-are-interconnected

Ulrich, B. (2009). Living in the World Risk Society. Economy and Society Journal, 35(3).

Compilation of References

Ulrich, B. (2009): Living in the World Risk Society. Economy and Society Journal, 35.

Ümarik, M., Loogma, K., & Tafel-Viia, K. (2014). Restructuring vocational schools as social innovation? *Journal of Educational Administration, 52*(1), 97–115. doi:10.1108/JEA-08-2012-0100

UN-DESA. (2022). *Total and urban population: UNCTAD Handbook of Statistics 2021.*United Nations Conference on Trade and Development (UNCTAD). https://hbs.unctad.org/total-and-urban-population/

UNESCO. (2009). *World Water Assessment Programme.* UNESDOC. WWW.UNESCO.ORG: https://unesdoc.unesco.org/ark:/48223/pf0000374903_spa?posInSet=1&queryId=N-EXPLORE-6f43efa9-f9da-451e-8948-31b90a8bd1e8

UNFCC. (2023). *Five Reasons Why Climate Action Needs Women.* UNFCC. https://unfccc.int/news/five-reasons-why-climate-action-needs women#:~:text=Particularly%20in%20developing%20countries%2C%20the,risk%20to%20their%20personal%20safety

United Nations Environment Programme (UNEP). (2023, March 31). *Panama to strengthen its sustainable finance framework: United Nations Environment – Finance initiative.* UNEP. https://www.unepfi.org/themes/climate-change/panama-to-strengthen-its-sustainable-finance-framework/*Urban law in Colombia.* (2018). UN-Habitat.

United Nations. (2006). *Adaptation to Climate Change in the Context of Sustainable Development: A Workshop to Strengthen Research and Understanding.* United Nations Department of Economic and Social Affairs. New Delhi,: The United Nations.

United Nations. (2019). *The Sustainable Development Goals Report 2019.* UN. https://www.un-ilibrary.org/content/books/9789210478878

Urry, J. (2015). Climate Change and Society. In J. Michie & C. L. Cooper (Eds.), *Why the Social Sciences Matter* (pp. 45–59). Palgrave Macmillan. doi:10.1057/9781137269928_4

USDA. (2011). Food Security. *U.S.department of Agriculture,* 38.

Varanasi, A. (2022, September 21). How colonialism spawned and continues to exacerbate the climate crisis. *State of the Planet, 23.* https://news.climate.columbia.edu/2022/09/21/how-colonialism-spawned-and-continues-to-exacerbate-the-climate-crisis/

Vavtar, L. (2014). Environmental lobby effectiveness–the case of Lithuania and the United Kingdom. *Socialinių mokslų studijos, 6*(2), 313–330. https://repository.mruni.eu/handle/007/13317

Vision 2030 Jamaica Sector Plans. (2008). Vision 2030. https://www.vision2030.gov.jm/vision-2030-jamaica-sector-plans/

Vision 2030. (2015). *Vision 2030: The national development strategy of Trinidad and Tobago.* Ministry of Planning and Development. https://www.planning.gov.tt/content/vision-2030

Vysotska, O. Y., & Vysotskyi, O. Y. (2022). Green consumer culture as a factor of sustainable development of society. Journal of Geology. *Geography and Geoecology, 31*(1), 171–185. doi:10.15421/112217

Wang, y. (2020). *Multidisciplinary perspectives on media fandom.*GlobalIGI.

Westhues, M. (2014). *Climate change and environmental documentary film; An analysis of "An Inconvenient Truth" and "The 11 Th Hour."* GRIN publishing.

Weston, P. (2019, May 23). *Sea Levels Could Rise More Than Two Meters by 2100.* Independent Arabia. https://www.independentarabia.com/node/27226

Weston, P. (2019, May 23). *Sea Levels Could Rise More Than Two Meters by 2100*. Independent Arabia. https://www.independentarabia.com: https://www.independentarabia.com/node/27226/مستويات_سطح_البحر_يمكن_أن_ترتفع_أكثر_من_مترين_بحلول_2100

Weston, P. (n.d.). *Sea Levels Coud Rise more than Two meters by 2100.*

Whyte, K. (2017). Indigenous climate change studies: Indigenizing futures, decolonizing the Anthropocene. *English Language Notes, 55*(1–2), 153–162. doi:10.1215/00138282-55.1-2.153

Whyte, K. (2018). Settler colonialism, ecology, and environmental injustice. *Environment and Society, 9*(1), 125–144. https://www.jstor.org/stable/26879582. doi:10.3167/ares.2018.090109

Wickstrom, S. (2003). The politics of development in Indigenous Panama. *Latin American Perspectives, 30*(4), 43–68. doi:10.1177/0094582X03030004006

Willoquet- Maricondi, P. (2010). *Framing the world: Explorations in Ecocriticism and film.* University of Virginia press Charlottesville and London.

Windvsky, R. S., & Douglas, P. G. (2008). Leadership Skills in Modern Organization, New York.

Windvsky, R. S., & Douglas, P. G. (2008). Leadership Skills in Modern Organization. New York.

World Bank. (2016). Odisha -Poverty, growth and inequality. India state briefs. Washington, D.C.: World Bank Group. https://documents.worldbank.org/curated/en/484521468197097972/Odisha-Poverty-growth-and-inequality

World Commission on Environment and Development. (1987). *Our common future.* Oxford University Press.

World Day to Combat Desertification and Drought, 17 June/Background. (n.d.). The United Nations. https://www.un.org/ar/observances/desertification-day/background

World Day to Combat Desertification and Drought. (n.d.). United Nations. https://www.un.org/en/observances/desertification-day

www,milletsodisha.com

Xaxa, V., & Ramanathan, U. (2014). *Report of the high level committee on socioeconomic, health and educational status of tribal communities of India.* Ministry of Tribal Affairs, Government of India. https://cjp.org.in/wp-content/uploads/2019/10/2014-Xaxa-Tribal-Committee-Report.pdf

Youssef, A. A. (1982). *Climatic Characteristics of the Heat Element in Egypt during the Twentieth Century: a Study in Climatic Geography.* Faculty of Arts, Ain Shams University.

Yuchi, Z, Y. (et al), (2022). Building social resilience in North Korea can mitigate the impacts of climate change on food security. *Nature food, 3*(714 99- 511)

Zagkar, R. (2015). Algerian youth between the fragility of psychological formation and the challenges of citizenship. [Unpublished research, Ammar Thaliji University, Algeria].

Zahran, Z. B. (2007). *Rains on the North African Coast, a Study in Climatic Geography.* Department of Geography, Faculty of Human Studies, Al-Azhar University.

Zayed, A. (2006). *Social capital among the professional segments of the middle class, Center for Research and Social Studies, Faculty of Arts* (1st ed.). Cairo University.

Zehr, S. (2013). The sociology of global climate change. *Wiley Interdisciplinary Reviews: Climate Change, 6*(2), 129–150. doi:10.1002/wcc.328

Compilation of References

Московский стандарт реновации. (2020). *City Projects*. Moscow government. https://www.mos.ru/city/projects/renovation/

Compilation of References

Related References

To continue our tradition of advancing academic research, we have compiled a list of recommended IGI Global readings. These references will provide additional information and guidance to further enrich your knowledge and assist you with your own research and future publications.

Abbasnejad, B., Moeinzadeh, S., Ahankoob, A., & Wong, P. S. (2021). The Role of Collaboration in the Implementation of BIM-Enabled Projects. In J. Underwood & M. Shelbourn (Eds.), *Handbook of Research on Driving Transformational Change in the Digital Built Environment* (pp. 27–62). IGI Global. https://doi.org/10.4018/978-1-7998-6600-8.ch002

Abdulrahman, K. O., Mahamood, R. M., & Akinlabi, E. T. (2022). Additive Manufacturing (AM): Processing Technique for Lightweight Alloys and Composite Material. In K. Kumar, B. Babu, & J. Davim (Ed.), *Handbook of Research on Advancements in the Processing, Characterization, and Application of Lightweight Materials* (pp. 27-48). IGI Global. https://doi.org/10.4018/978-1-7998-7864-3.ch002

Agrawal, R., Sharma, P., & Saxena, A. (2021). A Diamond Cut Leather Substrate Antenna for BAN (Body Area Network) Application. In V. Singh, V. Dubey, A. Saxena, R. Tiwari, & H. Sharma (Eds.), *Emerging Materials and Advanced Designs for Wearable Antennas* (pp. 54–59). IGI Global. https://doi.org/10.4018/978-1-7998-7611-3.ch004

Ahmad, F., Al-Ammar, E. A., & Alsaidan, I. (2022). Battery Swapping Station: A Potential Solution to Address the Limitations of EV Charging Infrastructure. In M. Alam, R. Pillai, & N. Murugesan (Eds.), *Developing Charging Infrastructure and Technologies for Electric Vehicles* (pp. 195–207). IGI Global. doi:10.4018/978-1-7998-6858-3.ch010

Aikhuele, D. (2018). A Study of Product Development Engineering and Design Reliability Concerns. *International Journal of Applied Industrial Engineering*, 5(1), 79–89. doi:10.4018/IJAIE.2018010105

Al-Khatri, H., & Al-Atrash, F. (2021). Occupants' Habits and Natural Ventilation in a Hot Arid Climate. In R. González-Lezcano (Ed.), *Advancements in Sustainable Architecture and Energy Efficiency* (pp. 146–168). IGI Global. https://doi.org/10.4018/978-1-7998-7023-4.ch007

Al-Shebeeb, O. A., Rangaswamy, S., Gopalakrishan, B., & Devaru, D. G. (2017). Evaluation and Indexing of Process Plans Based on Electrical Demand and Energy Consumption. *International Journal of Manufacturing, Materials, and Mechanical Engineering, 7*(3), 1–19. doi:10.4018/IJMMME.2017070101

Amuda, M. O., Lawal, T. F., & Akinlabi, E. T. (2017). Research Progress on Rheological Behavior of AA7075 Aluminum Alloy During Hot Deformation. *International Journal of Materials Forming and Machining Processes, 4*(1), 53–96. doi:10.4018/IJMFMP.2017010104

Amuda, M. O., Lawal, T. F., & Mridha, S. (2021). Microstructure and Mechanical Properties of Silicon Carbide-Treated Ferritic Stainless Steel Welds. In L. Burstein (Ed.), *Handbook of Research on Advancements in Manufacturing, Materials, and Mechanical Engineering* (pp. 395–411). IGI Global. https://doi.org/10.4018/978-1-7998-4939-1.ch019

Anikeev, V., Gasem, K. A., & Fan, M. (2021). Application of Supercritical Technologies in Clean Energy Production: A Review. In L. Chen (Ed.), *Handbook of Research on Advancements in Supercritical Fluids Applications for Sustainable Energy Systems* (pp. 792–821). IGI Global. https://doi.org/10.4018/978-1-7998-5796-9.ch022

Arafat, M. Y., Saleem, I., & Devi, T. P. (2022). Drivers of EV Charging Infrastructure Entrepreneurship in India. In M. Alam, R. Pillai, & N. Murugesan (Eds.), *Developing Charging Infrastructure and Technologies for Electric Vehicles* (pp. 208–219). IGI Global. https://doi.org/10.4018/978-1-7998-6858-3.ch011

Araujo, A., & Manninen, H. (2022). Contribution of Project-Based Learning on Social Skills Development: An Industrial Engineer Perspective. In A. Alves & N. van Hattum-Janssen (Eds.), *Training Engineering Students for Modern Technological Advancement* (pp. 119–145). IGI Global. https://doi.org/10.4018/978-1-7998-8816-1.ch006

Armutlu, H. (2018). Intelligent Biomedical Engineering Operations by Cloud Computing Technologies. In U. Kose, G. Guraksin, & O. Deperlioglu (Eds.), *Nature-Inspired Intelligent Techniques for Solving Biomedical Engineering Problems* (pp. 297–317). Hershey, PA: IGI Global. doi:10.4018/978-1-5225-4769-3.ch015

Atik, M., Sadek, M., & Shahrour, I. (2017). Single-Run Adaptive Pushover Procedure for Shear Wall Structures. In V. Plevris, G. Kremmyda, & Y. Fahjan (Eds.), *Performance-Based Seismic Design of Concrete Structures and Infrastructures* (pp. 59–83). Hershey, PA: IGI Global. doi:10.4018/978-1-5225-2089-4.ch003

Attia, H. (2021). Smart Power Microgrid Impact on Sustainable Building. In R. González-Lezcano (Ed.), *Advancements in Sustainable Architecture and Energy Efficiency* (pp. 169–194). IGI Global. https://doi.org/10.4018/978-1-7998-7023-4.ch008

Aydin, A., Akyol, E., Gungor, M., Kaya, A., & Tasdelen, S. (2018). Geophysical Surveys in Engineering Geology Investigations With Field Examples. In N. Ceryan (Ed.), *Handbook of Research on Trends and Digital Advances in Engineering Geology* (pp. 257–280). Hershey, PA: IGI Global. doi:10.4018/978-1-5225-2709-1.ch007

Related References

Ayoobkhan, M. U. D., Y., A., J., Easwaran, B., & R., T. (2021). Smart Connected Digital Products and IoT Platform With the Digital Twin. In P. Vasant, G. Weber, & W. Punurai (Ed.), Research Advancements in Smart Technology, Optimization, and Renewable Energy (pp. 330-350). IGI Global. https:// doi.org/ doi:10.4018/978-1-7998-3970-5.ch016

Baeza Moyano, D., & González Lezcano, R. A. (2021). The Importance of Light in Our Lives: Towards New Lighting in Schools. In R. González-Lezcano (Ed.), *Advancements in Sustainable Architecture and Energy Efficiency* (pp. 239–256). IGI Global. https://doi.org/10.4018/978-1-7998-7023-4.ch011

Bagdadee, A. H. (2021). A Brief Assessment of the Energy Sector of Bangladesh. *International Journal of Energy Optimization and Engineering, 10*(1), 36–55. doi:10.4018/IJEOE.2021010103

Baklezos, A. T., & Hadjigeorgiou, N. G. (2021). Magnetic Sensors for Space Applications and Magnetic Cleanliness Considerations. In C. Nikolopoulos (Ed.), *Recent Trends on Electromagnetic Environmental Effects for Aeronautics and Space Applications* (pp. 147–185). IGI Global. https://doi.org/10.4018/978-1-7998-4879-0.ch006

Bas, T. G. (2017). Nutraceutical Industry with the Collaboration of Biotechnology and Nutrigenomics Engineering: The Significance of Intellectual Property in the Entrepreneurship and Scientific Research Ecosystems. In T. Bas & J. Zhao (Eds.), *Comparative Approaches to Biotechnology Development and Use in Developed and Emerging Nations* (pp. 1–17). Hershey, PA: IGI Global. doi:10.4018/978-1-5225-1040-6.ch001

Bazeer Ahamed, B., & Periakaruppan, S. (2021). Taxonomy of Influence Maximization Techniques in Unknown Social Networks. In P. Vasant, G. Weber, & W. Punurai (Eds.), *Research Advancements in Smart Technology, Optimization, and Renewable Energy* (pp. 351-363). IGI Global. https://doi.org/10.4018/978-1-7998-3970-5.ch017

Beale, R., & André, J. (2017). *Design Solutions and Innovations in Temporary Structures.* Hershey, PA: IGI Global. doi:10.4018/978-1-5225-2199-0

Behnam, B. (2017). Simulating Post-Earthquake Fire Loading in Conventional RC Structures. In P. Samui, S. Chakraborty, & D. Kim (Eds.), *Modeling and Simulation Techniques in Structural Engineering* (pp. 425–444). Hershey, PA: IGI Global. doi:10.4018/978-1-5225-0588-4.ch015

Ben Hamida, I., Salah, S. B., Msahli, F., & Mimouni, M. F. (2018). Distribution Network Reconfiguration Using SPEA2 for Power Loss Minimization and Reliability Improvement. *International Journal of Energy Optimization and Engineering, 7*(1), 50–65. doi:10.4018/IJEOE.2018010103

Bentarzi, H. (2021). Fault Tree-Based Root Cause Analysis Used to Study Mal-Operation of a Protective Relay in a Smart Grid. In A. Recioui & H. Bentarzi (Eds.), *Optimizing and Measuring Smart Grid Operation and Control* (pp. 289–308). IGI Global. https://doi.org/10.4018/978-1-7998-4027-5.ch012

Beysens, D. A., Garrabos, Y., & Zappoli, B. (2021). Thermal Effects in Near-Critical Fluids: Piston Effect and Related Phenomena. In L. Chen (Ed.), *Handbook of Research on Advancements in Supercritical Fluids Applications for Sustainable Energy Systems* (pp. 1–31). IGI Global. https://doi.org/10.4018/978-1-7998-5796-9.ch001

Bhaskar, S. V., & Kudal, H. N. (2017). Effect of TiCN and AlCrN Coating on Tribological Behaviour of Plasma-nitrided AISI 4140 Steel. *International Journal of Surface Engineering and Interdisciplinary Materials Science, 5*(2), 1–17. doi:10.4018/IJSEIMS.2017070101

Bhuyan, D. (2018). Designing of a Twin Tube Shock Absorber: A Study in Reverse Engineering. In K. Kumar & J. Davim (Eds.), *Design and Optimization of Mechanical Engineering Products* (pp. 83–104). Hershey, PA: IGI Global. doi:10.4018/978-1-5225-3401-3.ch005

Blumberg, G. (2021). Blockchains for Use in Construction and Engineering Projects. In J. Underwood & M. Shelbourn (Eds.), *Handbook of Research on Driving Transformational Change in the Digital Built Environment* (pp. 179–208). IGI Global. https://doi.org/10.4018/978-1-7998-6600-8.ch008

Bolboaca, A. M. (2021). Considerations Regarding the Use of Fuel Cells in Combined Heat and Power for Stationary Applications. In G. Badea, R. Felseghi, & I. Aşchilean (Eds.), *Hydrogen Fuel Cell Technology for Stationary Applications* (pp. 239–275). IGI Global. https://doi.org/10.4018/978-1-7998-4945-2.ch010

Burstein, L. (2021). Simulation Tool for Cable Design. In L. Burstein (Ed.), *Handbook of Research on Advancements in Manufacturing, Materials, and Mechanical Engineering* (pp. 54–74). IGI Global. https://doi.org/10.4018/978-1-7998-4939-1.ch003

Calderon, F. A., Giolo, E. G., Frau, C. D., Rengel, M. G., Rodriguez, H., Tornello, M., ... Gallucci, R. (2018). Seismic Microzonation and Site Effects Detection Through Microtremors Measures: A Review. In N. Ceryan (Ed.), *Handbook of Research on Trends and Digital Advances in Engineering Geology* (pp. 326–349). Hershey, PA: IGI Global. doi:10.4018/978-1-5225-2709-1.ch009

Ceryan, N., & Can, N. K. (2018). Prediction of The Uniaxial Compressive Strength of Rocks Materials. In N. Ceryan (Ed.), *Handbook of Research on Trends and Digital Advances in Engineering Geology* (pp. 31–96). Hershey, PA: IGI Global. doi:10.4018/978-1-5225-2709-1.ch002

Ceryan, S. (2018). Weathering Indices Used in Evaluation of the Weathering State of Rock Material. In N. Ceryan (Ed.), *Handbook of Research on Trends and Digital Advances in Engineering Geology* (pp. 132–186). Hershey, PA: IGI Global. doi:10.4018/978-1-5225-2709-1.ch004

Chen, H., Padilla, R. V., & Besarati, S. (2017). Supercritical Fluids and Their Applications in Power Generation. In L. Chen & Y. Iwamoto (Eds.), *Advanced Applications of Supercritical Fluids in Energy Systems* (pp. 369–402). Hershey, PA: IGI Global. doi:10.4018/978-1-5225-2047-4.ch012

Chen, H., Padilla, R. V., & Besarati, S. (2021). Supercritical Fluids and Their Applications in Power Generation. In L. Chen (Ed.), *Handbook of Research on Advancements in Supercritical Fluids Applications for Sustainable Energy Systems* (pp. 566–599). IGI Global. https://doi.org/10.4018/978-1-7998-5796-9.ch016

Chen, L. (2017). Principles, Experiments, and Numerical Studies of Supercritical Fluid Natural Circulation System. In L. Chen & Y. Iwamoto (Eds.), *Advanced Applications of Supercritical Fluids in Energy Systems* (pp. 136–187). Hershey, PA: IGI Global. doi:10.4018/978-1-5225-2047-4.ch005

Related References

Chen, L. (2021). Principles, Experiments, and Numerical Studies of Supercritical Fluid Natural Circulation System. In L. Chen (Ed.), *Handbook of Research on Advancements in Supercritical Fluids Applications for Sustainable Energy Systems* (pp. 219–269). IGI Global. https://doi.org/10.4018/978-1-7998-5796-9.ch007

Chiba, Y., Marif, Y., Henini, N., & Tlemcani, A. (2021). Modeling of Magnetic Refrigeration Device by Using Artificial Neural Networks Approach. *International Journal of Energy Optimization and Engineering*, *10*(4), 68–76. https://doi.org/10.4018/IJEOE.2021100105

Clementi, F., Di Sciascio, G., Di Sciascio, S., & Lenci, S. (2017). Influence of the Shear-Bending Interaction on the Global Capacity of Reinforced Concrete Frames: A Brief Overview of the New Perspectives. In V. Plevris, G. Kremmyda, & Y. Fahjan (Eds.), *Performance-Based Seismic Design of Concrete Structures and Infrastructures* (pp. 84–111). Hershey, PA: IGI Global. doi:10.4018/978-1-5225-2089-4.ch004

Codinhoto, R., Fialho, B. C., Pinti, L., & Fabricio, M. M. (2021). BIM and IoT for Facilities Management: Understanding Key Maintenance Issues. In J. Underwood & M. Shelbourn (Eds.), *Handbook of Research on Driving Transformational Change in the Digital Built Environment* (pp. 209–231). IGI Global. doi:10.4018/978-1-7998-6600-8.ch009

Cortés-Polo, D., Calle-Cancho, J., Carmona-Murillo, J., & González-Sánchez, J. (2017). Future Trends in Mobile-Fixed Integration for Next Generation Networks: Classification and Analysis. *International Journal of Vehicular Telematics and Infotainment Systems*, *1*(1), 33–53. doi:10.4018/IJVTIS.2017010103

Costa, H. G., Sheremetieff, F. H., & Araújo, E. A. (2022). Influence of Game-Based Methods in Developing Engineering Competences. In A. Alves & N. van Hattum-Janssen (Eds.), *Training Engineering Students for Modern Technological Advancement* (pp. 69–88). IGI Global. https://doi.org/10.4018/978-1-7998-8816-1.ch004

Cui, X., Zeng, S., Li, Z., Zheng, Q., Yu, X., & Han, B. (2018). Advanced Composites for Civil Engineering Infrastructures. In K. Kumar & J. Davim (Eds.), *Composites and Advanced Materials for Industrial Applications* (pp. 212–248). Hershey, PA: IGI Global. doi:10.4018/978-1-5225-5216-1.ch010

Dalgıç, S., & Kuşku, İ. (2018). Geological and Geotechnical Investigations in Tunneling. In N. Ceryan (Ed.), *Handbook of Research on Trends and Digital Advances in Engineering Geology* (pp. 482–529). Hershey, PA: IGI Global. doi:10.4018/978-1-5225-2709-1.ch014

Dang, C., & Hihara, E. (2021). Study on Cooling Heat Transfer of Supercritical Carbon Dioxide Applied to Transcritical Carbon Dioxide Heat Pump. In L. Chen (Ed.), *Handbook of Research on Advancements in Supercritical Fluids Applications for Sustainable Energy Systems* (pp. 451–493). IGI Global. https://doi.org/10.4018/978-1-7998-5796-9.ch013

Daus, Y., Kharchenko, V., & Yudaev, I. (2021). Research of Solar Energy Potential of Photovoltaic Installations on Enclosing Structures of Buildings. *International Journal of Energy Optimization and Engineering*, *10*(4), 18–34. https://doi.org/10.4018/IJEOE.2021100102

Daus, Y., Kharchenko, V., & Yudaev, I. (2021). Optimizing Layout of Distributed Generation Sources of Power Supply System of Agricultural Object. *International Journal of Energy Optimization and Engineering*, *10*(3), 70–84. https://doi.org/10.4018/IJEOE.2021070104

de la Varga, D., Soto, M., Arias, C. A., van Oirschot, D., Kilian, R., Pascual, A., & Álvarez, J. A. (2017). Constructed Wetlands for Industrial Wastewater Treatment and Removal of Nutrients. In Á. Val del Río, J. Campos Gómez, & A. Mosquera Corral (Eds.), *Technologies for the Treatment and Recovery of Nutrients from Industrial Wastewater* (pp. 202–230). Hershey, PA: IGI Global. doi:10.4018/978-1-5225-1037-6.ch008

Deb, S., Ammar, E. A., AlRajhi, H., Alsaidan, I., & Shariff, S. M. (2022). V2G Pilot Projects: Review and Lessons Learnt. In M. Alam, R. Pillai, & N. Murugesan (Eds.), *Developing Charging Infrastructure and Technologies for Electric Vehicles* (pp. 252–267). IGI Global. https://doi.org/10.4018/978-1-7998-6858-3.ch014

Dekhandji, F. Z., & Rais, M. C. (2021). A Comparative Study of Power Quality Monitoring Using Various Techniques. In A. Recioui & H. Bentarzi (Eds.), *Optimizing and Measuring Smart Grid Operation and Control* (pp. 259–288). IGI Global. https://doi.org/10.4018/978-1-7998-4027-5.ch011

Deperlioglu, O. (2018). Intelligent Techniques Inspired by Nature and Used in Biomedical Engineering. In U. Kose, G. Guraksin, & O. Deperlioglu (Eds.), *Nature-Inspired Intelligent Techniques for Solving Biomedical Engineering Problems* (pp. 51–77). Hershey, PA: IGI Global. doi:10.4018/978-1-5225-4769-3.ch003

Dhurpate, P. R., & Tang, H. (2021). Quantitative Analysis of the Impact of Inter-Line Conveyor Capacity for Throughput of Manufacturing Systems. *International Journal of Manufacturing, Materials, and Mechanical Engineering, 11*(1), 1–17. https://doi.org/10.4018/IJMMME.2021010101

Dinkar, S., & Deep, K. (2021). A Survey of Recent Variants and Applications of Antlion Optimizer. *International Journal of Energy Optimization and Engineering, 10*(2), 48–73. doi:10.4018/IJEOE.2021040103

Dixit, A. (2018). Application of Silica-Gel-Reinforced Aluminium Composite on the Piston of Internal Combustion Engine: Comparative Study of Silica-Gel-Reinforced Aluminium Composite Piston With Aluminium Alloy Piston. In K. Kumar & J. Davim (Eds.), *Composites and Advanced Materials for Industrial Applications* (pp. 63–98). Hershey, PA: IGI Global. doi:10.4018/978-1-5225-5216-1.ch004

Drabecki, M. P., & Kułak, K. B. (2021). Global Pandemics on European Electrical Energy Markets: Lessons Learned From the COVID-19 Outbreak. *International Journal of Energy Optimization and Engineering, 10*(3), 24–46. https://doi.org/10.4018/IJEOE.2021070102

Dutta, M. M. (2021). Nanomaterials for Food and Agriculture. In M. Bhat, I. Wani, & S. Ashraf (Eds.), *Applications of Nanomaterials in Agriculture, Food Science, and Medicine* (pp. 75–97). IGI Global. doi:10.4018/978-1-7998-5563-7.ch004

Dutta, M. M., & Goswami, M. (2021). Coating Materials: Nano-Materials. In S. Roy & G. Bose (Eds.), *Advanced Surface Coating Techniques for Modern Industrial Applications* (pp. 1–30). IGI Global. doi:10.4018/978-1-7998-4870-7.ch001

Elsayed, A. M., Dakkama, H. J., Mahmoud, S., Al-Dadah, R., & Kaialy, W. (2017). Sustainable Cooling Research Using Activated Carbon Adsorbents and Their Environmental Impact. In T. Kobayashi (Ed.), *Applied Environmental Materials Science for Sustainability* (pp. 186–221). Hershey, PA: IGI Global. doi:10.4018/978-1-5225-1971-3.ch009

Related References

Ercanoglu, M., & Sonmez, H. (2018). General Trends and New Perspectives on Landslide Mapping and Assessment Methods. In N. Ceryan (Ed.), *Handbook of Research on Trends and Digital Advances in Engineering Geology* (pp. 350–379). Hershey, PA: IGI Global. doi:10.4018/978-1-5225-2709-1.ch010

Faroz, S. A., Pujari, N. N., Rastogi, R., & Ghosh, S. (2017). Risk Analysis of Structural Engineering Systems Using Bayesian Inference. In P. Samui, S. Chakraborty, & D. Kim (Eds.), *Modeling and Simulation Techniques in Structural Engineering* (pp. 390–424). Hershey, PA: IGI Global. doi:10.4018/978-1-5225-0588-4.ch014

Fekik, A., Hamida, M. L., Denoun, H., Azar, A. T., Kamal, N. A., Vaidyanathan, S., Bousbaine, A., & Benamrouche, N. (2022). Multilevel Inverter for Hybrid Fuel Cell/PV Energy Conversion System. In A. Fekik & N. Benamrouche (Eds.), *Modeling and Control of Static Converters for Hybrid Storage Systems* (pp. 233–270). IGI Global. https://doi.org/10.4018/978-1-7998-7447-8.ch009

Fekik, A., Hamida, M. L., Houassine, H., Azar, A. T., Kamal, N. A., Denoun, H., Vaidyanathan, S., & Sambas, A. (2022). Power Quality Improvement for Grid-Connected Photovoltaic Panels Using Direct Power Control. In A. Fekik & N. Benamrouche (Eds.), *Modeling and Control of Static Converters for Hybrid Storage Systems* (pp. 107–142). IGI Global. https://doi.org/10.4018/978-1-7998-7447-8.ch005

Fernando, P. R., Hamigah, T., Disne, S., Wickramasingha, G. G., & Sutharshan, A. (2018). The Evaluation of Engineering Properties of Low Cost Concrete Blocks by Partial Doping of Sand with Sawdust: Low Cost Sawdust Concrete Block. *International Journal of Strategic Engineering*, 1(2), 26–42. doi:10.4018/IJoSE.2018070103

Ferro, G., Minciardi, R., Parodi, L., & Robba, M. (2022). Optimal Charging Management of Microgrid-Integrated Electric Vehicles. In M. Alam, R. Pillai, & N. Murugesan (Eds.), *Developing Charging Infrastructure and Technologies for Electric Vehicles* (pp. 133–155). IGI Global. https://doi.org/10.4018/978-1-7998-6858-3.ch007

Flumerfelt, S., & Green, C. (2022). Graduate Lean Leadership Education: A Case Study of a Program. In A. Alves & N. van Hattum-Janssen (Eds.), *Training Engineering Students for Modern Technological Advancement* (pp. 202–224). IGI Global. https://doi.org/10.4018/978-1-7998-8816-1.ch010

Galli, B. J. (2021). Implications of Economic Decision Making to the Project Manager. *International Journal of Strategic Engineering*, 4(1), 19–32. https://doi.org/10.4018/IJoSE.2021010102

Gento, A. M., Pimentel, C., & Pascual, J. A. (2022). Teaching Circular Economy and Lean Management in a Learning Factory. In A. Alves & N. van Hattum-Janssen (Eds.), *Training Engineering Students for Modern Technological Advancement* (pp. 183–201). IGI Global. https://doi.org/10.4018/978-1-7998-8816-1.ch009

Ghosh, S., Mitra, S., Ghosh, S., & Chakraborty, S. (2017). Seismic Reliability Analysis in the Framework of Metamodelling Based Monte Carlo Simulation. In P. Samui, S. Chakraborty, & D. Kim (Eds.), *Modeling and Simulation Techniques in Structural Engineering* (pp. 192–208). Hershey, PA: IGI Global. doi:10.4018/978-1-5225-0588-4.ch006

Gil, M., & Otero, B. (2017). Learning Engineering Skills through Creativity and Collaboration: A Game-Based Proposal. In R. Alexandre Peixoto de Queirós & M. Pinto (Eds.), *Gamification-Based E-Learning Strategies for Computer Programming Education* (pp. 14–29). Hershey, PA: IGI Global. doi:10.4018/978-1-5225-1034-5.ch002

Gill, J., Ayre, M., & Mills, J. (2017). Revisioning the Engineering Profession: How to Make It Happen! In M. Gray & K. Thomas (Eds.), *Strategies for Increasing Diversity in Engineering Majors and Careers* (pp. 156–175). Hershey, PA: IGI Global. doi:10.4018/978-1-5225-2212-6.ch008

Godzhaev, Z., Senkevich, S., Kuzmin, V., & Melikov, I. (2021). Use of the Neural Network Controller of Sprung Mass to Reduce Vibrations From Road Irregularities. In P. Vasant, G. Weber, & W. Punurai (Ed.), *Research Advancements in Smart Technology, Optimization, and Renewable Energy* (pp. 69-87). IGI Global. https://doi.org/10.4018/978-1-7998-3970-5.ch005

Gomes de Gusmão, C. M. (2022). Digital Competencies and Transformation in Higher Education: Upskilling With Extension Actions. In A. Alves & N. van Hattum-Janssen (Eds.), *Training Engineering Students for Modern Technological Advancement* (pp. 313–328). IGI Global. https://doi.org/10.4018/978-1-7998-8816-1.ch015A

Goyal, N., Ram, M., & Kumar, P. (2017). Welding Process under Fault Coverage Approach for Reliability and MTTF. In M. Ram & J. Davim (Eds.), *Mathematical Concepts and Applications in Mechanical Engineering and Mechatronics* (pp. 222–245). Hershey, PA: IGI Global. doi:10.4018/978-1-5225-1639-2.ch011

Gray, M., & Lundy, C. (2017). Engineering Study Abroad: High Impact Strategy for Increasing Access. In M. Gray & K. Thomas (Eds.), *Strategies for Increasing Diversity in Engineering Majors and Careers* (pp. 42–59). Hershey, PA: IGI Global. doi:10.4018/978-1-5225-2212-6.ch003

Güler, O., & Varol, T. (2021). Fabrication of Functionally Graded Metal and Ceramic Powders Synthesized by Electroless Deposition. In S. Roy & G. Bose (Eds.), *Advanced Surface Coating Techniques for Modern Industrial Applications* (pp. 150–187). IGI Global. https://doi.org/10.4018/978-1-7998-4870-7.ch007

Guraksin, G. E. (2018). Internet of Things and Nature-Inspired Intelligent Techniques for the Future of Biomedical Engineering. In U. Kose, G. Guraksin, & O. Deperlioglu (Eds.), *Nature-Inspired Intelligent Techniques for Solving Biomedical Engineering Problems* (pp. 263–282). Hershey, PA: IGI Global. doi:10.4018/978-1-5225-4769-3.ch013

Hamida, M. L., Fekik, A., Denoun, H., Ardjal, A., & Bokhtache, A. A. (2022). Flying Capacitor Inverter Integration in a Renewable Energy System. In A. Fekik & N. Benamrouche (Eds.), *Modeling and Control of Static Converters for Hybrid Storage Systems* (pp. 287–306). IGI Global. https://doi.org/10.4018/978-1-7998-7447-8.ch011

Hasegawa, N., & Takahashi, Y. (2021). Control of Soap Bubble Ejection Robot Using Facial Expressions. *International Journal of Manufacturing, Materials, and Mechanical Engineering, 11*(2), 1–16. https://doi.org/10.4018/IJMMME.2021040101

Related References

Hejazi, T., & Akbari, L. (2017). A Multiresponse Optimization Model for Statistical Design of Processes with Discrete Variables. In M. Ram & J. Davim (Eds.), *Mathematical Concepts and Applications in Mechanical Engineering and Mechatronics* (pp. 17–37). Hershey, PA: IGI Global. doi:10.4018/978-1-5225-1639-2.ch002

Hejazi, T., & Hejazi, A. (2017). Monte Carlo Simulation for Reliability-Based Design of Automotive Complex Subsystems. In M. Ram & J. Davim (Eds.), *Mathematical Concepts and Applications in Mechanical Engineering and Mechatronics* (pp. 177–200). Hershey, PA: IGI Global. doi:10.4018/978-1-5225-1639-2.ch009

Hejazi, T., & Poursabbagh, H. (2017). Reliability Analysis of Engineering Systems: An Accelerated Life Testing for Boiler Tubes. In M. Ram & J. Davim (Eds.), *Mathematical Concepts and Applications in Mechanical Engineering and Mechatronics* (pp. 154–176). Hershey, PA: IGI Global. doi:10.4018/978-1-5225-1639-2.ch008

Henao, J., Poblano-Salas, C. A., Vargas, F., Giraldo-Betancur, A. L., Corona-Castuera, J., & Sotelo-Mazón, O. (2021). Principles and Applications of Thermal Spray Coatings. In S. Roy & G. Bose (Eds.), *Advanced Surface Coating Techniques for Modern Industrial Applications* (pp. 31–70). IGI Global. https://doi.org/10.4018/978-1-7998-4870-7.ch002

Henao, J., & Sotelo, O. (2018). Surface Engineering at High Temperature: Thermal Cycling and Corrosion Resistance. In A. Pakseresht (Ed.), *Production, Properties, and Applications of High Temperature Coatings* (pp. 131–159). Hershey, PA: IGI Global. doi:10.4018/978-1-5225-4194-3.ch006

Hrnčič, M. K., Cör, D., & Knez, Ž. (2021). Supercritical Fluids as a Tool for Green Energy and Chemicals. In L. Chen (Ed.), *Handbook of Research on Advancements in Supercritical Fluids Applications for Sustainable Energy Systems* (pp. 761–791). IGI Global. doi:10.4018/978-1-7998-5796-9.ch021

Ibrahim, O., Erdem, S., & Gurbuz, E. (2021). Studying Physical and Chemical Properties of Graphene Oxide and Reduced Graphene Oxide and Their Applications in Sustainable Building Materials. In R. González-Lezcano (Ed.), *Advancements in Sustainable Architecture and Energy Efficiency* (pp. 221–238). IGI Global. https://doi.org/10.4018/978-1-7998-7023-4.ch010

Ihianle, I. K., Islam, S., Naeem, U., & Ebenuwa, S. H. (2021). Exploiting Patterns of Object Use for Human Activity Recognition. In A. Nwajana & I. Ihianle (Eds.), *Handbook of Research on 5G Networks and Advancements in Computing, Electronics, and Electrical Engineering* (pp. 382–401). IGI Global. https://doi.org/10.4018/978-1-7998-6992-4.ch015

Ijemaru, G. K., Ngharamike, E. T., Oleka, E. U., & Nwajana, A. O. (2021). An Energy-Efficient Model for Opportunistic Data Collection in IoV-Enabled SC Waste Management. In A. Nwajana & I. Ihianle (Eds.), *Handbook of Research on 5G Networks and Advancements in Computing, Electronics, and Electrical Engineering* (pp. 1–19). IGI Global. https://doi.org/10.4018/978-1-7998-6992-4.ch001

Ilori, O. O., Adetan, D. A., & Umoru, L. E. (2017). Effect of Cutting Parameters on the Surface Residual Stress of Face-Milled Pearlitic Ductile Iron. *International Journal of Materials Forming and Machining Processes*, 4(1), 38–52. doi:10.4018/IJMFMP.2017010103

Imam, M. H., Tasadduq, I. A., Ahmad, A., Aldosari, F., & Khan, H. (2017). Automated Generation of Course Improvement Plans Using Expert System. *International Journal of Quality Assurance in Engineering and Technology Education, 6*(1), 1–12. doi:10.4018/IJQAETE.2017010101

Injeti, S. K., & Kumar, T. V. (2018). A WDO Framework for Optimal Deployment of DGs and DSCs in a Radial Distribution System Under Daily Load Pattern to Improve Techno-Economic Benefits. *International Journal of Energy Optimization and Engineering, 7*(2), 1–38. doi:10.4018/IJEOE.2018040101

Ishii, N., Anami, K., & Knisely, C. W. (2018). *Dynamic Stability of Hydraulic Gates and Engineering for Flood Prevention.* Hershey, PA: IGI Global. doi:10.4018/978-1-5225-3079-4

Iwamoto, Y., & Yamaguchi, H. (2021). Application of Supercritical Carbon Dioxide for Solar Water Heater. In L. Chen (Ed.), *Handbook of Research on Advancements in Supercritical Fluids Applications for Sustainable Energy Systems* (pp. 370–387). IGI Global. https://doi.org/10.4018/978-1-7998-5796-9.ch010

Jayapalan, S. (2018). A Review of Chemical Treatments on Natural Fibers-Based Hybrid Composites for Engineering Applications. In K. Kumar & J. Davim (Eds.), *Composites and Advanced Materials for Industrial Applications* (pp. 16–37). Hershey, PA: IGI Global. doi:10.4018/978-1-5225-5216-1.ch002

Kapetanakis, T. N., Vardiambasis, I. O., Ioannidou, M. P., & Konstantaras, A. I. (2021). Modeling Antenna Radiation Using Artificial Intelligence Techniques: The Case of a Circular Loop Antenna. In C. Nikolopoulos (Ed.), *Recent Trends on Electromagnetic Environmental Effects for Aeronautics and Space Applications* (pp. 186–225). IGI Global. https://doi.org/10.4018/978-1-7998-4879-0.ch007

Karkalos, N. E., Markopoulos, A. P., & Dossis, M. F. (2017). Optimal Model Parameters of Inverse Kinematics Solution of a 3R Robotic Manipulator Using ANN Models. *International Journal of Manufacturing, Materials, and Mechanical Engineering, 7*(3), 20–40. doi:10.4018/IJMMME.2017070102

Kelly, M., Costello, M., Nicholson, G., & O'Connor, J. (2021). The Evolving Integration of BIM Into Built Environment Programmes in a Higher Education Institute. In J. Underwood & M. Shelbourn (Eds.), *Handbook of Research on Driving Transformational Change in the Digital Built Environment* (pp. 294–326). IGI Global. https://doi.org/10.4018/978-1-7998-6600-8.ch012

Kesimal, A., Karaman, K., Cihangir, F., & Ercikdi, B. (2018). Excavatability Assessment of Rock Masses for Geotechnical Studies. In N. Ceryan (Ed.), *Handbook of Research on Trends and Digital Advances in Engineering Geology* (pp. 231–256). Hershey, PA: IGI Global. doi:10.4018/978-1-5225-2709-1.ch006

Knoflacher, H. (2017). The Role of Engineers and Their Tools in the Transport Sector after Paradigm Change: From Assumptions and Extrapolations to Science. In H. Knoflacher & E. Ocalir-Akunal (Eds.), *Engineering Tools and Solutions for Sustainable Transportation Planning* (pp. 1–29). Hershey, PA: IGI Global. doi:10.4018/978-1-5225-2116-7.ch001

Kose, U. (2018). Towards an Intelligent Biomedical Engineering With Nature-Inspired Artificial Intelligence Techniques. In U. Kose, G. Guraksin, & O. Deperlioglu (Eds.), *Nature-Inspired Intelligent Techniques for Solving Biomedical Engineering Problems* (pp. 1–26). Hershey, PA: IGI Global. doi:10.4018/978-1-5225-4769-3.ch001

Related References

Kostić, S. (2018). A Review on Enhanced Stability Analyses of Soil Slopes Using Statistical Design. In N. Ceryan (Ed.), *Handbook of Research on Trends and Digital Advances in Engineering Geology* (pp. 446–481). Hershey, PA: IGI Global. doi:10.4018/978-1-5225-2709-1.ch013

Kumar, A., Patil, P. P., & Prajapati, Y. K. (2018). *Advanced Numerical Simulations in Mechanical Engineering*. Hershey, PA: IGI Global. doi:10.4018/978-1-5225-3722-9

Kumar, G. R., Rajyalakshmi, G., & Manupati, V. K. (2017). Surface Micro Patterning of Aluminium Reinforced Composite through Laser Peening. *International Journal of Manufacturing, Materials, and Mechanical Engineering, 7*(4), 15–27. doi:10.4018/IJMMME.2017100102

Kumar, N., Basu, D. N., & Chen, L. (2021). Effect of Flow Acceleration and Buoyancy on Thermalhydraulics of sCO2 in Mini/Micro-Channel. In L. Chen (Ed.), *Handbook of Research on Advancements in Supercritical Fluids Applications for Sustainable Energy Systems* (pp. 161–182). IGI Global. doi:10.4018/978-1-7998-5796-9.ch005

Kumari, N., & Kumar, K. (2018). Fabrication of Orthotic Calipers With Epoxy-Based Green Composite. In K. Kumar & J. Davim (Eds.), *Composites and Advanced Materials for Industrial Applications* (pp. 157–176). Hershey, PA: IGI Global. doi:10.4018/978-1-5225-5216-1.ch008

Kuppusamy, R. R. (2018). Development of Aerospace Composite Structures Through Vacuum-Enhanced Resin Transfer Moulding Technology (VERTMTy): Vacuum-Enhanced Resin Transfer Moulding. In K. Kumar & J. Davim (Eds.), *Composites and Advanced Materials for Industrial Applications* (pp. 99–111). Hershey, PA: IGI Global. doi:10.4018/978-1-5225-5216-1.ch005

Kurganov, V. A., Zeigarnik, Y. A., & Maslakova, I. V. (2021). Normal and Deteriorated Heat Transfer Under Heating Turbulent Supercritical Pressure Coolants Flows in Round Tubes. In L. Chen (Ed.), *Handbook of Research on Advancements in Supercritical Fluids Applications for Sustainable Energy Systems* (pp. 494–532). IGI Global. https://doi.org/10.4018/978-1-7998-5796-9.ch014

Li, H., & Zhang, Y. (2021). Heat Transfer and Fluid Flow Modeling for Supercritical Fluids in Advanced Energy Systems. In L. Chen (Ed.), *Handbook of Research on Advancements in Supercritical Fluids Applications for Sustainable Energy Systems* (pp. 388–422). IGI Global. https://doi.org/10.4018/978-1-7998-5796-9.ch011

Loy, J., Howell, S., & Cooper, R. (2017). Engineering Teams: Supporting Diversity in Engineering Education. In M. Gray & K. Thomas (Eds.), *Strategies for Increasing Diversity in Engineering Majors and Careers* (pp. 106–129). Hershey, PA: IGI Global. doi:10.4018/978-1-5225-2212-6.ch006

Macher, G., Armengaud, E., Kreiner, C., Brenner, E., Schmittner, C., Ma, Z., ... Krammer, M. (2018). Integration of Security in the Development Lifecycle of Dependable Automotive CPS. In N. Druml, A. Genser, A. Krieg, M. Menghin, & A. Hoeller (Eds.), *Solutions for Cyber-Physical Systems Ubiquity* (pp. 383–423). Hershey, PA: IGI Global. doi:10.4018/978-1-5225-2845-6.ch015

Madhu, M. N., Singh, J. G., Mohan, V., & Ongsakul, W. (2021). Transmission Risk Optimization in Interconnected Systems: Risk-Adjusted Available Transfer Capability. In P. Vasant, G. Weber, & W. Punurai (Ed.), *Research Advancements in Smart Technology, Optimization, and Renewable Energy* (pp. 183-199). IGI Global. https://doi.org/10.4018/978-1-7998-3970-5.ch010

Mahendramani, G., & Lakshmana Swamy, N. (2018). Effect of Weld Groove Area on Distortion of Butt Welded Joints in Submerged Arc Welding. *International Journal of Manufacturing, Materials, and Mechanical Engineering, 8*(2), 33–44. doi:10.4018/IJMMME.2018040103

Makropoulos, G., Koumaras, H., Setaki, F., Filis, K., Lutz, T., Montowtt, P., Tomaszewski, L., Dybiec, P., & Järvet, T. (2021). 5G and Unmanned Aerial Vehicles (UAVs) Use Cases: Analysis of the Ecosystem, Architecture, and Applications. In A. Nwajana & I. Ihianle (Eds.), *Handbook of Research on 5G Networks and Advancements in Computing, Electronics, and Electrical Engineering* (pp. 36–69). IGI Global. https://doi.org/10.4018/978-1-7998-6992-4.ch003

Meric, E. M., Erdem, S., & Gurbuz, E. (2021). Application of Phase Change Materials in Construction Materials for Thermal Energy Storage Systems in Buildings. In R. González-Lezcano (Ed.), *Advancements in Sustainable Architecture and Energy Efficiency* (pp. 1–20). IGI Global. https://doi.org/10.4018/978-1-7998-7023-4.ch001

Mihret, E. T., & Yitayih, K. A. (2021). Operation of VANET Communications: The Convergence of UAV System With LTE/4G and WAVE Technologies. *International Journal of Smart Vehicles and Smart Transportation, 4*(1), 29–51. https://doi.org/10.4018/IJSVST.2021010103

Mir, M. A., Bhat, B. A., Sheikh, B. A., Rather, G. A., Mehraj, S., & Mir, W. R. (2021). Nanomedicine in Human Health Therapeutics and Drug Delivery: Nanobiotechnology and Nanobiomedicine. In M. Bhat, I. Wani, & S. Ashraf (Eds.), *Applications of Nanomaterials in Agriculture, Food Science, and Medicine* (pp. 229–251). IGI Global. doi:10.4018/978-1-7998-5563-7.ch013

Mohammadzadeh, S., & Kim, Y. (2017). Nonlinear System Identification of Smart Buildings. In P. Samui, S. Chakraborty, & D. Kim (Eds.), *Modeling and Simulation Techniques in Structural Engineering* (pp. 328–347). Hershey, PA: IGI Global. doi:10.4018/978-1-5225-0588-4.ch011

Molina, G. J., Aktaruzzaman, F., Soloiu, V., & Rahman, M. (2017). Design and Testing of a Jet-Impingement Instrument to Study Surface-Modification Effects by Nanofluids. *International Journal of Surface Engineering and Interdisciplinary Materials Science, 5*(2), 43–61. doi:10.4018/IJSEIMS.2017070104

Moreno-Rangel, A., & Carrillo, G. (2021). Energy-Efficient Homes: A Heaven for Respiratory Illnesses. In R. González-Lezcano (Ed.), *Advancements in Sustainable Architecture and Energy Efficiency* (pp. 49–71). IGI Global. https://doi.org/10.4018/978-1-7998-7023-4.ch003

Msomi, V., & Jantjies, B. T. (2021). Correlative Analysis Between Tensile Properties and Tool Rotational Speeds of Friction Stir Welded Similar Aluminium Alloy Joints. *International Journal of Surface Engineering and Interdisciplinary Materials Science, 9*(2), 58–78. https://doi.org/10.4018/IJSEIMS.2021070104

Muigai, M. N., Mwema, F. M., Akinlabi, E. T., & Obiko, J. O. (2021). Surface Engineering of Materials Through Weld-Based Technologies: An Overview. In S. Roy & G. Bose (Eds.), *Advanced Surface Coating Techniques for Modern Industrial Applications* (pp. 247–260). IGI Global. doi:10.4018/978-1-7998-4870-7.ch011

Related References

Mukherjee, A., Saeed, R. A., Dutta, S., & Naskar, M. K. (2017). Fault Tracking Framework for Software-Defined Networking (SDN). In C. Singhal & S. De (Eds.), *Resource Allocation in Next-Generation Broadband Wireless Access Networks* (pp. 247–272). Hershey, PA: IGI Global. doi:10.4018/978-1-5225-2023-8.ch011

Mukhopadhyay, A., Barman, T. K., & Sahoo, P. (2018). Electroless Nickel Coatings for High Temperature Applications. In K. Kumar & J. Davim (Eds.), *Composites and Advanced Materials for Industrial Applications* (pp. 297–331). Hershey, PA: IGI Global. doi:10.4018/978-1-5225-5216-1.ch013

Mwema, F. M., & Wambua, J. M. (2022). Machining of Poly Methyl Methacrylate (PMMA) and Other Olymeric Materials: A Review. In K. Kumar, B. Babu, & J. Davim (Eds.), *Handbook of Research on Advancements in the Processing, Characterization, and Application of Lightweight Materials* (pp. 363–379). IGI Global. https://doi.org/10.4018/978-1-7998-7864-3.ch016

Mykhailyshyn, R., Savkiv, V., Boyko, I., Prada, E., & Virgala, I. (2021). Substantiation of Parameters of Friction Elements of Bernoulli Grippers With a Cylindrical Nozzle. *International Journal of Manufacturing, Materials, and Mechanical Engineering, 11*(2), 17–39. https://doi.org/10.4018/IJMMME.2021040102

Náprstek, J., & Fischer, C. (2017). Dynamic Stability and Post-Critical Processes of Slender Auto-Parametric Systems. In V. Plevris, G. Kremmyda, & Y. Fahjan (Eds.), *Performance-Based Seismic Design of Concrete Structures and Infrastructures* (pp. 128–171). Hershey, PA: IGI Global. doi:10.4018/978-1-5225-2089-4.ch006

Nautiyal, L., Shivach, P., & Ram, M. (2018). Optimal Designs by Means of Genetic Algorithms. In M. Ram & J. Davim (Eds.), *Soft Computing Techniques and Applications in Mechanical Engineering* (pp. 151–161). Hershey, PA: IGI Global. doi:10.4018/978-1-5225-3035-0.ch007

Nazir, R. (2017). Advanced Nanomaterials for Water Engineering and Treatment: Nano-Metal Oxides and Their Nanocomposites. In T. Saleh (Ed.), *Advanced Nanomaterials for Water Engineering, Treatment, and Hydraulics* (pp. 84–126). Hershey, PA: IGI Global. doi:10.4018/978-1-5225-2136-5.ch005

Nikolopoulos, C. D. (2021). Recent Advances on Measuring and Modeling ELF-Radiated Emissions for Space Applications. In C. Nikolopoulos (Ed.), *Recent Trends on Electromagnetic Environmental Effects for Aeronautics and Space Applications* (pp. 1–38). IGI Global. https://doi.org/10.4018/978-1-7998-4879-0.ch001

Nogueira, A. F., Ribeiro, J. C., Fernández de Vega, F., & Zenha-Rela, M. A. (2018). Evolutionary Approaches to Test Data Generation for Object-Oriented Software: Overview of Techniques and Tools. In M. Khosrow-Pour, D.B.A. (Ed.), Incorporating Nature-Inspired Paradigms in Computational Applications (pp. 162-194). Hershey, PA: IGI Global. https://doi.org/ doi:10.4018/978-1-5225-5020-4.ch006

Nwajana, A. O., Obi, E. R., Ijemaru, G. K., Oleka, E. U., & Anthony, D. C. (2021). Fundamentals of RF/Microwave Bandpass Filter Design. In A. Nwajana & I. Ihianle (Eds.), *Handbook of Research on 5G Networks and Advancements in Computing, Electronics, and Electrical Engineering* (pp. 149–164). IGI Global. https://doi.org/10.4018/978-1-7998-6992-4.ch005

Ogbodo, E. A. (2021). Comparative Study of Transmission Line Junction vs. Asynchronously Coupled Junction Diplexers. In A. Nwajana & I. Ihianle (Eds.), *Handbook of Research on 5G Networks and Advancements in Computing, Electronics, and Electrical Engineering* (pp. 326–336). IGI Global. https://doi.org/10.4018/978-1-7998-6992-4.ch013

Orosa, J. A., Vergara, D., Fraguela, F., & Masdías-Bonome, A. (2021). Statistical Understanding and Optimization of Building Energy Consumption and Climate Change Consequences. In R. González-Lezcano (Ed.), *Advancements in Sustainable Architecture and Energy Efficiency* (pp. 195–220). IGI Global. https://doi.org/10.4018/978-1-7998-7023-4.ch009

Osho, M. B. (2018). Industrial Enzyme Technology: Potential Applications. In S. Bharati & P. Chaurasia (Eds.), *Research Advancements in Pharmaceutical, Nutritional, and Industrial Enzymology* (pp. 375–394). Hershey, PA: IGI Global. doi:10.4018/978-1-5225-5237-6.ch017

Ouadi, A., & Zitouni, A. (2021). Phasor Measurement Improvement Using Digital Filter in a Smart Grid. In A. Recioui & H. Bentarzi (Eds.), *Optimizing and Measuring Smart Grid Operation and Control* (pp. 100–117). IGI Global. https://doi.org/10.4018/978-1-7998-4027-5.ch005

Padmaja, P., & Marutheswar, G. (2017). Certain Investigation on Secured Data Transmission in Wireless Sensor Networks. *International Journal of Mobile Computing and Multimedia Communications*, 8(1), 48–61. doi:10.4018/IJMCMC.2017010104

Palmer, S., & Hall, W. (2017). An Evaluation of Group Work in First-Year Engineering Design Education. In R. Tucker (Ed.), *Collaboration and Student Engagement in Design Education* (pp. 145–168). Hershey, PA: IGI Global. doi:10.4018/978-1-5225-0726-0.ch007

Panchenko, V. (2021). Prospects for Energy Supply of the Arctic Zone Objects of Russia Using Frost-Resistant Solar Modules. In P. Vasant, G. Weber, & W. Punurai (Eds.), *Research Advancements in Smart Technology, Optimization, and Renewable Energy* (pp. 149-169). IGI Global. https://doi.org/10.4018/978-1-7998-3970-5.ch008

Panchenko, V. (2021). Photovoltaic Thermal Module With Paraboloid Type Solar Concentrators. *International Journal of Energy Optimization and Engineering*, 10(2), 1–23. https://doi.org/10.4018/IJEOE.2021040101

Pandey, K., & Datta, S. (2021). Dry Machining of Inconel 825 Superalloys: Performance of Tool Inserts (Carbide, Cermet, and SiAlON). *International Journal of Manufacturing, Materials, and Mechanical Engineering*, 11(4), 26–39. doi:10.4018/IJMMME.2021100102

Panneer, R. (2017). Effect of Composition of Fibers on Properties of Hybrid Composites. *International Journal of Manufacturing, Materials, and Mechanical Engineering*, 7(4), 28–43. doi:10.4018/IJMMME.2017100103

Pany, C. (2021). Estimation of Correct Long-Seam Mismatch Using FEA to Compare the Measured Strain in a Non-Destructive Testing of a Pressurant Tank: A Reverse Problem. *International Journal of Smart Vehicles and Smart Transportation*, 4(1), 16–28. doi:10.4018/IJSVST.2021010102

Related References

Paul, S., & Roy, P. (2018). Optimal Design of Power System Stabilizer Using a Novel Evolutionary Algorithm. *International Journal of Energy Optimization and Engineering, 7*(3), 24–46. doi:10.4018/IJEOE.2018070102

Paul, S., & Roy, P. K. (2021). Oppositional Differential Search Algorithm for the Optimal Tuning of Both Single Input and Dual Input Power System Stabilizer. In P. Vasant, G. Weber, & W. Punurai (Eds.), *Research Advancements in Smart Technology, Optimization, and Renewable Energy* (pp. 256-282). IGI Global. https://doi.org/10.4018/978-1-7998-3970-5.ch013

Pavaloiu, A. (2018). Artificial Intelligence Ethics in Biomedical-Engineering-Oriented Problems. In U. Kose, G. Guraksin, & O. Deperlioglu (Eds.), *Nature-Inspired Intelligent Techniques for Solving Biomedical Engineering Problems* (pp. 219–231). Hershey, PA: IGI Global. doi:10.4018/978-1-5225-4769-3.ch010

Pioro, I., Mahdi, M., & Popov, R. (2017). Application of Supercritical Pressures in Power Engineering. In L. Chen & Y. Iwamoto (Eds.), *Advanced Applications of Supercritical Fluids in Energy Systems* (pp. 404–457). Hershey, PA: IGI Global. doi:10.4018/978-1-5225-2047-4.ch013

Plaksina, T., & Gildin, E. (2017). Rigorous Integrated Evolutionary Workflow for Optimal Exploitation of Unconventional Gas Assets. *International Journal of Energy Optimization and Engineering, 6*(1), 101–122. doi:10.4018/IJEOE.2017010106

Popat, J., Kakadiya, H., Tak, L., Singh, N. K., Majeed, M. A., & Mahajan, V. (2021). Reliability of Smart Grid Including Cyber Impact: A Case Study. In R. Singh, A. Singh, A. Dwivedi, & P. Nagabhushan (Eds.), *Computational Methodologies for Electrical and Electronics Engineers* (pp. 163–174). IGI Global. https://doi.org/10.4018/978-1-7998-3327-7.ch013

Quiza, R., La Fé-Perdomo, I., Rivas, M., & Ramtahalsing, V. (2021). Triple Bottom Line-Focused Optimization of Oblique Turning Processes Based on Hybrid Modeling: A Study Case on AISI 1045 Steel Turning. In L. Burstein (Ed.), *Handbook of Research on Advancements in Manufacturing, Materials, and Mechanical Engineering* (pp. 215–241). IGI Global. https://doi.org/10.4018/978-1-7998-4939-1.ch010

Rahmani, M. K. (2022). Blockchain Technology: Principles and Algorithms. In S. Khan, M. Syed, R. Hammad, & A. Bushager (Eds.), *Blockchain Technology and Computational Excellence for Society 5.0* (pp. 16–27). IGI Global. https://doi.org/10.4018/978-1-7998-8382-1.ch002

Ramdani, N., & Azibi, M. (2018). Polymer Composite Materials for Microelectronics Packaging Applications: Composites for Microelectronics Packaging. In K. Kumar & J. Davim (Eds.), *Composites and Advanced Materials for Industrial Applications* (pp. 177–211). Hershey, PA: IGI Global. doi:10.4018/978-1-5225-5216-1.ch009

Ramesh, M., Garg, R., & Subrahmanyam, G. V. (2017). Investigation of Influence of Quenching and Annealing on the Plane Fracture Toughness and Brittle to Ductile Transition Temperature of the Zinc Coated Structural Steel Materials. *International Journal of Surface Engineering and Interdisciplinary Materials Science, 5*(2), 33–42. doi:10.4018/IJSEIMS.2017070103

Robinson, J., & Beneroso, D. (2022). Project-Based Learning in Chemical Engineering: Curriculum and Assessment, Culture and Learning Spaces. In A. Alves & N. van Hattum-Janssen (Eds.), *Training Engineering Students for Modern Technological Advancement* (pp. 1–19). IGI Global. https://doi.org/10.4018/978-1-7998-8816-1.ch001

Rondon, B. (2021). Experimental Characterization of Admittance Meter With Crude Oil Emulsions. *International Journal of Electronics, Communications, and Measurement Engineering, 10*(2), 51–59. https://doi.org/10.4018/IJECME.2021070104

Rudolf, S., Biryuk, V. V., & Volov, V. (2018). Vortex Effect, Vortex Power: Technology of Vortex Power Engineering. In V. Kharchenko & P. Vasant (Eds.), *Handbook of Research on Renewable Energy and Electric Resources for Sustainable Rural Development* (pp. 500–533). Hershey, PA: IGI Global. doi:10.4018/978-1-5225-3867-7.ch021

Sah, A., Bhadula, S. J., Dumka, A., & Rawat, S. (2018). A Software Engineering Perspective for Development of Enterprise Applications. In A. Elçi (Ed.), *Handbook of Research on Contemporary Perspectives on Web-Based Systems* (pp. 1–23). Hershey, PA: IGI Global. doi:10.4018/978-1-5225-5384-7.ch001

Sahli, Y., Zitouni, B., & Hocine, B. M. (2021). Three-Dimensional Numerical Study of Overheating of Two Intermediate Temperature P-AS-SOFC Geometrical Configurations. In G. Badea, R. Felseghi, & I. Aşchilean (Eds.), *Hydrogen Fuel Cell Technology for Stationary Applications* (pp. 186–222). IGI Global. https://doi.org/10.4018/978-1-7998-4945-2.ch008

Sahoo, P., & Roy, S. (2017). Tribological Behavior of Electroless Ni-P, Ni-P-W and Ni-P-Cu Coatings: A Comparison. *International Journal of Surface Engineering and Interdisciplinary Materials Science, 5*(1), 1–15. doi:10.4018/IJSEIMS.2017010101

Sahoo, S. (2018). Laminated Composite Hypar Shells as Roofing Units: Static and Dynamic Behavior. In K. Kumar & J. Davim (Eds.), *Composites and Advanced Materials for Industrial Applications* (pp. 249–269). Hershey, PA: IGI Global. doi:10.4018/978-1-5225-5216-1.ch011

Sahu, H., & Hungyo, M. (2018). Introduction to SDN and NFV. In A. Dumka (Ed.), *Innovations in Software-Defined Networking and Network Functions Virtualization* (pp. 1–25). Hershey, PA: IGI Global. doi:10.4018/978-1-5225-3640-6.ch001

Salem, A. M., & Shmelova, T. (2018). Intelligent Expert Decision Support Systems: Methodologies, Applications, and Challenges. In T. Shmelova, Y. Sikirda, N. Rizun, A. Salem, & Y. Kovalyov (Eds.), *Socio-Technical Decision Support in Air Navigation Systems: Emerging Research and Opportunities* (pp. 215–242). Hershey, PA: IGI Global. doi:10.4018/978-1-5225-3108-1.ch007

Samal, M. (2017). FE Analysis and Experimental Investigation of Cracked and Un-Cracked Thin-Walled Tubular Components to Evaluate Mechanical and Fracture Properties. In P. Samui, S. Chakraborty, & D. Kim (Eds.), *Modeling and Simulation Techniques in Structural Engineering* (pp. 266–293). Hershey, PA: IGI Global. doi:10.4018/978-1-5225-0588-4.ch009

Samal, M., & Balakrishnan, K. (2017). Experiments on a Ring Tension Setup and FE Analysis to Evaluate Transverse Mechanical Properties of Tubular Components. In P. Samui, S. Chakraborty, & D. Kim (Eds.), *Modeling and Simulation Techniques in Structural Engineering* (pp. 91–115). Hershey, PA: IGI Global. doi:10.4018/978-1-5225-0588-4.ch004

Samarasinghe, D. A., & Wood, E. (2021). Innovative Digital Technologies. In J. Underwood & M. Shelbourn (Eds.), *Handbook of Research on Driving Transformational Change in the Digital Built Environment* (pp. 142–163). IGI Global. https://doi.org/10.4018/978-1-7998-6600-8.ch006

Related References

Sawant, S. (2018). Deep Learning and Biomedical Engineering. In U. Kose, G. Guraksin, & O. Deperlioglu (Eds.), *Nature-Inspired Intelligent Techniques for Solving Biomedical Engineering Problems* (pp. 283–296). Hershey, PA: IGI Global. doi:10.4018/978-1-5225-4769-3.ch014

Schulenberg, T. (2021). Energy Conversion Using the Supercritical Steam Cycle. In L. Chen (Ed.), *Handbook of Research on Advancements in Supercritical Fluids Applications for Sustainable Energy Systems* (pp. 659–681). IGI Global. doi:10.4018/978-1-7998-5796-9.ch018

Sezgin, H., & Berkalp, O. B. (2018). Textile-Reinforced Composites for the Automotive Industry. In K. Kumar & J. Davim (Eds.), *Composites and Advanced Materials for Industrial Applications* (pp. 129–156). Hershey, PA: IGI Global. doi:10.4018/978-1-5225-5216-1.ch007

Shaaban, A. A., & Shehata, O. M. (2021). Combining Response Surface Method and Metaheuristic Algorithms for Optimizing SPIF Process. *International Journal of Manufacturing, Materials, and Mechanical Engineering, 11*(4), 1–25. https://doi.org/10.4018/IJMMME.2021100101

Shafaati Shemami, M., & Sefid, M. (2022). Implementation and Demonstration of Electric Vehicle-to-Home (V2H) Application: A Case Study. In M. Alam, R. Pillai, & N. Murugesan (Eds.), *Developing Charging Infrastructure and Technologies for Electric Vehicles* (pp. 268–293). IGI Global. https://doi.org/10.4018/978-1-7998-6858-3.ch015

Shah, M. Z., Gazder, U., Bhatti, M. S., & Hussain, M. (2018). Comparative Performance Evaluation of Effects of Modifier in Asphaltic Concrete Mix. *International Journal of Strategic Engineering, 1*(2), 13–25. doi:10.4018/IJoSE.2018070102

Sharma, N., & Kumar, K. (2018). Fabrication of Porous NiTi Alloy Using Organic Binders. In K. Kumar & J. Davim (Eds.), *Composites and Advanced Materials for Industrial Applications* (pp. 38–62). Hershey, PA: IGI Global. doi:10.4018/978-1-5225-5216-1.ch003

Shivach, P., Nautiyal, L., & Ram, M. (2018). Applying Multi-Objective Optimization Algorithms to Mechanical Engineering. In M. Ram & J. Davim (Eds.), *Soft Computing Techniques and Applications in Mechanical Engineering* (pp. 287–301). Hershey, PA: IGI Global. doi:10.4018/978-1-5225-3035-0.ch014

Shmelova, T. (2018). Stochastic Methods for Estimation and Problem Solving in Engineering: Stochastic Methods of Decision Making in Aviation. In S. Kadry (Ed.), *Stochastic Methods for Estimation and Problem Solving in Engineering* (pp. 139–160). Hershey, PA: IGI Global. doi:10.4018/978-1-5225-5045-7.ch006

Siero González, L. R., & Romo Vázquez, A. (2017). Didactic Sequences Teaching Mathematics for Engineers With Focus on Differential Equations. In M. Ramírez-Montoya (Ed.), *Handbook of Research on Driving STEM Learning With Educational Technologies* (pp. 129–151). Hershey, PA: IGI Global. doi:10.4018/978-1-5225-2026-9.ch007

Sim, M. S., You, K. Y., Esa, F., & Chan, Y. L. (2021). Nanostructured Electromagnetic Metamaterials for Sensing Applications. In M. Bhat, I. Wani, & S. Ashraf (Eds.), *Applications of Nanomaterials in Agriculture, Food Science, and Medicine* (pp. 141–164). IGI Global. https://doi.org/10.4018/978-1-7998-5563-7.ch009

Singh, R., & Dutta, S. (2018). Visible Light Active Nanocomposites for Photocatalytic Applications. In K. Kumar & J. Davim (Eds.), *Composites and Advanced Materials for Industrial Applications* (pp. 270–296). Hershey, PA: IGI Global. doi:10.4018/978-1-5225-5216-1.ch012

Skripov, P. V., Yampol'skiy, A. D., & Rutin, S. B. (2021). High-Power Heat Transfer in Supercritical Fluids: Microscale Times and Sizes. In L. Chen (Ed.), *Handbook of Research on Advancements in Supercritical Fluids Applications for Sustainable Energy Systems* (pp. 424–450). IGI Global. https://doi.org/10.4018/978-1-7998-5796-9.ch012

Sözbilir, H., Özkaymak, Ç., Uzel, B., & Sümer, Ö. (2018). Criteria for Surface Rupture Microzonation of Active Faults for Earthquake Hazards in Urban Areas. In N. Ceryan (Ed.), *Handbook of Research on Trends and Digital Advances in Engineering Geology* (pp. 187–230). Hershey, PA: IGI Global. doi:10.4018/978-1-5225-2709-1.ch005

Stanciu, I. (2018). Stochastic Methods in Microsystems Engineering. In S. Kadry (Ed.), *Stochastic Methods for Estimation and Problem Solving in Engineering* (pp. 161–176). Hershey, PA: IGI Global. doi:10.4018/978-1-5225-5045-7.ch007

Strebkov, D., Nekrasov, A., Trubnikov, V., & Nekrasov, A. (2018). Single-Wire Resonant Electric Power Systems for Renewable-Based Electric Grid. In V. Kharchenko & P. Vasant (Eds.), *Handbook of Research on Renewable Energy and Electric Resources for Sustainable Rural Development* (pp. 449–474). Hershey, PA: IGI Global. doi:10.4018/978-1-5225-3867-7.ch019

Sukhyy, K., Belyanovskaya, E., & Sukhyy, M. (2021). *Basic Principles for Substantiation of Working Pair Choice*. IGI Global. doi:10.4018/978-1-7998-4432-7.ch002

Suri, M. S., & Kaliyaperumal, D. (2022). Extension of Aspiration Level Model for Optimal Planning of Fast Charging Stations. In A. Fekik & N. Benamrouche (Eds.), *Modeling and Control of Static Converters for Hybrid Storage Systems* (pp. 91–106). IGI Global. https://doi.org/10.4018/978-1-7998-7447-8.ch004

Tallet, E., Gledson, B., Rogage, K., Thompson, A., & Wiggett, D. (2021). Digitally-Enabled Design Management. In J. Underwood & M. Shelbourn (Eds.), *Handbook of Research on Driving Transformational Change in the Digital Built Environment* (pp. 63–89). IGI Global. https://doi.org/10.4018/978-1-7998-6600-8.ch003

Terki, A., & Boubertakh, H. (2021). A New Hybrid Binary-Real Coded Cuckoo Search and Tabu Search Algorithm for Solving the Unit-Commitment Problem. *International Journal of Energy Optimization and Engineering, 10*(2), 104–119. https://doi.org/10.4018/IJEOE.2021040105

Tüdeş, Ş., Kumlu, K. B., & Ceryan, S. (2018). Integration Between Urban Planning and Natural Hazards For Resilient City. In N. Ceryan (Ed.), *Handbook of Research on Trends and Digital Advances in Engineering Geology* (pp. 591–630). Hershey, PA: IGI Global. doi:10.4018/978-1-5225-2709-1.ch017

Ulamis, K. (2018). Soil Liquefaction Assessment by Anisotropic Cyclic Triaxial Test. In N. Ceryan (Ed.), *Handbook of Research on Trends and Digital Advances in Engineering Geology* (pp. 631–664). Hershey, PA: IGI Global. doi:10.4018/978-1-5225-2709-1.ch018

Related References

Valente, M., & Milani, G. (2017). Seismic Assessment and Retrofitting of an Under-Designed RC Frame Through a Displacement-Based Approach. In V. Plevris, G. Kremmyda, & Y. Fahjan (Eds.), *Performance-Based Seismic Design of Concrete Structures and Infrastructures* (pp. 36–58). Hershey, PA: IGI Global. doi:10.4018/978-1-5225-2089-4.ch002

Vargas-Bernal, R. (2021). Advances in Electromagnetic Environmental Shielding for Aeronautics and Space Applications. In C. Nikolopoulos (Ed.), *Recent Trends on Electromagnetic Environmental Effects for Aeronautics and Space Applications* (pp. 80–96). IGI Global. https://doi.org/10.4018/978-1-7998-4879-0.ch003

Vasant, P. (2018). A General Medical Diagnosis System Formed by Artificial Neural Networks and Swarm Intelligence Techniques. In U. Kose, G. Guraksin, & O. Deperlioglu (Eds.), *Nature-Inspired Intelligent Techniques for Solving Biomedical Engineering Problems* (pp. 130–145). Hershey, PA: IGI Global. doi:10.4018/978-1-5225-4769-3.ch006

Verner, C. M., & Sarwar, D. (2021). Avoiding Project Failure and Achieving Project Success in NHS IT System Projects in the United Kingdom. *International Journal of Strategic Engineering*, *4*(1), 33–54. https://doi.org/10.4018/IJoSE.2021010103

Verrollot, J., Tolonen, A., Harkonen, J., & Haapasalo, H. J. (2018). Challenges and Enablers for Rapid Product Development. *International Journal of Applied Industrial Engineering*, *5*(1), 25–49. doi:10.4018/IJAIE.2018010102

Wan, A. C., Zulu, S. L., & Khosrow-Shahi, F. (2021). Industry Views on BIM for Site Safety in Hong Kong. In J. Underwood & M. Shelbourn (Eds.), *Handbook of Research on Driving Transformational Change in the Digital Built Environment* (pp. 120–140). IGI Global. https://doi.org/10.4018/978-1-7998-6600-8.ch005

Yardimci, A. G., & Karpuz, C. (2018). Fuzzy Rock Mass Rating: Soft-Computing-Aided Preliminary Stability Analysis of Weak Rock Slopes. In N. Ceryan (Ed.), *Handbook of Research on Trends and Digital Advances in Engineering Geology* (pp. 97–131). Hershey, PA: IGI Global. doi:10.4018/978-1-5225-2709-1.ch003

You, K. Y. (2021). Development Electronic Design Automation for RF/Microwave Antenna Using MATLAB GUI. In A. Nwajana & I. Ihianle (Eds.), *Handbook of Research on 5G Networks and Advancements in Computing, Electronics, and Electrical Engineering* (pp. 70–148). IGI Global. https://doi.org/10.4018/978-1-7998-6992-4.ch004

Yousefi, Y., Gratton, P., & Sarwar, D. (2021). Investigating the Opportunities to Improve the Thermal Performance of a Case Study Building in London. *International Journal of Strategic Engineering*, *4*(1), 1–18. https://doi.org/10.4018/IJoSE.2021010101

Zindani, D., & Kumar, K. (2018). Industrial Applications of Polymer Composite Materials. In K. Kumar & J. Davim (Eds.), *Composites and Advanced Materials for Industrial Applications* (pp. 1–15). Hershey, PA: IGI Global. doi:10.4018/978-1-5225-5216-1.ch001

Zindani, D., Maity, S. R., & Bhowmik, S. (2018). A Decision-Making Approach for Material Selection of Polymeric Composite Bumper Beam. In K. Kumar & J. Davim (Eds.), *Composites and Advanced Materials for Industrial Applications* (pp. 112–128). Hershey, PA: IGI Global. doi:10.4018/978-1-5225-5216-1.ch006

About the Contributors

Islam Abdel Wareth holds a PhD in Education, specializing in mental health. He is a member of the Egyptian Association for Psychological Studies, a member of the Association of Psychologists, a member of the Syndicate of Egyptian Scholars, a research associate in psychological and strategic studies at local, regional and international studies centers. Scientific Experiments SPSS (Capacity Development Center of Faculty Members at Alexandria University).

Shar-Lee E. Amori is a PhD student in the School of Planning, Faculty of Engineering, McGill University. She holds a Master's Degree in International Public and Development Management (IPDM), and a Bachelor's in International Relations and Public Policy, both from the University of West Indies, Mona. Her research focuses on urban planning, policy and design, and areas of interconnection with sustainable development, smart cities, climate adaptation, socioeconomic resilience, urban economics, urban informality and participatory governance.

Valentina Burkšienė (Valentina Burksiene), born in 1963, Ph.D. in Management and Administration in 2012. Member of editorial board of scientific journals. Scientific interests: Sustainable Development, Sustainable Organizations, Strategic Management, Tourism and Recreation, Public Administration, Regional Development.

Jaroslav Dvorak, born in 1974, Ph.D. in Political Sciences in 2011. He was visiting researcher at Uppsala University (2017), Institute of Russian and Eurasian Studies, Sweden, and visiting professor at Bialystok Technical University (2017), Poland. Jaroslav Dvorak is involved in the editorial board of international scientific journals. Jaroslav Dvorak is a professor and head of the Department of Public Administration and Political Sciences at Klaipeda University, Lithuania. He has executive and expert experience in national and international institutions. He was the Klaipeda University representative on the Klaipeda regional development council (2019-2021). He is a member of the Research Board at People Powered.

Arij Elbadrawy is an Associate Professor of Sociology (2017) and winner of the Arab Women's Organization (League of Arab States) award in 2007. She participated in many local and international conferences and forums in: (Egypt, Lebanon, Algeria, Morocco and Jordon). Some of her research papers have been published in international journals in: (Egypt, UK, Lebanon and Algeria) and she participated in workshops, panel discussions and cultural seminars as well. Her works include: (Research Methods between Traditionalism and Contemporary, 2019), (Sociology of Culture: A Theoretical Vision and

Applied Studies - 2018). It is worth mentioning that there are copies of her books in university libraries of Columbia, Princeton, Pennsylvania and Stanford (Feminist creativity in Egypt, Sociological analysis of the biographies of creative women - 2012) and there are copies of this book in the library of the American University in Cairo (AUC) and 15 university libraries between Canada and the United States of America, including Harvard University. She lectured and supervised the Department of Sociology of Linguistics at the Higher Institute of Languages in Mansoura. She was also assigned to teach in many institutes affiliated with the Academy of Arts. She lectured twice at Blida 2 University in Algeria

Nieaz Muhammad Fattah is a lecturer at the University of Halabja located in sulaimanyah-Iraqi Kurdistan and the author of five books: Education in the era of globalization and technology, For the health of our generations, the relationship between you and the universe, Kindergarten as an introduction to education, and a message to those who suffer from staying up late. Also, she translated a book from Arabic to Kurdish language named: The Trial of Socrates. The author was born and raised in Sulaimanyah-Iraqi kurdistan, and earned her bachelor's degree at the University of Sulaimanyah. However, she had the opportunity to complete her master's and PHD degrees at Alexandria University in Egypt. Not to mention that right now she has her manuscript under publication. The main aim of her studies and research is to underline the factors that caused the loss of cultural identity amongst the youth in the Eastern societies in a globalized world.

Bdor M. Osama has a Bachelor's degree from the Department of Geography and Geographic Information Systems, Division of Surveying and Maps, Class of 2007, Grade Good. - Master's degree in maps entitled "Spatial Analysis of Traffic Problems in the City of Alexandria" (February 2013). - PhD from the Institute of Mediterranean Studies Division of Environmental Studies in the Mediterranean Basin, entitled "Geographical Evaluation of Some Oil and Natural Gas Ports in the Mediterranean Basin" (January 2018). Participation in seminars: - Participation in the first cultural season of the Institute for Mediterranean Studies under the title "Mediterranean Basin Countries and Their Relationship Through the Ages" in the third lecture of this season with a lecture entitled "Future Development of Oil and Natural Gas Ports in the Mediterranean Basin: Future Vision 2030" on 6/15/2021 . - Participation in the first scientific seminar of the Institute for Mediterranean Studies, entitled "Mediterranean Environment: Opportunities and Challenges", with a lecture entitled "The Blue Economy and its Role in Sustainable Development in the Mediterranean Basin", on 4/6/2022 Participation in the second scientific seminar of the Institute for Mediterranean Studies, entitled "The Mediterranean: Region and the World," with a research titled "The Impact of the Covid-19 Pandemic on the Mediterranean Environment," from November 28-29, 2022. - Participation in the third scientific seminar of the Institute for Mediterranean Studies, entitled "Mediterranean Environment: Opportunities and Challenges", with a lecture entitled "Plastic Waste and its Impact on the Mediterranean Environment", on 10/6/2023. Scientific publication: - "Spatial Analysis of Traffic Problems in the City of Alexandria", Journal of the Second Conference for Postgraduate Students at the Faculty of Arts - University of Alexandria, March 2017 - "The environmental geostrategic importance of natural gas production and transportation in the eastern Mediterranean basin" Journal of Human and Literary Studies - Faculty of Arts - Kafr El-Sheikh University, scientific journal, twenty-third issue, June 2020. - "Environmental Assessment of East Port Said Port", Assiut University Journal of Environmental Research, Center for Environmental Studies and Research at Assiut University, October 2021 issue. Presentation of a Master's thesis entitled "Spatial Analysis of Traffic Problems in the City of Alexandria" in the Journal of the Faculty of Arts - University of Alexandria, Issue No. 108,

About the Contributors

April 2022. Presentation of a doctoral dissertation entitled "Geographical Evaluation of Some Oil and Natural Gas Ports in the Mediterranean Basin" in the Journal of the Faculty of Arts - Alexandria University, Issue No. 109, July 2022.

Prageetha G. Raju is MBA PhD from Osmania University with specialization in Human Resource Management. She possesses 21 years of MBA, Executive MBA, Masters in Law and PhD programs in various premier B-Schools in India. Since 1.5 years, she has turned into management research and consulting and has started her own enterprise, Rainbow Management Research Consultants, at Hyderabad city in Telangana state, in India

Marwa Tawfiq has more than 16 years of experience with focus on Short and long-time therapeutic services as well as counseling for Children, Adolescents, Adult, family therapy, dealing with differences in relationships (couple therapy). These services come in the form of play therapy, individual therapy and group therapy. • Specializes in Educational and Clinical work having a multidisciplinary approach such as the psychosocial, spiritual rehabilitation, life coaching, personal & social Skills training program based on Cognitive-Behavioral Therapy and Gestalt Therapy. • My areas of interest include (anxiety disorder, generalized anxiety disorder, social anxiety disorder, panic disorder, agoraphobia, post-traumatic stress disorder, Grief issues, depression, Bipolar disorder, thought disorder, Sleep disorders, Sexual disorders, and Various behavioral addictions, Life Skills, Dependency, self-confidence and self-esteem. • Experience focus on the development of children in all areas, Behavior Modification, understanding of the child's function, in terms of intellect and scholastic abilities. This is done through therapy, counseling, observations as well as parent and teacher guidance. • Responsible for coordinating the role which involves planning, implementing and evaluating educational and therapeutic interventions for each individual child and handling problem early. • Preparing programs and educational plans related to helping children with special needs to integrate with normal children

Index

2030 1, 18, 27, 32, 38, 44, 54, 90, 98, 105, 113, 115-116, 118, 123, 126, 128-129, 156-159, 170, 223

A

active citizen 195

Adaptation 1-6, 9-10, 12, 15-17, 19, 28, 45, 49-50, 61-62, 103, 105, 108, 114, 117, 119-121, 123, 128, 158, 162-166, 168-169, 172, 175-177, 180-181, 185, 187, 189, 213, 215, 220-221, 223, 226

air pollution 55, 63, 90-92, 94, 100, 102-104, 106, 166

Anthropocene 54, 61, 65, 109, 125, 129

Autopoiesis 115, 129

awareness 4, 6, 13-17, 20, 28, 50, 57, 61, 76, 81, 85, 119, 132, 139, 150, 159-160, 163-164, 186-187, 190-192, 194-195, 198, 203-208, 211, 213, 218, 226

B

Building Community Resilience 1-3, 5, 12, 15-17

C

Caribbean 108, 114, 120, 125, 128, 132

city sinking 1

Climate change 1-30, 36-38, 40-57, 60-66, 68-75, 77-78, 87-95, 98, 100-111, 113-114, 118, 123-132, 150-151, 153-155, 157-166, 168-172, 175, 177, 180-181, 185-191, 193-194, 198, 203, 207, 209-214, 218-230

Climate changes 5, 8-9, 11, 14-15, 17, 19-21, 25, 27, 29-30, 32, 35-38, 40-43, 49-50, 52-53, 65-66, 131-133, 135, 150, 164, 175, 181, 184, 186-188, 191, 203, 207, 210 214, 218 221, 223, 225 227, 229-230

Climate Changes Risks 17

climate innovation 153

Climate Justice 71, 108-109, 117

Climate-resilience 81

commercial advertisements 190, 200-201

communication management 90-91, 93, 95-98, 106

Consumer Culture 190-193, 196-199, 207, 209

controlling climate change 153

Crisis management 106, 174-179, 185, 188-189

Cultural Imperialism 110, 121, 129

D

Dominant Culture Narrative 129

E

economics 18, 49-51, 68, 88, 106, 123, 125, 128, 151, 153, 184, 229-230

economics of carbon 153

environment 2, 5, 10, 12-14, 18, 24, 28, 30, 32, 36, 42-43, 45, 49, 53-57, 60, 63-64, 66, 90, 92, 94-95, 98, 101, 105-107, 109, 114, 117, 124, 127, 129, 140, 154-157, 159-161, 166, 170, 176-177, 180, 189, 191, 194, 196, 207-208, 214-215, 219, 221, 226, 228-229

Environmental Cinema 52, 60

Environmental Transformation 52, 62

extreme climatic 21, 36, 174-176, 182, 184-185, 187-188, 212

F

Food Security 3, 10, 17, 20, 38, 40-41, 43, 50, 54, 64, 70-71, 73, 75-77, 79, 84, 87, 89, 132, 151, 180, 188-193, 195, 198, 207-209, 226

G

generations 7, 16, 18, 27-28, 53, 56, 111, 118, 157, 165, 192

GIS 18-19, 29

green economy 153, 160

Index

H

Hyperlocalism 117, 130

I

Immigration 1, 10, 12
Indigenous Rights 108
Industrial development 190

L

Latin America 108, 114, 120, 125-128
Liquid Life 194, 196, 207-208
Livelihood 2, 70, 75-76, 84, 86, 162-163, 193

M

Millets 70-72, 75-77, 80-88
Mitigation of climate change risks 1-2

N

Natural Disasters 3, 52-53, 63-64, 66, 88, 132, 154, 162, 164, 179, 188
Nutritional Deficiencies 70

O

Odisha Millets Mission 70-72, 75, 79, 85-86, 88

P

Participatory Governance 108-109, 119, 124
Plantationocene 109-110, 130
politics 2, 17-18, 51, 69, 87, 95, 101, 103-104, 125-129, 162, 171, 195
Poshansakhis 70, 84
Psychological fragility 131, 133-136, 139-140, 145-151
Psychosomatic disorders 131-136, 142-152

S

Science Fiction Films 52

SDGs 18, 109, 112, 119, 124, 153, 157
Self-defeat 131-138, 145-150
smart green tech 153
Social Inclusion 2, 5-6, 9, 11, 16-17, 108-109, 119-120, 122-124
Social Morphology 53
Social resilience 3, 17, 174-176, 182, 185
Social Responsibility 4, 6, 15-17, 189, 210-214, 218-219, 221-223, 225, 227-230
Socio-Climate Justice 108-109, 115, 119, 124, 130
Sociocybernetics 109, 115-117, 120-124, 126, 130
socioeconomic aspects 18
sustainability 2, 4, 6, 15, 27, 38, 56, 67, 86, 88, 90, 94, 103-104, 106, 154, 156-158, 160, 167, 169-170, 189
sustainable development 1, 12, 15, 18, 20-21, 27-30, 38, 41-44, 48-51, 56, 73, 90, 92, 94-95, 104, 107, 109, 119, 121, 124, 127, 153, 156-159, 186, 209, 211, 213, 220, 230
sustainable port city 90
Sustainable Urbanization 108-109, 115

T

technological development 160
Third Sector Institutions 210-212, 218, 224, 228-229
Tribal Population 70, 72, 74, 76

U

University students 131, 134-135, 145-148, 151
Urban Imperialism 116, 130
Urban planning 11, 109-110, 113, 115, 119-120, 122, 124, 128, 130, 168-169

W

Women Empowerment 70, 72-73, 77

Recommended Reference Books

IGI Global's reference books are available in three unique pricing formats:
Print Only, E-Book Only, or Print + E-Book.

Order direct through IGI Global's Online Bookstore at
www.igi-global.com or through your preferred provider.

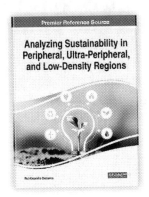

Analyzing Sustainability in Peripheral, Ultra-Peripheral, and Low-Density Regions

ISBN: 9781668445488
EISBN: 9781668445501
© 2022; 334 pp.
List Price: US$ 240

Frameworks for Sustainable Development Goals to Manage Economic, Social, and Environmental Shocks and Disasters

ISBN: 9781668467503
EISBN: 9781668467527
© 2022; 296 pp.
List Price: US$ 240

Disruptive Technologies and Eco-Innovation for Sustainable Development

ISBN: 9781799889007
EISBN: 9781799889021
© 2022; 340 pp.
List Price: US$ 215

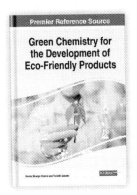

Green Chemistry for the Development of Eco-Friendly Products

ISBN: 9781799898511
EISBN: 9781799898535
© 2022; 276 pp.
List Price: US$ 250

Leadership Approaches to the Science of Water and Sustainability

ISBN: 9781799896913
EISBN: 9781799896937
© 2022; 219 pp.
List Price: US$ 215

Research Anthology on Environmental and Societal Impacts of Climate Change

ISBN: 9781668436868
EISBN: 9781668436875
© 2022; 2,064 pp.
List Price: US$ 2,085

Do you want to stay current on the latest research trends, product announcements, news, and special offers?
Join IGI Global's mailing list to receive customized recommendations, exclusive discounts, and more.
Sign up at: www.igi-global.com/newsletters.

Publisher of Timely, Peer-Reviewed Inclusive Research Since 1988

www.igi-global.com   Sign up at www.igi-global.com/newsletters   facebook.com/igiglobal   twitter.com/igiglobal   linkedin.com/igiglobal

Ensure Quality Research is Introduced to the Academic Community

Become an Evaluator for IGI Global Authored Book Projects

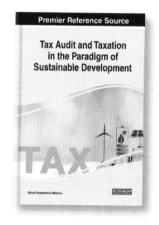

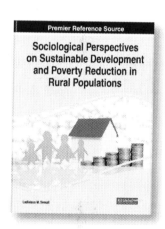

The overall success of an authored book project is dependent on quality and timely manuscript evaluations.

Applications and Inquiries may be sent to:
development@igi-global.com

Applicants must have a doctorate (or equivalent degree) as well as publishing, research, and reviewing experience. Authored Book Evaluators are appointed for one-year terms and are expected to complete at least three evaluations per term. Upon successful completion of this term, evaluators can be considered for an additional term.

If you have a colleague that may be interested in this opportunity, we encourage you to share this information with them.

Easily Identify, Acquire, and Utilize Published
Peer-Reviewed Findings in Support of Your Current Research

IGI Global OnDemand

Purchase Individual IGI Global OnDemand Book Chapters and Journal Articles

For More Information:
www.igi-global.com/e-resources/ondemand/

Browse through 150,000+ Articles and Chapters!

Find specific research related to your current studies and projects that have been contributed by international researchers from prestigious institutions, including:

- Accurate and Advanced Search
- Affordably Acquire Research
- Instantly Access Your Content
- Benefit from the InfoSci Platform Features

"*It really provides* an excellent entry into the research literature of the field. *It presents a manageable number of* highly relevant sources *on topics of interest to a wide range of researchers. The sources are* scholarly, but also accessible *to 'practitioners'.*"

- Ms. Lisa Stimatz, MLS, University of North Carolina at Chapel Hill, USA

Interested in Additional Savings?

Subscribe to
IGI Global OnDemand *Plus*

Learn More

Acquire content from over 128,000+ research-focused book chapters and 33,000+ scholarly journal articles for as low as US$ 5 per article/chapter (original retail price for an article/chapter: US$ 37.50).

7,300+ E-BOOKS. ADVANCED RESEARCH. INCLUSIVE & AFFORDABLE.

IGI Global e-Book Collection

- **Flexible Purchasing Options** (Perpetual, Subscription, EBA, etc.)
- Multi-Year Agreements with **No Price Increases** Guaranteed
- **No Additional Charge** for Multi-User Licensing
- No Maintenance, Hosting, or Archiving Fees
- Continually Enhanced & Innovated **Accessibility Compliance Features** (WCAG)

Handbook of Research on Digital Transformation, Industry Use Cases, and the Impact of Disruptive Technologies
ISBN: 9781799877127
EISBN: 9781799877141

Handbook of Research on New Investigations in Artificial Life, AI, and Machine Learning
ISBN: 9781799886860
EISBN: 9781799886877

Handbook of Research on Future of Work and Education
ISBN: 9781799882756
EISBN: 9781799882770

Research Anthology on Physical and Intellectual Disabilities in an Inclusive Society (4 Vols.)
ISBN: 9781668435427
EISBN: 9781668435434

Innovative Economic, Social, and Environmental Practices for Progressing Future Sustainability
ISBN: 9781799895909
EISBN: 9781799895923

Applied Guide for Event Study Research in Supply Chain Management
ISBN: 9781799889694
EISBN: 9781799889717

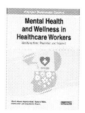

Mental Health and Wellness in Healthcare Workers
ISBN: 9781799888130
EISBN: 9781799888147

Clean Technologies and Sustainable Development in Civil Engineering
ISBN: 9781799898108
EISBN: 9781799898122

Request More Information, or Recommend the IGI Global e-Book Collection to Your Institution's Librarian

For More Information or to Request a Free Trial, Contact IGI Global's e-Collections Team: eresources@igi-global.com | 1-866-342-6657 ext. 100 | 717-533-8845 ext. 100

Are You Ready to Publish Your Research

IGI Global — PUBLISHER of TIMELY KNOWLEDGE

IGI Global offers book authorship and editorship opportunities across 11 subject areas, including business, computer science, education, science and engineering, social sciences, and more!

Benefits of Publishing with IGI Global:

- Free one-on-one editorial and promotional support.
- Expedited publishing timelines that can take your book from start to finish in less than one (1) year.
- Choose from a variety of formats, including Edited and Authored References, Handbooks of Research, Encyclopedias, and Research Insights.
- Utilize IGI Global's eEditorial Discovery® submission system in support of conducting the submission and double-blind peer review process.
- IGI Global maintains a strict adherence to ethical practices due in part to our full membership with the Committee on Publication Ethics (COPE).
- Indexing potential in prestigious indices such as Scopus®, Web of Science™, PsycINFO®, and ERIC – Education Resources Information Center.
- Ability to connect your ORCID iD to your IGI Global publications.
- Earn honorariums and royalties on your full book publications as well as complimentary content and exclusive discounts.

Join Your Colleagues from Prestigious Institutions, Including:

 Australian National University

 Massachusetts Institute of Technology

 Johns Hopkins University

 Tsinghua University

 Harvard University

 Columbia University IN THE CITY OF NEW YORK

Learn More at: www.igi-global.com/publish
or Contact IGI Global's Aquisitions Team at: acquisition@igi-global.com

Individual Article & Chapter Downloads
US$ 29.50/each

Easily Identify, Acquire, and Utilize Published Peer-Reviewed Findings in Support of Your Current Research

- Browse Over **170,000+ Articles & Chapters**
- **Accurate & Advanced** Search
- Affordably Acquire **International Research**
- **Instantly Access** Your Content
- Benefit from the **InfoSci® Platform Features**

THE UNIVERSITY *of* NORTH CAROLINA *at* CHAPEL HILL

" *It really provides* an excellent entry into the research literature of the field. *It presents a manageable number of* highly relevant sources *on topics of interest to a wide range of researchers. The sources are* scholarly, but also accessible *to 'practitioners'.* "

- Ms. Lisa Stimatz, MLS, University of North Carolina at Chapel Hill, USA

Interested in Additional Savings?

Subscribe to
IGI Global OnDemand *Plus*

Learn More

Acquire content from over 137,000+ research-focused book chapters and 33,000+ scholarly journal articles for as low as US$ 5 per article/chapter (original retail price for an article/chapter: US$ 29.50).

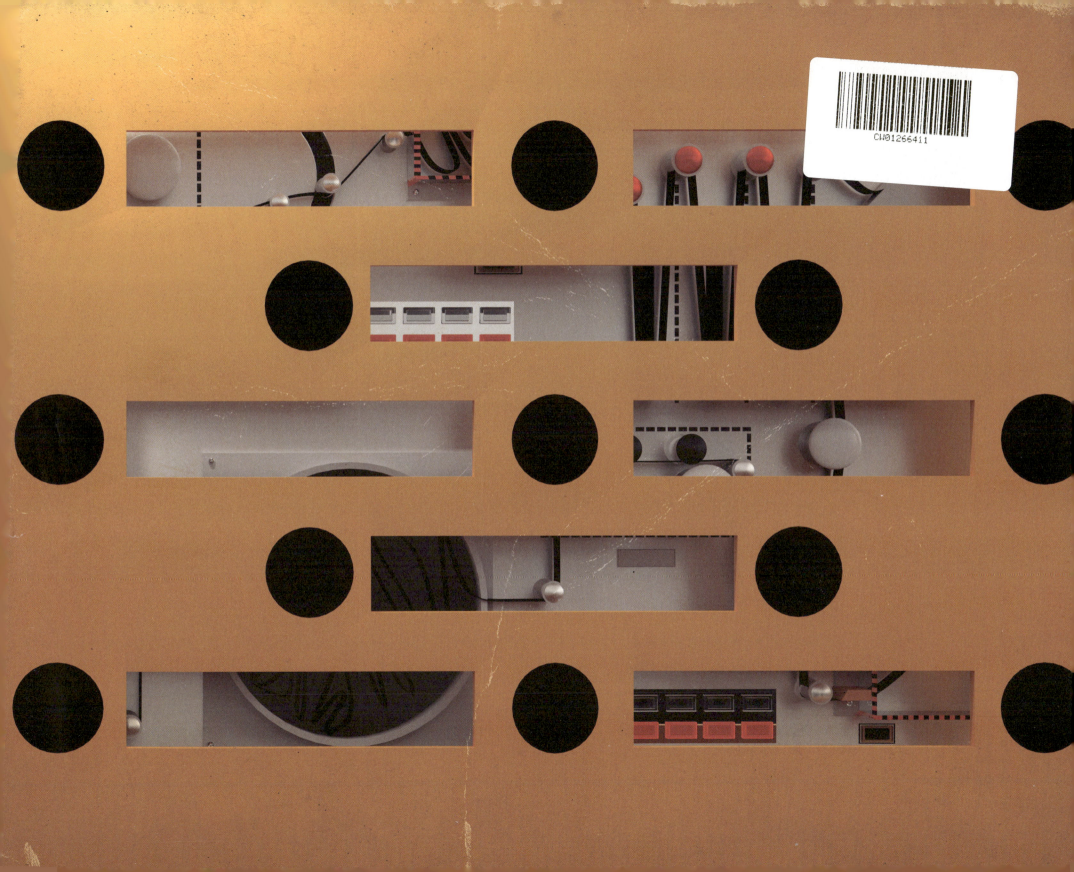

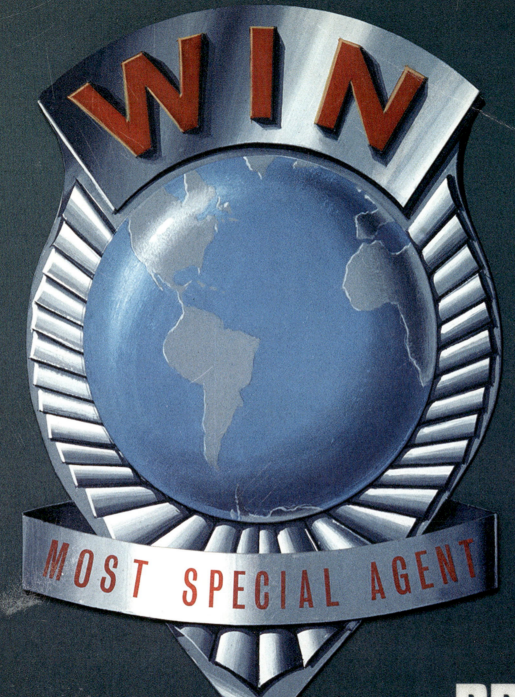

PROJECT 90
TECHNICAL OPERATIONS MANUAL

CONTENTS

Introduction by Shane Weston 4

The History of WIN 6

WIN London 10

SECTION 1: PROJECT 90

Introduction 16

Key Personnel 18

The McClaine Residence 22

The BIG RAT 24

The BIG RAT Brain Pattern
Tape Library 30

SECTION 2: EQUIPMENT

Introduction 34

Agent's Case 36

Glasses 38

Sidearm 40

WIN Communicator 42

Jetpack 44

Proton Lance 45

Recording Devices 46

Badge and ID Card 47

SECTION 3: APPROVED VEHICLES

Introduction 50

Jet Air Car 52

WIN Company Car 58

Stealth Hovercraft 62

DSRV Athena 66

Fast Attack Vehicle A14 70

OTC Triton OC2 Rocket 74

U59 Wildcat 78

VG 104 82

F116 86

SECTION 4: CASE STUDIES

Introduction 92

Hijacked 94

Splashdown 98

Business Holiday 102

Test Flight 106

The Professional 110

Big Fish 114

Runaway 118

Murder on the Moon 122

SECTION 5: CONCLUSION

Exit Strategy and Conclusion 126

INTRODUCTION

WELCOME TO WIN

To be clear, in an ideal situation, this book will never see the
light of day. In the event of my disappearance or death, I have made
arrangements for this collection of documents to be passed on to you,
my successor. Its contents are not to be stored or shared within
regular World Intelligence Network channels, and if its contents are
at risk of coming to light then you are to destroy the book and leave
no trace of WIN London's most secret operation, Project 90.

I'd also like to state that the questionable ethics involved in an
operation such as this are not lost on me. When my chief operative,
Sam Loover, walked into my office and told me of the existence of the
BIG RAT, a machine that could record the brain pattern of a person and
transfer all their knowledge and experience to another, I knew such a
device should never be made public as its potential for abuse could
be catastrophic. In my mind, the compromise we reached with Professor
McClaine and his son Joe remains the only way for such a machine to
exist and to achieve some much needed good in the world.

Again, I am well aware that asking someone as young as Joe to
effectively give up their childhood and potentially their life in the
pursuit of world peace is a decision that no one would normally
consider in good conscience, but I would ask you to withhold your
judgement until you have reviewed this account of the operation.

In this book you will find a breakdown of Project 90, its key
players, technology, the resources at its disposal, and accounts of
some of the successful missions that Joe has conducted and the good
that we have hopefully brought to the world.

Most importantly, you will be able to ask yourself, would you have
done the same?

From now on, the future of Project 90 is up to you.

Good luck.

Shane Weston

HISTORY OF THE WORLD INTELLIGENCE NETWORK

Like any intelligence organisation, the workings of the World Intelligence Network (WIN) are largely a mystery to those outside its walls. To the casual observer, WIN is merely a series of office buildings where hapless employees struggle with red tape and internal politics, their movements never telegraphed and their achievements never celebrated.

Members of the public who visit any of the headquarters of WIN will be told the story of the formation of the organisation: the brainchild of a successful politician and diplomat who went on to win the Nobel Peace Prize for his efforts in ending the Cold War and uniting the governments of Russia, the United Kingdom and the United States of America. This is, however, just the tip of the iceberg of the real truth.

In 1975, the heads of the CIA, the KGB and MI5 met in the basement level of a hotel in Geneva, Switzerland without the knowledge of their respective governments. This potentially treasonous meeting had been prompted by a string of nuclear crises that had strained East-West relations to the point that the world was dangerously close to armageddon.

A plan was hatched to exchange information and to shift their individual organisations' policies from attempting to gain advantages for their respective countries to making sure the balance of power would stay equal across all three nations. A secret network was set up, sharing intelligence between the territories and coordinating their agents in ways that would allow them to achieve their goals.

This huge risk relied on all three entities not abusing the system for their own ends, and the leaking of classified intelligence to rival governments not being discovered by their own side. Fortunately, all three were able to function and within five years tensions had eased to the point that American, British and Russian agents could cooperate openly on assignments involving international crime or external nations.

This new era of cooperation, coupled with a growing coalition of rival countries forming in Eastern Europe and Asia, prompted open talks to unify the various branches of each nation under a single banner, to share resources and remove the possibility of Russia and the West slipping back into a Cold War. In 1985, the newly ratified World Government was appointed and their first act was to begin the development of a World Army, World Navy, World Air Force and World Intelligence Network – the branch which had already been secretly operating in the shadows for some time.

International Reach

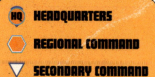

- **HQ** HEADQUARTERS
- **⬢** REGIONAL COMMAND
- **▽** SECONDARY COMMAND

Today, WIN has integrated the intelligence agencies of 18 countries and operates from three main headquarters – London, Moscow and Washington – with Thor Jurgens serving as the organisation's overall commander. WIN can also be invited to assist countries in situations where lives or freedom is at risk.

THOR JURGENS
Swedish national, Thor Jurgens is the incumbent Supreme Commander of WIN, controlling the organisation from its Washington headquarters. He succeeded former US Defence Secretary James Mahone Calloway after questions were raised over Calloway's neutrality and the fear of preferential treatment from Moscow.

SHANE WESTON AND IVAN BRENSKI
WIN has two secondary headquarters with a degree of independence from the main Washington branch. WIN London is run by former CIA agent, Shane Weston, alongside his second-in-command, Samuel Loover. The Moscow branch is commanded by ex-KGB director, Ivan Brenski, assisted by Petrana Kopski. The structure and operations of the three branches are managed by their respective chiefs, but they must also cooperate with one another and share vital information.

Supported by an army of analysts, technicians and support staff, WIN's field agents are considered to be the most exceptional in the world. Recruited from a wide variety of backgrounds, these agents are trained to survive under the most hostile of conditions or assume deep cover in the field for years at a time. Maintaining the secret identities of these agents while providing them with continuous support in the field is one of WIN's most complex tasks.

It should be stressed that forming WIN has presented many obstacles. The sheer size of the organisation has meant that, at times, resources have been thinly spread or difficult to requisition from stockpiles up to half a world away. Corruption has been uncovered within the organisation which has showed the screening process to be flawed, and old rivalries from certain sectors that once were enemies have hampered the success of some operations.

The Eastern Alliance has been able to create its own intelligence organisation that operates with ruthless efficiency, and it would appear organised crime has evolved to challenge WIN.

Ultimately, WIN will continue its mission to make sure the balance of power favours no single nation and to protect world peace at any cost.

WIN LONDON

Located in Whitehall, the headquarters of the WIN London office is a surprisingly unremarkable building from the outside, appearing to be only three stories tall with a small bronze plaque beside the door denoting its identity. Dominated by Shane Weston's generously-sized office, as well as a labyrinth of smaller offices, function rooms and lecture theatres, the building is designed to invoke the impression of an impenetrable wall of bureaucracy that hides the real face of WIN.

Accessible from a secret reception area located behind the public reception is the entrance to the underground portion of the building, stretching four levels below the surface and filled with workshops, operations centres and endless rows of computer banks. This is the real heart of WIN, monitoring all the various ongoing operations across Europe and Africa and connecting to its counterparts in the West and the East. The basement of the building provides access to a secret underground railway line, adjacent to the London Tube network, that allows VIPs to transfer quickly and seamlessly between the UK command of the World Armed Forces and the old MI6 buildings. A concealed helipad on the roof provides a third means of access.

WIN ACQUISITIONS DEPARTMENT
WIN is infamous for its ability to develop bespoke equipment swiftly. The acquisitions department on-site, run by Steve Johnson, will often receive unusual briefs for mission-specific equipment that will either need to be purchased or constructed on a tight deadline. This makes them a key resource for Project 90.

MONITORED
While the office is on a public street, the entire area is constantly monitored to identify anyone who might be spying on the building. In instances where enemy agents have been detected, personnel may be advised to use the Tube network.

WIN LABORATORY
Various laboratories and workshops are housed on-site. While a number of these will be used to develop equipment, most of them are used for analytics. A whole lab is set aside for voice print identification, while others are used for encryption and data sharing.

SHANE WESTON'S OFFICE

The nerve centre of WIN London is Shane Weston's office, located on the upper floor at the front of the building. It is a versatile space that can be transformed quickly by staff to become a briefing room or a room for high-level functions. For the most part, it is maintained in a comfortable yet efficient state with Shane Weston's desk acting as the focal point of the room with two wraparound sofas allowing for visiting guests to be briefed at their ease. Individual chairs are also available when only two or three people are present.

Putting the office of the most important person of the local WIN hierarchy in view of the street might seem unconventional but this actually leans into the more bullish aspect of Shane Weston's character. He will deliberately leave windows open and give the overall impression of lax security in order to lure in particularly susceptible enemy agents so that they can be monitored and fed false information. In reality, the office is constantly swept for bugs and the neighbouring street and buildings are closely monitored by WIN's security team.

The office decor is dictated by the commander-in-residence and Shane Weston has chosen to decorate the room with busts of the Greek gods Ares, Eirene and Themis – the gods of war, peace and justice, respectively.

A	Shane Weston's desk
B	Seating area
C	Windows to street
D	Secret exit behind curtain
E	Function table (hidden under floor)
F	BIG RAT antenna Installed by Professor McClaine following the inception of Project 90, this antenna has been used to record the brain patterns of multiple WIN agents and visiting dignitaries.
G	Retractable screen During briefings a projector screen can be raised from the floor. An even larger screen is located behind one of the paintings on the adjacent wall.
H	Entrance The primary entrance and exit to the room is through a padded door isolating the room from the rest of the building.

PROJECT 90

SECTION 1
PROJECT 90

INTRODUCTION

Only four people are aware of the existence of Project 90: Shane Weston, Supreme Commander WIN London; Sam Loover, Deputy Chief WIN London; Professor Ian McClaine, external contractor and inventor; and Joe McClaine, the Professor's son and now Special Agent Joe 90.

The project focuses on an invention developed by Professor McClaine in late 2012: the Brain Impulse Galvanoscope Record and Transfer or the "BIG RAT" as it has become known. This machine has the ability to record the complete knowledge and experience of a person and temporarily transfer them to another.

It became immediately apparent that such a machine could become one of the most important inventions ever developed, and crucially one of the most dangerous. The value of such a device would represent a considerable advantage in the hands of an intelligence organisation. It would remove the need for interrogation, specialised training and background screening, and allow WIN to outstrip its opponents in every way by using their own knowledge and experience against them. However, such a device could easily be misused. Any rival intelligence agency could take advantage of its existence if it was ever revealed, develop similar technology for themselves and quickly gain access to the knowledge and experience of their competitors – or sell the skills of the world's finest experts to the highest bidder.

So, the decision was made to offer Professor McClaine a unique proposal: to purchase access to his invention and only use it for top secret missions that the Professor would actively participate in. Furthermore, the only person who would benefit from the BIG RAT's capabilities would be his son, Joe, who could be "brained up" and deployed on missions with the knowledge and experience of an agent but with the cover of being a nine-year-old boy.

PROJECT 90 PARAMETERS

After an intense debate, a proposal was made with the following conditions:

- Joe and the BIG RAT will only be used in missions where people's lives and livelihoods are under threat.

- Joe will never be used to directly assassinate a target, and will only be expected to retaliate in self-defence.

- Recorded brain patterns will be stored securely and not be used to extract personal information that isn't vital to a mission in progress.

- Professor McClaine will have the power to veto any mission he deems to be too risky or that breaks any previous conditions.

- Professor McClaine and Joe have the power to end their participation in the operation as and when they see fit.

With these conditions agreed, Joe was (unofficially) welcomed into WIN as the 90th London field agent and given the designation "Joe 90".

Misdirection

As Project 90 continues, Shane Weston and Sam Loover have set up a series of more plausible cover stories to misdirect anyone who might try to delve deeper into the operation. Files relating to Josephine Stewart, a fictional espionage agent, have been included in official WIN documentation to account for any leaks surrounding Joe 90's involvement, and Professor McClaine has been designated as an external advisor reporting directly to Weston as a way of explaining his frequent visits to headquarters.

KEY PERSONNEL

NAME:
JOE McCLAINE

BORN: August 7, 2003
PLACE OF BIRTH: Hampstead, London
POSITION: WIN's Most Special Agent

Joe McClaine is the adopted son of Professor McClaine, and carries the affectionate title of WIN's "Most Special Agent". Little is known of Joe's birth parents who both died in a car accident when he was just 12 months old. With no living relatives to care for him, Joe was sent to Caxton Manor Orphanage in East London where he spent his early years nurtured in a community with other children.

Shortly after his fifth birthday, Joe met his future adoptive parents, Ian and Mary McClaine. It was during a day out in central London that the young boy's keen interest in cars drew him to the advanced, jet-powered vehicle built by Professor McClaine while it was parked conspicuously on Fleet Street. Joe had clambered into the unlocked car and tucked himself behind the seats to admire its impressive twin aero turbines. Unaware of Joe's presence when he returned to the vehicle, the Professor inadvertently drove all the way home to Dorset with the youngster as a stowaway.

When Joe revealed himself to the McClaines at the end of the journey, Ian and Mary instantly bonded with their guest and visited him regularly at the orphanage over the next few months. Joe was also given permission to stay at the McClaines' home in Culver Bay during weekends. The eventual end result was Joe's adoption into the McClaine family, giving him a permanent home and a sense of belonging which he had been lacking during his early years. Sadly, just one year later, Joe's life was struck by further tragedy when his adoptive mother was killed in a high-speed motorway accident during a visit to London. Six-year-old Joe was aided through the grief by forming an even closer bond with his adoptive father as they secretly worked together on a project dedicated to Mary's memory – the BIG RAT.

Joe enthusiastically played a vital role in the progress of the BIG RAT, as his developing mind made him the ideal candidate for testing the electronic transfer of brain patterns. The machine was fine-tuned to comfortably seat Joe, and it was during a demonstration of the BIG RAT's capabilities to Samuel Loover, a family friend of the McClaines, that Joe's potential for a role in WIN was realised.

With the express permission and assistance of Professor McClaine, Joe 90 undertakes field operations for WIN which no other agent could carry out. The BIG RAT supplies Joe with the brain patterns of experts from across the globe, granting him the knowledge and experience to complete any assignment with a high chance of success. While "brained up" and undertaking a mission, Joe maintains a professional and mature attitude to his work. He also demonstrates great bravery and dedication, often drawing upon his embedded expertise to calculate the odds in a dangerous situation and push forward to victory.

Joe inherently possesses great athletic ability, regularly accomplishing physical feats which an adult agent would be unable to achieve. WIN ensures that Joe undergoes routine medical and psychiatric check-ups, as well as counselling, to ensure that the effects of the BIG RAT do not impact on his future life. It has been confirmed within all reasonable parameters that once Joe's special glasses have been removed, he returns to his own mindset and personality whilst retaining some memory of the missions he has undertaken.

Joe enjoys a number of hobbies in his spare time including fishing, camping, playing football, listening to music and reading comic books. He is currently attending Culver County Junior School, and has shown a greater inclination towards adventure and outdoor pursuits rather than academic studies.

NAME:
IAN MCCLAINE

BORN: June 24, 1965
PLACE OF BIRTH: Baron's Court, London
POSITION: Electronics Engineer

Adoptive father to Joe and inventor of the BIG RAT, Professor McClaine serves WIN as an unofficial scientific advisor and a capable field operative. Son of physicist Robert McClaine and novelist Katherine McClaine, Ian – nicknamed "Mac" – was educated at Canford Public School in Dorset where he excelled academically. Mac went on to study electronics and aerodynamics at St. Lucifer's College, Cambridge. Supplementing his studies, he became fascinated with the art of ballooning, and within three days had single-handedly constructed and flown his first balloon over Cambridge using his own bespoke navigational equipment.

Upon graduating, Mac was awarded a scholarship to Stanford University in California for the following academic year, leaving him with 12 months to further his passion for ballooning. Mac joined London's Aerodynamic Academy and conducted several balloon flights around the world – flying over the Alps, along the Mississippi River and across the full expanse of Australia.

At Stanford, Mac obtained a further degree in electronics alongside his friend and future colleague, Samuel Loover. During a summer vacation in Sam's hometown of Flagstaff, Arizona, Mac's life was endangered by a balloon crash which left him stranded in the desert for five days with a broken leg. Following an instinct, Sam was able to rescue Mac who, after a four-month stay in hospital, fully recovered from the ordeal.

Mac returned to England after graduating and rejoined the Aerodynamic Academy to conquer new feats with his balloons. He also accepted a position as an electronics controller for a private computer firm. After eight years and a promotion to general manager, Mac had developed a number of great computing advancements. At the age of 36, he returned to Dorset and bought a picturesque cottage in Culver Bay, converting the tunnels underneath it into a vast laboratory. Mac continued his late father's work and turned his expertise towards teaching and researching brain patterns and their capacity for computerisation, earning him a professorship at Cambridge University. Echoing his mother's literary achievements, Professor McClaine caught the writing bug and published his first book to great acclaim. The celebrity writer became embroiled in his unfinished manuscripts and decided to hire an assistant – Mary Reed was the first to apply. The Professor and Mary were old friends from their days at the Aerodynamic Academy and soon rekindled their relationship. They were married within three months.

The McClaines retreated to a leisurely life at the cottage allowing the Professor to experiment and write from his underground laboratory. He also had a special project underway in the cottage's garage – a unique vehicle powered by jet turbines which could travel on roads, sea and in the air. One day after a publisher's meeting in London, Profesor McClaine returned home and discovered his special car was carrying an unusual cargo – a five-year-old orphan named Joe. The McClaines bonded with the boy over the following months and subsequently adopted him as their son. Joe adored his new parents and the boy became their pride and joy. Unfortunately, a year of familial bliss ended tragically when Mary died in a car accident. Professor McClaine's grief was channelled into a determination to reach the pinnacle of his brain pattern research by completing work on the BIG RAT.

Initial hesitation gave way to Professor McClaine allowing his invention, and his son, to play a part in maintaining world peace. Encouraged by Sam Loover, now WIN London's Deputy Chief, the Professor's role in WIN is two-fold. Primarily, he maintains the BIG RAT and its library of brain pattern recordings, enabling Joe to gain the knowledge and experience required to undertake missions. The Professor also assists Joe 90 in the field, often providing support from his multi-purpose vehicle. Since Mary's death, the Professor has maintained a bachelor's lifestyle with aid from his housekeeper, Mrs Harris. The secrecy required to protect the BIG RAT has necessitated the Professor stepping out of the limelight, and devoting himself to his fatherly duties and to WIN. In his spare time, however, the Professor maintains connections within the artistic and scientific community, and enjoys ballooning, fishing, classical music and engineering.

NAME:
SAMUEL WILLIAM LOOVER

BORN: December 15, 1969
PLACE OF BIRTH: Flagstaff, Arizona, USA
POSITION: Deputy Chief WIN London

Samuel Loover holds a long career dedicated to maintaining world security, and currently works as second-in-command to Shane Weston at the WIN London branch. Sam was born in the small town of Flagstaff, Arizona in the heart of the desert, and enjoyed a childhood in the outdoors – hiking, hunting and fishing. His father, geologist William Loover, encouraged Sam's determination for a career in the field of engineering. Willie funded Sam's education at Arizona's top technical school in Phoenix, and aged 10, the boy began eight years of studies specialising in electronic engineering. Sam felt stifled by big city life but was set on achieving his goal of reaching Stanford University. The only interruption to his work came at the age of 13 when Sam's mother, Jacqueline, died tragically in a freak sand storm destroying half of the Loovers' hometown.

Completing his training with top results in electronics, engineering and dynamics, Sam received a scholarship grant to study at Stanford. Devastated for years by the loss of Jacqueline, the Loover family were revitalised by Sam's achievement. The news had a miraculous effect on Willie, who was seemingly cured of a mysterious illness which had threatened his life. Sam moved to Stanford and formed a lasting friendship with Ian "Mac" McClaine, a fellow electronics student. The Englishman shared Sam's spirit of adventure, and Mac joined him for summer vacations at the Loover homestead in Flagstaff. One such visit led to disaster, however, when Mac was left injured and alone in the Arizona desert following a balloon crash. Sam, knowledgeable of the unforgiving landscape, managed to track down and rescue Mac after five days of searching.

Mac recovered, eternally grateful to Sam for saving his life. The pair graduated from Stanford with perfect grades in electronics. While Mac returned to England, Sam was immediately offered a position at Cape Kennedy working on the base's electronic security systems. Within five years, Sam had climbed the ladder to the position of chief security officer. His prowess for electronics combined with an inherent security instinct had made Sam an invaluable asset to the American government. He was soon promoted to security controller for the entire Western block of the US, and then again to the prestigious role of Chief Security Adviser to the US Secretary for Defense.

It was during this time that the CIA merged with other Western secret service organisations to form the World Intelligence Network. Sam was chosen to join the service as a leading field operative, allowing him the opportunity to combine his security expertise with his rugged love of outdoor adventuring. Sam's new career took him around the world and into some of the most dangerous territories imaginable. Throughout his first 15 years of active duty, Sam eliminated many notorious international espionage and terrorist networks. He was decorated for his service and awarded a promotion to WIN's London office as Deputy Chief. Sam greatly respects his commander, Shane Weston, and oversees all agent operations while streamlining the service for peak efficiency.

Sam's crowning achievement was the induction of Joe 90 and the BIG RAT into WIN. Having reunited with his former Stanford colleague, Sam saw the potential in Mac's invention, and in his son, Joe, for maintaining world peace. "Uncle Sam" (Joe's nickname for his direct superior) takes command of all operations completed by Joe 90, and often returns to the field to assist in missions as required. He has assumed responsibility for Joe's personal safety and maintains the security of the BIG RAT so that only key personnel are aware of the invention's very existence.

In his spare time, Sam's hobbies see him return to his spiritual home in nature for activities including fishing, rock climbing and kayaking. He also enjoys exercising his mind with logic puzzles, and devising complex and imaginative training assignments for new recruits.

NAME:
SHANE WESTON

BORN: July 10, 1968
PLACE OF BIRTH: Carson City, Nevada, USA
POSITION: WIN Deputy Controller, Supreme Commander of WIN London

Answering only to Thor Jurgens, Shane Weston is the incumbent chief of London-based operations for WIN, and deputy controller of the organisation overall. The life of WIN's brightest star is rooted in a turbulent upbringing on the rural outskirts of Carson City, Nevada. Shane's childhood was marred by the death of his mother when the boy was seven-years-old. The family's grief manifested in the breakdown of Shane's relationship with his father, which eventually led to him abandoning the Westons' ranch to travel east at the age of 12. He lived rough and alone, enjoying a oneness with nature synonymous with his Cherokee ancestry. With little education, Shane's talents resided in outdoor endurance and superhuman physical feats.

Aged 16, Shane joined the United States Army for eight years as a marine, rising quickly from private to corporal. Shane led his own platoon into conflicts across Asia where the US was aiming to achieve world peace. Corporal Weston's heroic deeds in battle earned him the distinctive order of valour, the Purple Heart, and promotion to captain. Returning to America after his successful campaign, Shane had to adapt his untamed bravery in the field towards a desk-job and the often tedious administration involved with maintaining the global harmony he had helped to achieve.

Dreading civilian life, Shane signed up for another 10 years of service in the Army Intelligence Network and achieved his dream of becoming a commander. Shane's new responsibilities saw him gain a fresh enthusiasm for work, now as an active intelligence agent attached to the CIA. Once again, Commander Weston was recognised for his action in conquering international terrorist groups worldwide, which culminated in a mission to save London from annihilation. While he was in England, Shane married a high-ranking naval officer, Vice Admiral Susan Denver.

Returning to the US with his new family, Shane transitioned with the rest of the CIA into WIN. Shane's life and career were almost brought to an end during an assignment in South America to eliminate the underworld assassin, Lomax Brunt. The armed confrontation took place in the Amazon jungle, and only one man would survive. Brunt received three gunshot wounds to the heart, while Shane was dealt a spinal injury. Shane proved to be the victor, outliving Brunt, but the battle had cost him any hope of continuing with the high demands and pressures of arduous field work, and led him to offer his resignation from active service after a two-year hospital stay.

Unwilling to lose his vast experience, WIN offered Shane the position of deputy chief of the Washington-based operations where he made a name for himself as a demanding teacher to incoming operatives and a brilliant strategist. During his five years of responsibility for the American secret service files, Shane finally shrugged off his loner mentality. He led his team like a family and positioned himself as a paternal figure to his agents – a far cry from the distant relationship Shane had endured with his own father. In recognition of Shane's dedication to WIN, he was awarded full command of the London office and deputy controller of the entire global organisation.

With his deputy, Sam Loover, Shane leads WIN London with a steely determination to continue maintaining world peace. He encourages new initiatives to outpace the technological and strategic capabilities of enemies threatening world security. One such initiative was onboarding nine-year-old Joe McClaine as WIN's "Most Special Agent" and harnessing the capabilities of the BIG RAT. Shane supervises many of Joe 90's assignments personally, either by formulating plans from his office in London or a base in the field.

Shane's restlessness often leads him to attempt challenging exploits across Europe from climbing in the Alps to windsurfing on the Riviera in his spare time. Shane and Sue Weston publicly sponsor several charities supporting military veterans, while maintaining Shane's cover as one of the foremost figures in global security.

THE MCCLAINE RESIDENCE

PROJECT 90

Atop a cliff in Culver Bay, along the coast in the county of Dorset, sits an old Tudor cottage. Over the centuries it has had many residents and is now the home of Professor McClaine, his son Joe and the nerve centre of the entire Project 90 operation.

The Professor bought the cottage in 2001 for several reasons: as a restoration project since the property had fallen into disrepair; its proximity to the sea and multiple popular fishing spots; and for the network of tunnels that run underneath which offered the opportunity to install an underground laboratory without disturbing the landscape above.

A garage, in a similar Tudor style, has been added and was used as a workshop while the underground caves were being converted and latterly to house Professor McClaine's Jet Air Car – the first major invention he devised at the cottage.

While Professor McClaine and Joe often undertake the three-hour drive (or half-hour flight) to be briefed in London, the cottage is considered the main staging site for Project 90. The Professor and Joe are commonly briefed by either Sam Loover or Shane Weston at the cottage in order to help keep Project 90 off WIN's books. Security has been stepped up accordingly around the grounds with extra improvements by Professor McClaine added in.

Interior

The interior of the cottage features a generous living room, dining room, study, kitchen and larder with a spiral staircase leading to three bedrooms on the upper level. A lift to the underground laboratory is located behind a traditional wooden door. Outwardly, the building is still in keeping with its Tudor roots, but a whole host of technology – for security and comfort – is hidden inside the various furnishings.

MRS HARRIS
Mrs Ada Harris, a resident from one of the local towns, has been employed for the last five years as the McClaines' housekeeper. She is responsible for keeping the cottage tidy and occasionally making food.

CLIFFSIDE VIEW
A path at the back of the cottage leads down to the base of the cliff and a series of caves which have been modified into an emergency exit from the laboratory in the event that the cottage – for whatever reason – is compromised. A makeshift dock has been added for fishing trips or for guests to arrive by boat. The proximity of the cottage to the sea has proved to be a security risk in the past, but a sonar sensor installed in a buoy just offshore now keeps track of all shipping in the immediate area.

THE BRAIN IMPULSE GALVANOSCOPE RECORD AND TRANSFER

PROJECT 90

The invention at the centre of the operation, the BIG RAT, was the culmination of almost two decades of work by Professor McClaine accelerated over the last few years following the death of his wife. Built in secret in the laboratory underneath the cottage, the BIG RAT is a marvel of modern technology and has the potential to change the world as we know it, for good or for worse.

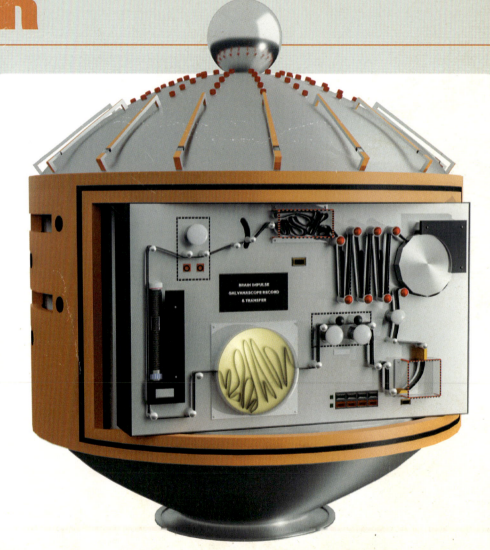

In the simplest terms, the BIG RAT is able to record a person's entire knowledge and experience (or brain pattern), transfer it onto a reel of magnetic tape and then imprint that brain pattern onto another person.

Primary Unit

Although all components of the machine are referred to collectively as the BIG RAT, the large cylindrical unit at the eastern side of the laboratory is the true BIG RAT computer, while the subsidiary equipment is used for projection and reflection. The sphere on top receives the incoming brain pattern signal which is interpreted by the machine. It is then processed and transferred to magnetic tape in a way that can be reinterpreted and projected by the machine later on. Once completed, the tape reels are transferred to a spool in the control room which allows them to be easily loaded and unloaded. When the BIG RAT is transferring a brain pattern into a new subject it reads the pre-recorded tape and projects the brain pattern signal through the floor and into the projector.

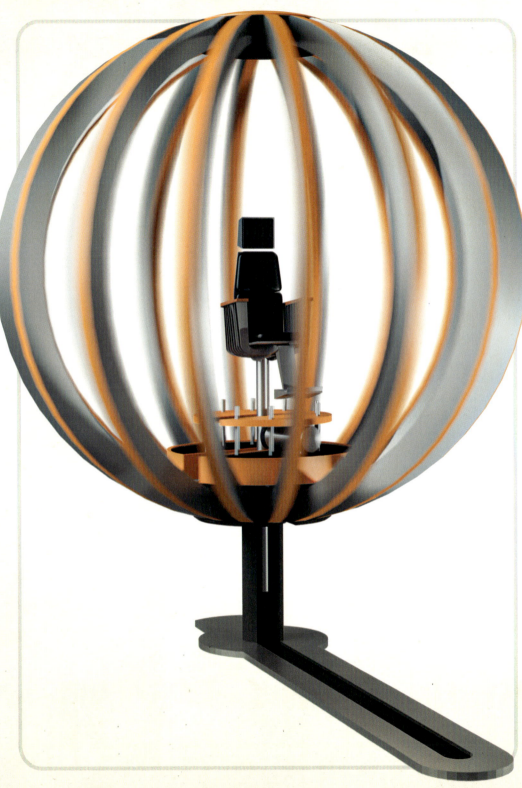

Rat Trap

Taking on the appearance of a segmented orange with a chair at the centre, the Rat Trap is in fact a focusing device that takes the brain pattern signal from the Projection Wall and directs it towards the head of the occupant from all directions. At the same time, electrodes attached to the temples stimulate as yet unused areas of the brain that are dedicated to memory. These electrodes allow the area to be written on and enable the occupant to access new knowledge while sitting in the chair. A portable version of the electrodes has been added to a pair of spectacles to allow the user to continue to access these memories when in the field. To revert to normal, a scan of the occupant taken before the procedure is used to remove the additional brain pattern from the subject's mind.

To gain entry into the Rat Trap, two of the rings pull apart allowing the chair to be lowered to the ground.

Projection Wall

Located at the back of the room, the Projection Wall collects the brain pattern signal from the Primary Unit and projects it towards the spinning rings of the Rat Trap. The control panel in front of the wall houses various settings to fine tune the projector as it needs to be calibrated for different users. The colourful banks of lights on either side of the projector, while seemingly decorative, are tied to the various components within the projector and the Rat Trap, and can be used to identify faults with the machine.

PROJECT 90

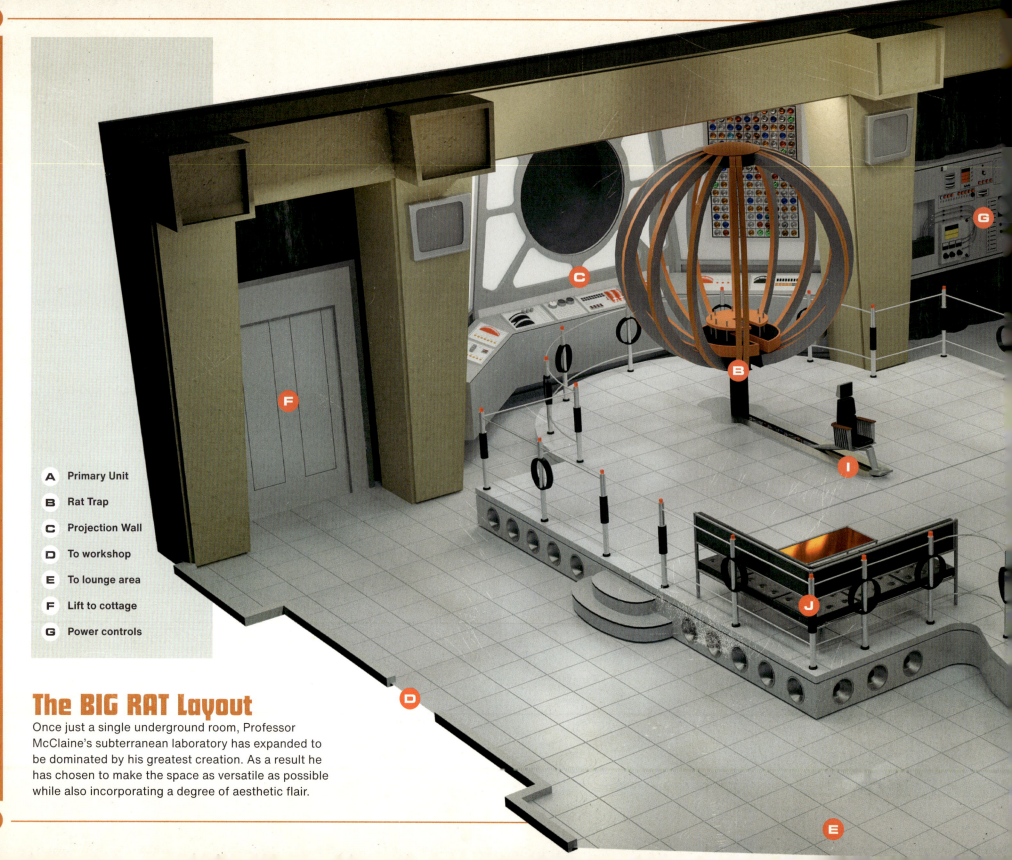

A — Primary Unit
B — Rat Trap
C — Projection Wall
D — To workshop
E — To lounge area
F — Lift to cottage
G — Power controls

The BIG RAT Layout

Once just a single underground room, Professor McClaine's subterranean laboratory has expanded to be dominated by his greatest creation. As a result he has chosen to make the space as versatile as possible while also incorporating a degree of aesthetic flair.

H **Control room**
The main control room for the machine is on a raised plinth allowing a good view of all the components. From here, the BIG RAT can be activated, calibrated and fed new brain patterns.

I **Chair**
When not in use, the chair is lowered to the ground to allow the user to access it. The electrodes can still be applied in this configuration allowing a brain pattern to be accessed.

J **Seating area**
As the BIG RAT requires such a large open space, Professor McClaine has added a seating area between the components for general conversation or briefings. As the brain pattern is projected behind the Rat Trap, this area is safe to use during operation of the machine.

K **Antenna monitoring station**
Built into a secret compartment in the roof of the cottage, a receiver dish can be deployed in order to receive brain patterns transmitted from a recorder. Professor McClaine has explained this away to his neighbours as an advanced aerial for picking up foreign television signals.

Workshop

With the main area of the laboratory now occupied by the BIG RAT, a new workshop had to be constructed. Partitioned away from the rest of the lab it houses a variety of equipment such as lathes, drills, grinders and anything else Professor McClaine might need for his inventions and developments.

This area also contains Professor McClaine's safe where family finances, documents and the plans for his multitude of inventions are kept.

Lounge Area

A more recent addition to the laboratory is an adjoining area with expanded seating and a television. Originally used as a recreational area on particularly cold nights when the cottage's heating was not up to standard, this area has become a makeshift staging area where Professor McClaine, Shane Weston and Sam Loover can stay in communication with Joe and direct missions in reasonable comfort.

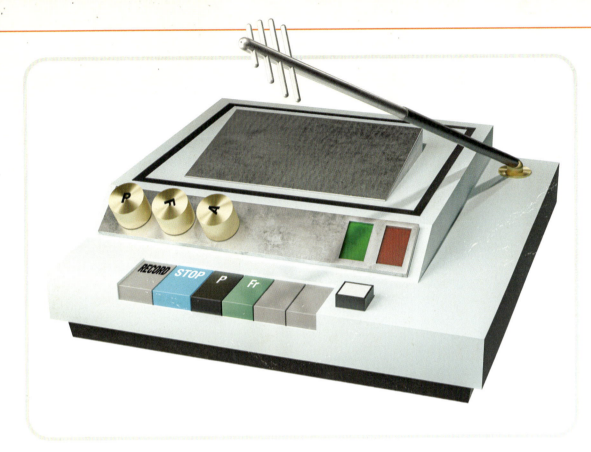

POTENTIAL DRAWBACKS

The BIG RAT is still relatively new technology, and while it has been declared safe to use we have started to see certain drawbacks in its design. On one mission, Joe was erroneously given the brain pattern of a double agent causing him to unwittingly veer off course and almost deliver top secret codes to enemy agents. On another occasion, Joe was given the brain pattern of a pilot who had suffered a memory blackout from a fear of landing. As a result, the landing procedure was not transferred over to Joe requiring him to be talked through it by the original pilot.

Going forward, a more advanced screening process will be used in order to evaluate potential brain patterns and determine if they have any underlying neurological conditions or ulterior motives.

Recording Equipment

While the BIG RAT is capable of receiving brain patterns, it requires a device in the field in order to actually scan and transmit them. Over the course of the first year of Project 90, multiple iterations of this device were developed with the final version able to be disguised within a briefcase that could be surreptitiously used to scan targets at a range of four metres for around three minutes. Depending on the distance from the cottage, a relay balloon can be deployed to boost the signal while the BIG RAT and cottage antenna can be activated remotely. Plans are currently underway to see if the booster signal can function as part of WIN's existing satellite network.

THE BIG RAT BRAIN PATTERN TAPE LIBRARY

PROJECT 90

The ever-expanding library of brain pattern recordings from the world's leading experts is stored on tapes inside the BIG RAT for use at a moment's notice. For cataloguing, each recording is assigned an alphanumeric code based upon its source, age, quality and effectiveness. When a recording is updated, its tape number is revised accordingly. What follows is a selection of the tapes which have been utilised by Joe 90 during WIN assignments.

TAPE #	LAST NAME	FIRST NAME	TITLE/RANK	OCCUPATION/SKILLS	SPECIAL NOTES
P4	Aston	Kenneth	Doctor	Archaeologist	Aztec Specialist
H6	Baxter	Howard	Professor	Chemical Engineer	
Q12	Brinker	Clive		Underground Explorer	
Q9	Brown	John	Professor	Metallurgist	Formerly Tape # A8
G10	Culshaw	Jonathan	Sir	Impressionist	
Q7	Davies	Angela		Safecracker	Caution: See Criminal File
Q15	Drake	Len		Accountant	
E6	Drayton	Charles	Colonel	Astronaut	
X9	Ealing	David	Lord	World Bank Vice President	
A16	Fernandez	Jessie		WIN Agent & Former Political Bodyguard	
V5	Flambeau	Guy		Olympic Bobsleigh Champion	Amateur Marksman & Helicopter Pilot
K13	Foley	Frank "Fearless"	Captain	VG 104 Pilot	
Q8	Fraser	Bill		Aquanaut	Formerly Tape # Q14
D19	Grant	Jim	Major	World Air Force Test Pilot	Caution: Key Knowledge Absent
C2	Hadid	Halley		Eastern Alliance Defence System Expert	
Q11	Harris	Graham	Doctor	Optician	

TAPE #	LAST NAME	FIRST NAME	TITLE/RANK	OCCUPATION/SKILLS	SPECIAL NOTES
S18	Hartley	Amelia	Colonel	World Army Driver	
Q14	Henderson	Shane	Colonel	Military Vehicle Expert	
Q16	Jackson	Malcolm	Doctor	Electrical Engineer	Formerly Tape # K10
E14	Johnson	Edward		WIN Agent	
P19	Johnson	Brad		Aeronautical Computer Engineer & Demolitions Expert	
Q1	Jones	Peter	Admiral	Submarine Officer	Formerly Tape # J12
T3	Jones	Lionel		Monorail Engineer	
A1	Jurgens	Thor	Secretary	Supreme Commander of WIN Washington	
Q3	Kados	Emil	Doctor	Neurosurgeon	
L7	Kelvin	William	Doctor	Undersea Explorer	Arctic Specialist
Q10	Kent	Tony		Railway Engineer	
G17	Laramie	Michael		WIN Agent	
A3	Loover	Samuel William		Deputy Chief of WIN London	Updated Monthly
S10	Malone	Lee		Trumpet Player	
Z1	McClaine	Ian	Professor	Electronics Expert & Balloonist	Updated Monthly. Used For Experimental Purposes
Z2	McClaine	Joe		WIN's Most Special Agent	Updated Monthly. Used For Experimental Purposes
Q6	Miller	Richard		Artist	
F26	Myers	Edwin	Professor	Tutor to Prince Kahib	
T12	Patel	Emily		Power Pack Specialist	
J11	Rodriguez	Martha		Financial Expert	Formerly Tape # M3
B3	Sanders	Gordon	Doctor	Explosives Expert	
Q4	Sherman	Hugh	Captain	Test Pilot	
Q5	Sladek	Igor		Concert Pianist & WIN Agent	
X3	Sloane	Harry		Head of WIN Special Courier Department	Caution: See Criminal File
X12	Summerfield	Henry George		Safecracker	Caution: See Criminal File
A8	Swan	Alex		WIN Agent & Expert Marksman	
Q2	Taylor	Charles	Major	Chief US Air Force Test Pilot & Small Arms Expert	Formerly Tape # A5
R14	Todd	Avery		WIN Agent	
Q13	Warner	James	Doctor	Geologist	
X18	Webb	Johnstone		Propellant and Thrusts Expert	
A2	Weston	Shane		Supreme Commander of WIN London	Updated Monthly
Q17	Wright	Peter		Solicitor	

SECTION 2
EQUIPMENT

EQUIPMENT 2

INTRODUCTION

In addition to having access to a huge stockpile of equipment, WIN's Research and Development (R&D) department has become world-renowned for creating highly advanced, mission-specific gear for WIN's agents, assisting in their objectives while also offering life-saving protection. Located beneath WIN London with similar departments found in Moscow and Washington, the R&D department is constantly testing new forms of field equipment while also finding innovative ways to conceal existing weapons, radios and tools.

As part of Project 90, Joe has been issued with specialised equipment, mostly miniaturised variants of WIN-issue gear disguised to look like toys.

It is impossible to predict what missions Joe may be needed for so Project 90 has full access to WIN's R&D and Acquisitions departments. Whether it is a spacesuit, a jet pack or a high-powered laser, anything required can be delivered to the cottage in Culver Bay within 12 hours of being ordered.

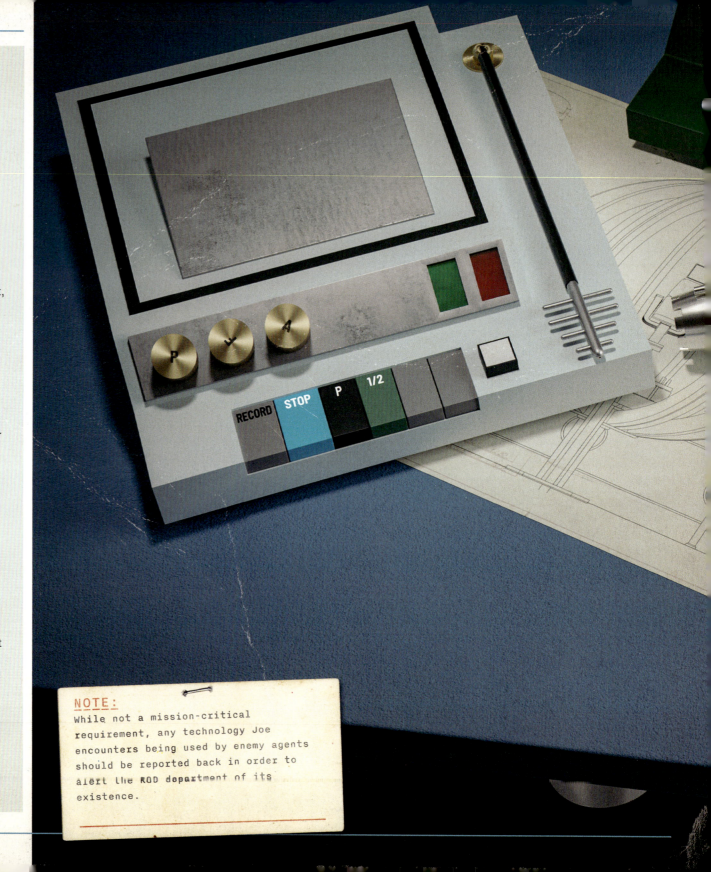

NOTE:
While not a mission-critical requirement, any technology Joe encounters being used by enemy agents should be reported back in order to alert the R&D department of its existence.

EQUIPMENT 2

AGENT'S CASE

Standard issue to all WIN field agents is either a case, satchel or backpack modified to secretly house their radios and weapons. No two of these carry items are the same so as not to immediately telegraph their identity to someone familiar with WIN's procedures.

Joe 90's case is a one-off design. Outwardly it is made of lightweight but durable plastic, sized appropriately to fit schoolbooks and stationery. The main compartment of the case opens traditionally and contains small bands for attaching pencils, rulers and a geometry set to the lid. As it looks like a child's case, it is unlikely to come under much scrutiny, but if searched then this compartment should be enough to alleviate any suspicion about the likely contents.

Compartment 1

Flipping the case over and rotating the right-hand foot reveals the first secret compartment: a flip-up lid containing Joe's special-issue sidearm and 2-way radio. The base of this compartment contains two ammunition clips, an agent-specific cypher book and technical specifications for the radio in case of damage.

Compartment 2

Rotating the left-hand foot reveals the other secret compartment containing Joe's Special Agent badge, electrode glasses and a suppressor for his sidearm. This base contains a wallet for alternate identification and travel documents.

It should be noted that the compartments are visible on X-ray, but the ingenious design of standard WIN-issue equipment means they are 'invisible'. Different methods might therefore need to be employed if moving non-standard items through airport security.

ADDITIONAL FEATURES

Another feature of Joe's case is a carbon-fibre plate that separates the front compartment from the back compartment. This plate is strong enough to stop armour-piercing rounds and allow the case to be used as a shield in extreme circumstances. Standard on all WIN carry cases is a self-destruct function. This has purposely been omitted from Joe's case so as not to put his life at any additional risk – a good faith gesture and reassurance for the McClaines.

EQUIPMENT

GLASSES

Without Joe's special glasses, he's just a boy. With them, he's a secret agent.

Transferring the entire knowledge and experience of one or more persons into a single mind can put an immense amount of strain on the recipient's brain. From the very beginning of his experiments, Professor McClaine realised that he would need a way to safeguard the receiver's mind in order to cut out the risk of potentially overwriting or permanently entangling a new brain pattern with the old. His solution was to code the pattern onto as yet unused parts of the brain, primarily the hippocampus and cerebellum, and electrically stimulate these parts to activate them. This way the brain pattern is kept separate from the original mind and lowers the possibility of major personality changes, deep trauma and residual memories being left behind after the new temporary brain patterns are wiped.

When connected to the BIG RAT, Joe uses electrodes attached to his temples to activate the new brain pattern. Once the electrodes are disconnected and Joe leaves the chair, his access to the brain pattern quickly fades. So, for Joe to be able to retain this information in the field, a portable solution was required.

OTHER HEADGEAR
There are many situations in which wearing spectacles is not convenient or possible. In these scenarios, the electrodes can be built into other forms of headgear such as swimming goggles or visors.

From specifications given by Professor McClaine, similar electrodes to the ones used on the BIG RAT were incorporated into a pair of spectacles that line up with Joe's temples and allow him to access the relevant brain pattern. The glasses, although lightweight and made of plastic, are incredibly durable and have been designed to provide Joe with an additional layer of protection. The lenses, which are not corrective, are polarised in order to dull the harsh light of muzzle flashes and explosions, while the material itself is toughened to resist shrapnel.

A Toughened lens
B Electrode contact point
C Power cells running up to the temples
D Softened pad arms
E Charging contacts that connect to a small battery in the case

MISSION CRUCIAL

While the spectacles provide Joe with a huge advantage on his missions, they are also his Achilles heel. If they are knocked off his face or forcibly removed then he loses all access to the brain pattern he has been given. Joe has been given field training to manage these situations but ultimately losing his glasses is considered a good reason for immediately terminating the mission and extracting him where possible.

EQUIPMENT 2

SIDEARM

For his own personal protection on missions, Joe has been issued with a customised variant of the WIN standard sidearm. Designed by Shane Weston himself, the WIN Special Issue Automatic Pistol marked a big departure from conventional firearms, with a focus on miniaturising the ammunition in order to allow the weapon to fire more rounds.

While maintaining the characteristics of a traditional automatic pistol, the sidearm uses a combination of combustion-based firing and electromagnets mounted in the barrel which propel specially miniaturised ammunition almost three times faster than a regular bullet. The extra speed offsets the reduced size of the smaller bullets and raises the effective range of the weapon. Sniper rifle elements such as a shoulder stock and a scope can be added to take advantage of this.

- **A** Trigger
- **B** Muzzle
- **C** Magazine
- **D** Thermal resistant grip
- **E** Suppressor

The magazine is loaded into the butt of the gun and can hold 200 rounds. Given the secretive nature of the ammunition, the gun does not eject rounds like a normal automatic weapon and instead cycles them back into the magazine for disposal later. In addition, the bullets tend to shatter on impact which makes them hard to detect.

It is important to note that the WIN sidearm is only rated for engaging infantry or unarmoured targets as bullets can ricochet off hard surfaces and potentially cause collateral damage. Another drawback, particularly in the case of Joe's model, is the potential for the weapon to overheat with successive rapid firing.

FIREARM TRAINING
While on missions Joe will be using the brain pattern of an experienced agent who is familiar with firearms. However, as part of his induction into WIN, Joe has been given firearms training. While considerably behind the marksmanship of other WIN agents, this knowledge gives Joe the ability to defend himself during emergency situations in which the brain pattern is insufficient or his glasses are lost.

Suppressor
A suppressor is included in Joe's case. It can be fitted to the front of the pistol and dampens the weapon's recoil, allowing for a more accurate shot as well as reducing the noise of the weapon by approximately 30 per cent. Overall, it offers a stealthy solution, although in many cases the objective is for Joe not to require his weapon at all.

EFFECTIVE IN THE FIELD
While in many ways the WIN sidearm could be considered over-designed and hard to mass-produce, it has proved its effectiveness in the field, with WIN agents being able to hold off large numbers of hostiles without having to stop to reload.

EQUIPMENT 2

STANDARD WIN COMMUNICATOR

Possibly the most important piece of equipment in any WIN agent's arsenal, the WIN communicator and its supporting infrastructure is one of the most costly projects the organisation has undertaken.

For an intelligence network now operating across three continents, it is vital for agents in the field to be able to communicate with their home base or with contacts on the other side of the planet in a manner which renders their signals secure and undetectable.

An expansive fleet of satellites were secretly launched in high orbit over the course of 10 years and now form a huge interconnected network with coverage that spans the entire globe. While the signal frequencies they operate within can be intercepted, they are encoded in such a way that they are impossible to decrypt by outsiders. In the unlikely case that a WIN communicator falls into enemy hands, fresh cyphers are distributed frequently so that WIN's communications network cannot be compromised for greater than 24 hours.

The communicator itself is of slim design, compact and stored in a titanium case which makes it extremely durable; even capable of deflecting bullets. A telescopic antenna transmits and receives information from the satellite network, while three buttons and a dial make up its interface. Two contact points on either side allow the device to be connected to earphones or its charging supply. Once charged, the communicator can remain active for two weeks of frequent use.

SUB-CHANNEL

Unlike Joe's other equipment, the communicator is not miniaturised as the item is already produced at an optimal size. The only key difference to the standard-issue model is that Joe's communicator has access to a special sub-channel that is only used by the four members of Project 90 to decrease the risk of the operation being exposed to the rest of the organisation.

A	Microphone/speaker	**E**	Volume button
B	Function dial	**F**	Frequency button
C	Antennae	**G**	Cypher button
D	Power light	**H**	Power connector (headphone connector on opposite side)

COMMUNICATOR `BLINDSPOTS`
In addition to global coverage, the communicator can continue to transmit and receive effectively up to a kilometre beneath the surface with only minimal interference. The only two blindspots occur when the communicator is shielded by lead or during atmospheric re-entry.

Satellite Network
The WIN satellites rotating the Earth operate in high orbit putting them out of the range of most competing nations. Publicly, they make up a meteorological system that can monitor the weather with high accuracy. The results are freely available which provides an incentive for the global community and thus prevents sabotage.

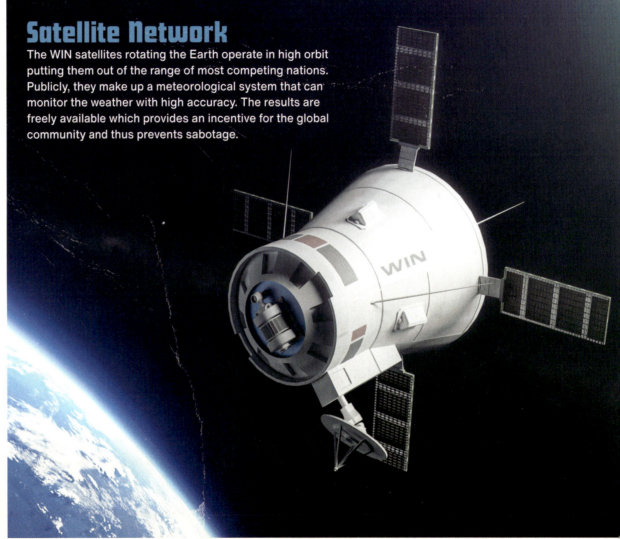

43

EQUIPMENT 2

JETPACK

Not an uncommon sight in the 2010s, jetpacks are a commercially available mode of personal transportation with pilots requiring a licence and specialised training in order to fly them. Four micro gas turbines – two pointed vertically for lift and two horizontally for forward momentum and steering – allow the wearer to move in three dimensions with ease. Currently available on the market are models used for scouting, construction, rescue and combat. However, these solutions are over-sized and imperfectly balanced for use by a nine-year-old boy.

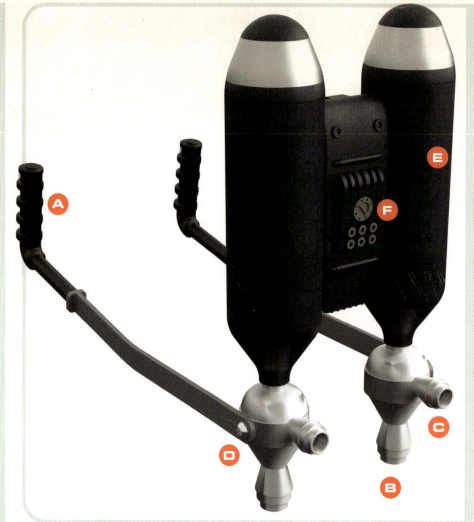

- **A** Control stick (left for vertical thrust, right for horizontal thrust)
- **B** Vertical micro gas turbine
- **C** Horizontal micro gas turbine
- **D** Control arm hinge
- **E** Fuel canister
- **F** Calibration panel

The special jetpack that Joe uses was developed by Professor McClaine himself using a miniaturised WIN field model adapted to accommodate Joe's size and weight.

Fortunately, as Joe requires less power to lift him, the smaller fuel capacity doesn't affect the performance of the device and allows for approximately 15 minutes of continuous flight time.

ADDITIONAL FEATURES
When not in use, the control arms of the jetpack can be folded away for ease of transport. The device can also be disassembled for concealment in luggage or secret compartments.

PROTON LANCE

An extremely specialised piece of equipment, the proton lance is often used in a laboratory environment for heating substances from a distance or slicing through dense materials. A variant of the device has been mounted to spacecraft and used to extract and recover Moon rock.

Able to heat its targets up to a temperature of 6000°F, the proton lance is considered highly dangerous, and its use is strictly prohibited to unlicensed persons or companies. However, elements of the criminal underworld have developed their own, unauthorised version of the lance. In particular, Henry George Summerfield specialised in using the lance in conjunction with a mounted X-ray scope to disable alarm systems and cut open maximum security safes. This black market variant of the proton lance is in the possession of WIN London, and the device now forms part of the inventory available for Project 90.

ADDITIONAL FEATURES

The proton lance can be disassembled and stowed in a small suitcase for ease of use. It can be attached to a small tripod for precision work, although a stock is also included to use the device freehand. It should be noted that because of the powerful potential of the proton lance, using it as an offensive weapon is outlawed under the revised Geneva Convention.

A Trigger
B Proton accelerator
C Emitter
D Stock
E X-ray scope
F Heat dissipation system

WIN LOG RECORDER

When operating in the field, every agent runs the risk of being incapacitated or killed while on mission. In order to guarantee that their work can be continued, WIN agents are required to carry a log recorder and update it regularly. Easily mistaken for a pen, the device is encased in a solid titanium shell. It is virtually indestructible and rated to survive the most catastrophic explosions intact. It is also impossible to access the recordings on it without using a specialised reader system only available to high-level WIN operatives.

ENCODER PEN

Many WIN agents under deep cover and constantly monitored by outside sources have been forced to use more avant-garde methods of relaying information back to headquarters. The encoder pen is a two-fold device. By looking like a conventional pen, it can be deployed inconspicuously by agents to secretly record up to two minutes of audio and transfer it onto extremely fine magnetic tape. The second function is its ability to transfer the magnetic tape onto paper, disguised as ink, as part of a signature or handwritten note. This ink can then be read by running a sensor over it.

BADGE AND WIN IDENTIFICATION CARD

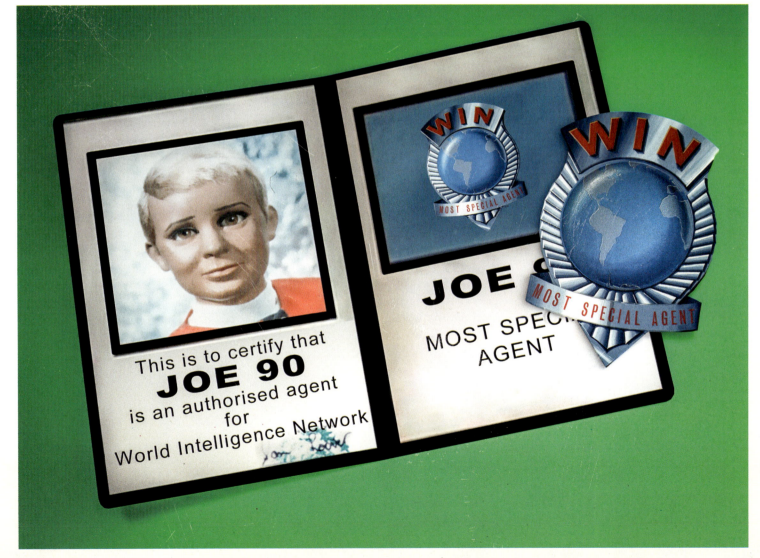

Many would question the logic in giving an undercover agent, particularly one as sensitive as Joe 90, a badge that outright identifies him as WIN's "Most Special Agent." However, this is yet another contributing factor to the air of plausible deniability about him. If captured or searched, much of Joe's equipment can easily be passed off as children's toys. Likewise, Joe introducing himself as an agent for WIN will lead any adult in the room to assume he's just a child playing a game. This enables Joe to continue conducting espionage-related activities in plain sight without question.

Joe's badge and WIN ID card are stored in one of the secret parts of his case.

SECTION 3
APPROVED VEHICLES

INTRODUCTION

APPROVED VEHICLES — 3

A vast organisation such as WIN requires access to a wide variety of vehicles which have either been developed internally for specific purposes, purchased from commercial suppliers or requisitioned from other branches of the World Armed Forces.

The following section will specifically identify and detail vehicles that the Project 90 outfit can access at short notice, and that Joe 90 can operate using readily sourced brain patterns from leading experts. As Project 90 develops, this volume will be expanded to reflect equipment added to WIN's inventory.

COMMERCIAL VEHICLES

In addition to the vehicles listed in this section, commercially available vehicles such as helijets, hovervans and boats can be easily sourced and made available to Joe, either through WIN or from private rental services.

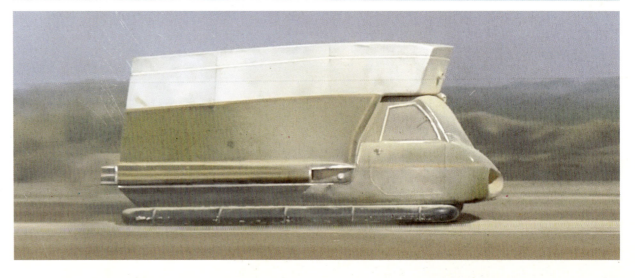

A	Spider (guard model)	2.2 m
B	Hovercraft	4.8 m
C	Sam Loover's car	5.5 m
D	Hovervan	6.0 m
E	Jet Air Car	8.2 m
F	Wildcat	9.3 m
G	Spider (riot control model)	10.4 m
H	A14	12.6 m
I	Recovery Submarine	15.4 m
J	U85	23.4 m
K	VG 104	34.1 m
L	F116	43.4 m
M	AV21 Passenger Jet	65.9 m
N	Orbital Glide Transport	97 m
O	OTC rocket	91.0 m

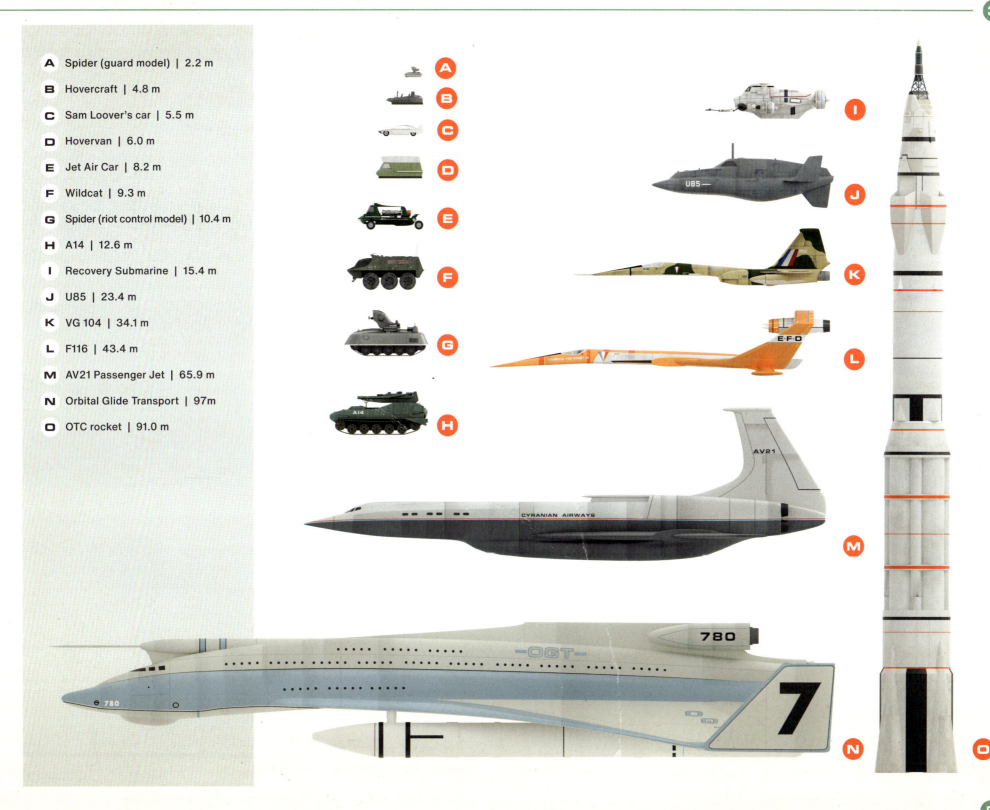

JET AIR CAR

Developed as a passion project by Professor McClaine and constructed in the garage of his cottage seven years ago, the Jet Air Car is a one-of-a-kind vehicle that can seamlessly transition between land, air and sea. Inspired by tales of a similar vehicle that operated in the US in the 1960s, this car is built purely for efficiency and practicality, and to anyone who wasn't an engineer it could be described as an aesthetic nightmare. It has, however, become a popular sight in the local community, and has harmlessly contributed to Professor McClaine's eccentric reputation.

Rather than using a traditional internal combustion engine, the Jet Air Car uses a twin-turbine Rolls Royce engine, more commonly found on jet aircraft, to propel the vehicle. In order to make the jet engine less fuel-intensive, Professor McClaine developed an experimental high-octane fuel that is refined using a special process back at the laboratory giving the vehicle a respectable 24 kilometres to the gallon when on the road. One unfortunate drawback to this large fuel tank is the specialised fuel required to fill it, which is only produced and available from Professor McClaine's cottage, thus limiting the vehicle's range, particularly on international trips. When required on a mission, WIN can airdrop fuel tanks in advance to strategic positions in order to allow Professor McClaine to use the vehicle.

The Jet Air Car's most impressive feature is its ability to transition between being a car, a plane or a boat in a matter of moments. Four VTOL jets on the underside of the vehicle provide lift as the car accelerates, its wings swing into position, and stability fins are deployed. In this configuration, it can accelerate to 482 kilometres per hour. To maximise fuel efficiency, the car is able to redirect the flow of air through its intakes and down through its VTOL, allowing the vehicle to travel on a cushion of air at low altitudes. For this reason, the Jet Air Car will most commonly be seen flying only a few metres above ground level.

Amphibious
The underside of the Jet Air Car resembles the hull of a boat and is watertight. This allows the craft to land on water and use its jet engines to propel it forward at around 70 knots. This would mostly be applicable as a fuel-efficient way to cross the English Channel rather than flying.

EQUIPMENT 3

ROAD LEGAL
The Jet Air Car is comparatively big and takes advantage of a nationwide road widening project from the 1990s to make the United Kingdom more accessible to larger transport vehicles. In order to navigate the traditional country roads around the cottage, the Jet Air Car will most often be forced to fly over them. Normally such an extravagant vehicle wouldn't be allowed on British roads or airspace, however, Professor McClaine has been able to negotiate special dispensation by volunteering his expertise to various institutions.

Now an almost eight-year-old design, the Jet Air Car has undergone multiple refits and iterations throughout its life and will continue to be improved going forward. Professor McClaine is currently trying to develop the high-octane fuel it runs on to be even more efficient, while also trying to develop a way to add water-tight ducts to the jet engine and ballast tanks to allow the craft to function underwater.

The Jet Air Car also holds an additional significance to the McClaine family, in that it was the contributing factor to Joe meeting his adoptive parents when he stowed away on board at the age of five. This degree of sentimentality has contributed to Professor McClaine's desire to refine the vehicle as opposed to building newer models.

CARGO CAGE

A small cargo area is located underneath the Jet Air Car which can be used to house a variety of peripherals such as a long-range brain pattern recorder and magnetic grabs. When not in use it can simply be used for storage.

MILITARY INTEREST

After seeing it in operation, interest in the vehicle's design has been expressed by independent companies and even the world military. However, given his current ties to WIN, Professor McClaine has opted to keep the design of the vehicle and its fuel system under close wraps for the time being.

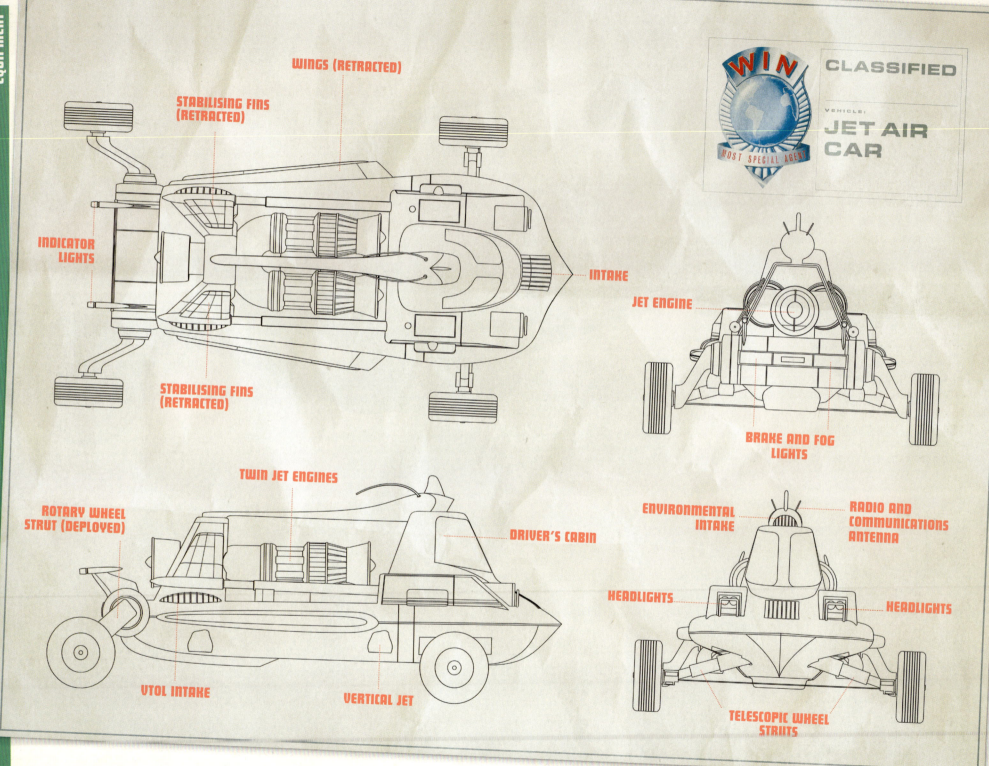

Interior

The Jet Air Car features a three-person cabin with access through the two-hinged windscreens at the front. The driver uses a simplified control system and can easily transition between the various modes of travel. A panoramic view takes advantage of the raised nature of the vehicle. For longer journeys, the third seat can be removed for additional storage.

A Driver's seat
B Passenger seat
C Aft compartment
D Steering wheel (retracts when in flight mode)
E Throttle controls
F Headlight and wiper controls
G Dashboard
H Joystick (retracts when in driving mode)
I Engine readouts
J Intake purge levers

AUTOPILOT
Professor McClaine has designed a simplistic autopilot system for the Jet Air Car that acts as cruise control on highways or allows the vehicle to automatically hover above a location enabling him to focus on other tasks such as operating installed peripherals or deploying the retractable rope ladder from the cockpit.

DASHBOARD
The vehicle's dashboard is dominated by a screen that acts as a rear-view mirror, as well as an aiming system for any additional equipment loaded into the vehicle. The passenger side features a generous compartment that can be used to store documents or tools.

WIN COMPANY CAR

As Deputy Chief of WIN London, Sam Loover requires a vehicle that will allow him to move around the country quickly and easily while also offering additional capabilities to help him conduct his more dangerous WIN duties. Much like the other high-ranking individuals in WIN, Sam was issued a specially developed company car built around an array of gadgets and designed to look like a luxurious yet sporty saloon.

The primary feature of the car is its gas turbine engine, allowing the vehicle to accelerate up to 200 kilometres per hour on a flat surface for speeding to emergency sites or when engaging in car chases. Two side-mounted intakes hold the car to the ground when travelling at this speed.

Among Sam's duties is the transportation of VIPs as well as the head of WIN London, Shane Weston, necessitating the car's armour-plated body and windows for the protection of the passengers. Additional safety measures are incorporated into the seats which are filled with liquid foam to aqua-magnetically hold passengers in the event of a collision. The vehicle's virtual indestructibility can be used defensively with the reinforced nose built in the shape of a point to maximise the ability of the car to ram targets or break through obstacles.

In particularly dangerous situations where the car is likely to not survive a collision, the sunroof can be jettisoned and the driver and passenger seats can be ejected from the vehicle.

RIDING SHOTGUN
Much like the other heads of WIN, Shane Weston is not allowed to drive unaccompanied and will travel either in his personalised armoured car with a chauffeur or with Sam. Most commonly the pair travel together to the cottage in Culver Bay to brief the McClaines on forthcoming missions.

STRIKING
Much like the Jet Air Car, Sam's car looks very conspicuous when parked in relatively quiet English villages. When building the vehicle, WIN engineers deliberately tried to produce something that would be used by a wealthy London businessman since Sam would most often be operating in the city.

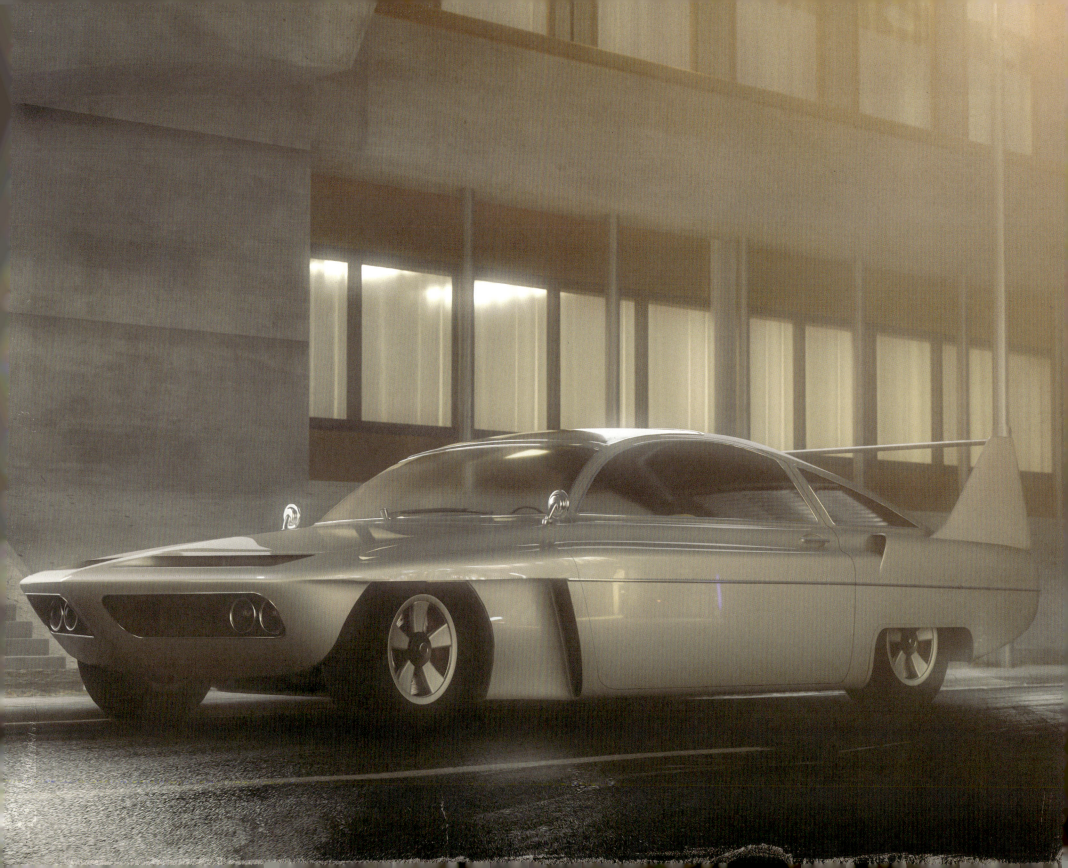

EQUIPMENT

MOBILE OFFICE
As Deputy Chief of WIN London, Sam's car has to effectively double as a mobile office and as such it is equipped with a phone, video screen, fax machine, recording equipment and surveillance cameras.

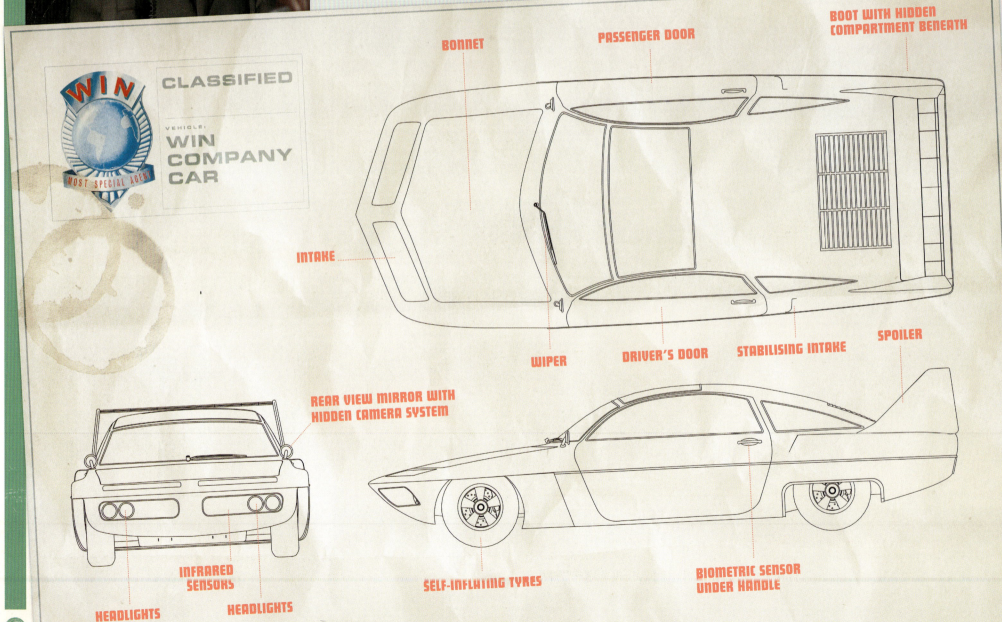

WIN — MOST SPECIAL AGENT
CLASSIFIED
VEHICLE: WIN COMPANY CAR

- BONNET
- PASSENGER DOOR
- BOOT WITH HIDDEN COMPARTMENT BENEATH
- INTAKE
- WIPER
- DRIVER'S DOOR
- STABILISING INTAKE
- SPOILER
- REAR VIEW MIRROR WITH HIDDEN CAMERA SYSTEM
- INFRARED SENSORS
- HEADLIGHTS
- HEADLIGHTS
- SELF-INFLATING TYRES
- BIOMETRIC SENSOR UNDER HANDLE

Sam's saloon is only one of the multiple variations of WIN company cars. Petrana Kopski, the Deputy Chief of WIN Moscow, drives a convertible "aqua car" that is equipped with limited amphibious ability, while Artur Caldeira, WIN's lead operative in Brazil, drives an armoured jeep with telescopic suspension allowing the vehicle to cover almost any terrain.

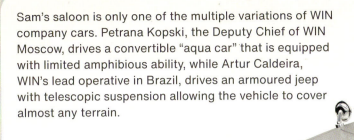

The Boot

The boot of Sam's saloon contains a false floor with a generous secret compartment beneath it, completely vacuum sealed to prevent it from being detected by sniffer dogs. This allows Sam to transport equipment or contraband across borders. An oxygen supply can be fitted in instances where a person may need to be smuggled past security checkpoints.

TEAM TRAVEL

As the Jet Air Car can only carry three people, Sam's car becomes the obvious choice when the whole team needs to travel to a location. Obviously, unlike the Jet Air Car, Sam's car can't fly so when travelling abroad, Sam will either load it onto a specially chartered WIN transporter, or otehwise utilise whatever resources are available on site.

STEALTH HOVERCRAFT

A specialised design for particular circumstances, the Stealth Hovercraft was originally developed by the World Navy as a means of deploying their marines ashore and into hiding under cover of darkness quickly. The hovercraft was chosen over more traditional dinghies as they can not only travel out of water but also quickly traverse obstacles on the beach. While the vehicle air-cushion is a vulnerable target, it is still better armoured than its counterpart.

The main obstacle towards designing a hovercraft built for stealth came in reducing the noise output of the vehicle's fans. Hovercraft are traditionally noisy due to the substantial airflow required to keep the skirt inflated while also propelling the vehicle forward. The solution was to use a single high-powered propellor mounted with specially designed noise-cancelling rotor blades while also enclosing the fan within a noise-dampening chassis. While not cutting out the noise completely, this does reduce the sound to a level comparable with most propeller-driven boats, allowing the craft to be almost silent at slow speeds when moving over solid ground. The chassis also functions to redirect the expelled air both vertically to lift the craft and horizontally to propel it forward.

While the World Navy used to operate six-man models of these vehicles in order to deploy large numbers of marines at once, several two-man models were developed for missions requiring smaller teams. A number of these models have been requisitioned by WIN for use in its more remote Asian and South American operations.

INTERIOR

The vehicle's interior is designed for a driver and a navigator with a single forward canopy to prevent the driver's view from being obscured. A tarpaulin can be pulled over in order to protect the occupants from rain or insects, but ultimately the vehicle's weak armour and lack of cover make it less than ideal for direct combat.

Original

The original larger transport version of the craft is very similar to its smaller counterpart with a wider chassis and larger fan in order to account for the greater passenger capacity. This version can be heavily modified in order to carry cargo or mounted machine guns.

CLASSIFIED

VEHICLE:
STEALTH HOVERCRAFT

STATS:
Crew complement: 2
Max speed: 75 kph
Range: 450 km

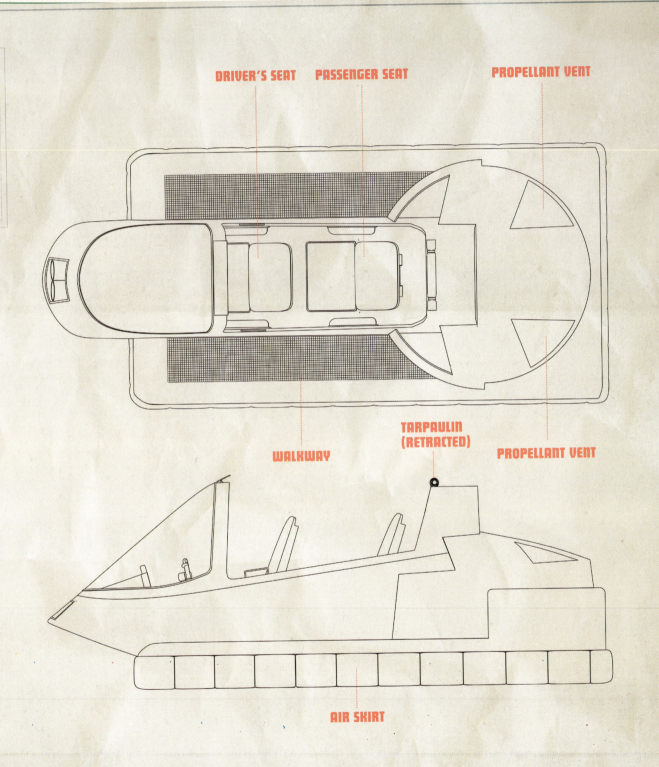

- DRIVER'S SEAT
- PASSENGER SEAT
- PROPELLANT VENT
- WALKWAY
- TARPAULIN (RETRACTED)
- PROPELLANT VENT
- AIR SKIRT

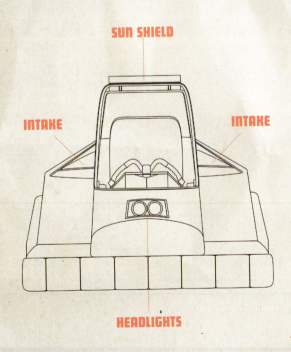

- SUN SHIELD
- INTAKE
- INTAKE
- HEADLIGHTS

Performance

In spite of reduced power necessitated by its small size, the hovercraft is fast, agile and can pull off some impressive manoeuvres when in the hands of an expert. It has been recorded achieving speeds of up to 75 kilometres per hour, and can even "jump" over small obstacles if the driver possesses the critical timing and skill.

PROJECT 90

When WIN Agent Flemming was captured and brutally interrogated to extract the location of a secret microfilm, Joe 90 was dispatched to rescue him armed with the brain pattern of Flemming's injured partner, Agent Laramie.

Laramie had concealed the hovercraft they used to enter the country in a boathouse several kilometres downstream from the facility Flemming was being held captive in. The craft's innate stealth characteristics and ability to glide over the hidden minefield guarding the fortress made it the perfect choice to sneak in and rescue Flemming.

DEFLATED

Once landed and deflated, the vehicle has a surprisingly small profile which allows it to be easily concealed and stowed away. The camouflage can be updated to assist with this to the point of fixing fake foliage and debris to the exterior of the vessel.

DEEP SEA RECOVERY VESSEL
– DSRV-3 ATHENA

Built as part of a run of three back in the 1980s, *Athena* and its sister vessels were initially commissioned by the British Navy for recovering hazardous materials from sunken ships and submarines. Their success in the field caught the eye of multiple scientific and archaeological research groups who used them frequently in expeditions including an extended mission to chart the seabed under the Arctic Circle by the famous explorer Dr William Kelvin.

The Deep Sea Recovery Vessel (DSRV) is designed around a sturdy central hull, constructed to survive the pressures of the depths for long periods of time, and four gimballed thrusters that can allow for precise movement or station-keeping. A forward-facing canopy can be modified to fit one or two aquanauts and a high-powered forward-facing light rig. Two major hard points on the front of the craft can be equipped with manipulator arms, sample capture devices, cutting equipment and other mission-specific peripherals. Smaller cameras, lights and sensors can be mounted at any part of the craft depending on the tasks required.

Since the dissolution of the Royal Navy and transfer to the World Navy, all three DSRVs were sold to private organisations. *Athena* is currently owned by the International Oceanic Exploration Organisation (IOEO), a company that WIN secretly owns a stake in allowing for the submarine and its parent vessel to be requisitioned and deployed at short notice while maintaining the cover of ocean exploration.

POLARSTERN
Athena is currently housed aboard the IOEO nuclear icebreaker *Polarstern*. Commissioned to operate in the arctic for long periods of time while also able to provide the crew with a reasonable amount of luxury, *Polarstern* is probably one of the most advanced vessels of its type in the world. With a displacement of 13,000 tonnes and able to cut through ice as deep as two metres at a speed of almost 10 knots, it is also one of the fastest.

COCKPIT

While traditional minisubs would have used a spherical canopy that was structurally more efficient, *Athena* uses specially designed glass to provide a canopy that can stand up to the same amount of pressure, while also allowing for a much wider field of view. The cockpit is a standard design and can be easily modified for one or two crew.

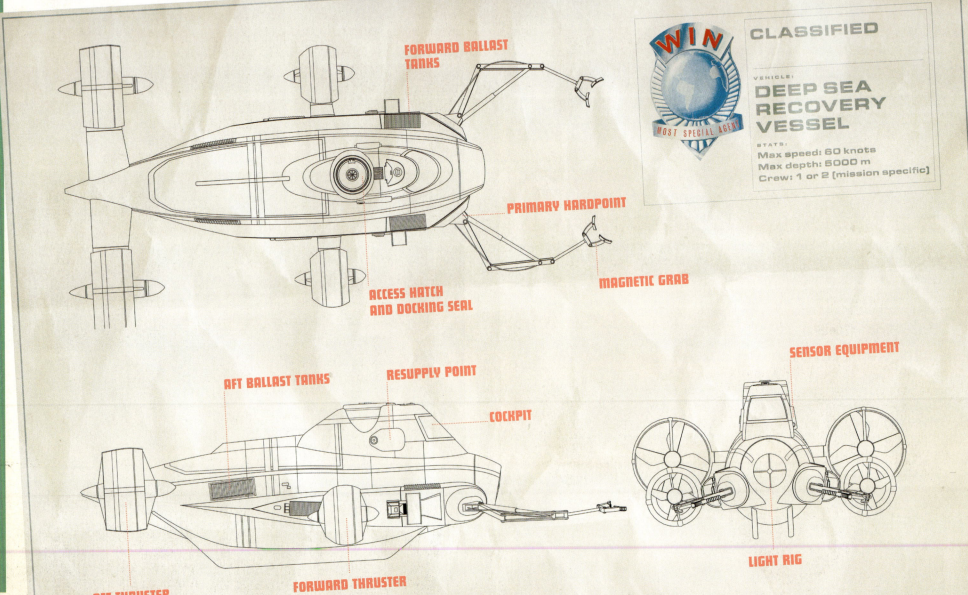

CLASSIFIED

VEHICLE:
DEEP SEA RECOVERY VESSEL

STATS:
Max speed: 60 knots
Max depth: 5000 m
Crew: 1 or 2 (mission specific)

- FORWARD BALLAST TANKS
- PRIMARY HARDPOINT
- MAGNETIC GRAB
- ACCESS HATCH AND DOCKING SEAL
- AFT BALLAST TANKS
- RESUPPLY POINT
- COCKPIT
- SENSOR EQUIPMENT
- LIGHT RIG
- AFT THRUSTER
- FORWARD THRUSTER

Modified

Over their lifetimes, each DSRV has been heavily modified from the original design in order to carry out the various missions they would perform. Although, at first glance, *Athena* appears antiquated to many of the newer and sleeker models in service today, its service record and the numerous accolades it has accumulated over the years have proven that the craft can still hold its own when in the hands of an experienced aquanaut.

PROJECT 90

A nuclear missile disappeared following the crash of a World Air Force transport plane in the Arctic Circle dangerously close to the Eastern Alliance sector. Fearing the missile punched through the ice and drifted into enemy territory, Shane Weston called in Joe 90. Equipped with the brain pattern of Dr William Kelvin, Joe used a DSRV minisub to follow the trail of debris into alliance territory where he discovered the bomb wedged in a crevice. Using his adopted experience, Joe was able to free the bomb and evade hunter-killer submarines dispatched from the nearby Vostula base, recovering the bomb and averting a diplomatic incident.

FAST ATTACK VEHICLE – A14

With large-scale battles and full-scale wars hopefully a thing of the past, the World Armed Forces have shifted their doctrine from mass-produced battle tanks and armoured vehicles designed to overwhelm targets and fortify defences, to smaller numbers of highly advanced surgical strike vehicles. The logic behind this strategy being that a single confrontation is much easier to address politically than a protracted assault that could escalate into a larger conflict.

Such vehicles are designed to strike without warning, break through defences, and destroy their targets as quickly as possible. The first of its kind to pass the prototype stage was the A14 Fast Attack Vehicle.

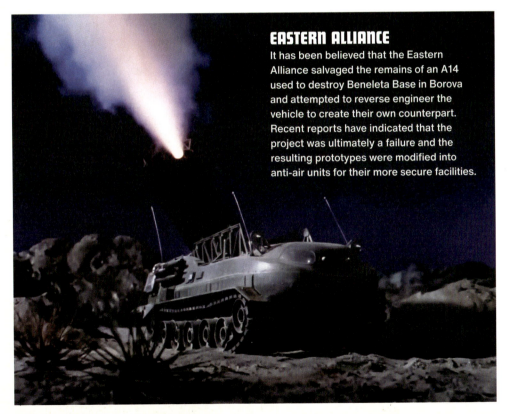

EASTERN ALLIANCE
It has been believed that the Eastern Alliance salvaged the remains of an A14 used to destroy Beneleta Base in Borova and attempted to reverse engineer the vehicle to create their own counterpart. Recent reports have indicated that the project was ultimately a failure and the resulting prototypes were modified into anti-air units for their more secure facilities.

AMPHIBIOUS
The amphibious nature of the A14 gives it a huge advantage in navigating terrain that would be impassable by regular vehicles. This also makes the craft harder to detect and intercept as it can easily traverse lakes, swamps, gorges and walls. In addition, the craft can be stowed aboard World Navy submarines and safely deployed up to three kilometres from shore.

APPROVED VEHICLES 3

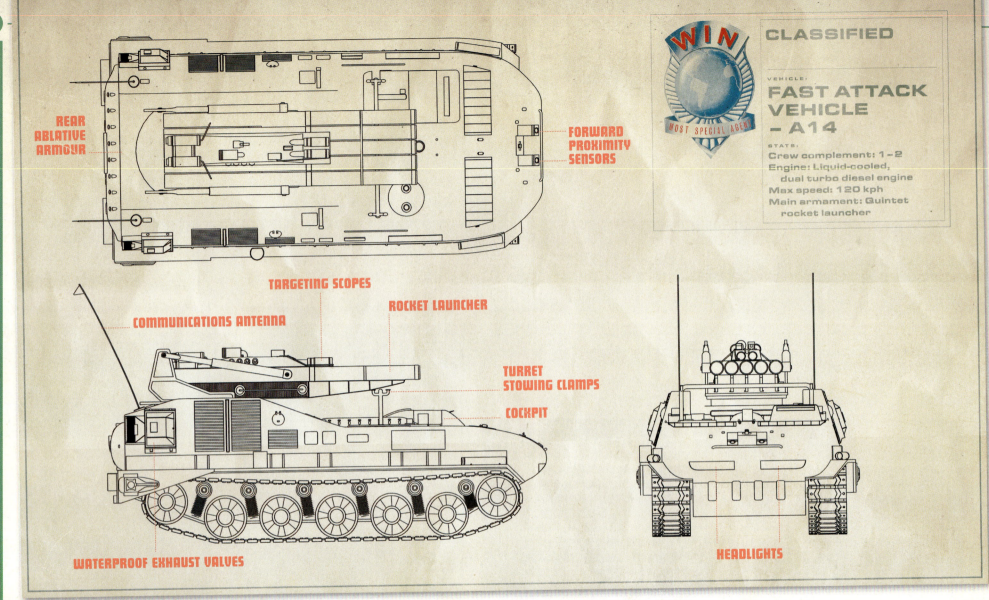

REAR ABLATIVE ARMOUR
FORWARD PROXIMITY SENSORS
TARGETING SCOPES
ROCKET LAUNCHER
COMMUNICATIONS ANTENNA
TURRET STOWING CLAMPS
COCKPIT
WATERPROOF EXHAUST VALVES
HEADLIGHTS

CLASSIFIED

VEHICLE:
FAST ATTACK VEHICLE – A14

STATS:
Crew complement: 1–2
Engine: Liquid-cooled, dual turbo diesel engine
Max speed: 120 kph
Main armament: Quintet rocket launcher

INTERIOR
The interior cockpit fits one driver but can be modified to carry a gunner or navigator as well. The only window is a small slit that the driver can see through so as to not compromise the armoured hull. Cameras located throughout the hull provide an almost 360-degree view assisting in navigation. These cameras also come in play at night if the vehicle has to move under the cover of darkness with its headlights off. Forward proximity sensors can help map out the environment in front of the vehicle, relaying an accurate picture to the driver.

EJECTOR SEAT
Should the hull of the A14 be compromised, an ejector seat is fitted as a means of escape. Powerful rockets blast the pilot clear of the vehicle and up to 200 metres in the air before deploying a chute. Toggles are provided to assist in a controlled descent. The cabin is offset to the side of the vehicle so that the pilot can eject safely without hitting the turret.

PROJECT 90

A coup d'etat in Borova resulted in the country reneging on its treaties with the World Government and demanding the immediate withdrawal of the World Army from its base at Beneleta. A swift withdrawal was promised on the condition that Borova would ethically disarm and dismantle the base. This agreement was not honoured by the new military government, and Borova was suddenly in possession of immense military power which greatly outmatched neighbouring countries. WIN was tasked with the mission to destroy the base and restore the balance of power in the region. The mission was set up but Colonel Henderson, who was due to undertake the attack, was injured in a training exercise.

Following a little bit of subterfuge in order to get Professor McClaine to agree to the mission, Joe 90 took the control of a specially adapted A14 packed with explosives with the aim of crashing the vehicle into the missile store and starting a chain reaction to destroy the base. Joe was easily able to dodge the base's defences and eject one kilometre before the vehicle collided with its target and utterly destroying Beneleta Base.

Performance

The high-speed, heavily-armoured A14 and its powerful forward-mounted rocket launcher represent a bold new direction in the World Army strategy for maintaining world peace. In combat, the vehicle can be deployed undercover via airdrop or submarine and wait until it receives clearance to attack. Once the go-ahead is given, the vehicle will make a direct run at the target ignoring anything in its path within reason. If enemy forces are able to mount a defence in time, the A14's hull is strong enough to deflect any shells or missiles used against it. Ultimately, when the A14's final target comes into range, the vehicle can unleash a barrage of rockets, either surgically or spread across an area that can obliterate most surface structures.

APPROVED VEHICLES 3

ORBITAL TRANSFER COMPANY
– TRITON OC2 ROCKET

The flagship vessel of the Orbital Transfer Company (OTC), the Triton OC2 Rocket is an adaptable, multi-stage rocket capable of lifting heavy payloads into orbit for the construction of large satellites and space stations. Launching from bases such as Cape Canaveral USA, Cornwall UK and Uchinoura Japan, the OTC fleet is responsible for building and maintaining multiple World Government-backed space projects.

Developed as a successor to the Saturn programme of rockets, the Triton series would attempt to increase the size of the payload while improving the fuel efficiency and thrust in order to keep the rocket a manageable size. After nearly 15 years of research and development, the Triton series entered active service in the late 1980s and remains in operation today.

While the Tritons have enjoyed many years of an almost flawless safety record, the age of the technology has become apparent in the early 2010s with the number of aborted launches and complete system failures beginning to rise. Years of an "if it's not broken, don't fix it attitude" at OTC has meant that the development of a replacement rocket has fallen behind. With the company forming a major part of WIN's satellite network, WIN has had to step in and push the company to rectify this issue in preparation for a potential satellite blackout in the future.

VEHICLE ASSEMBLY BUILDING
Most OTC facilities are large enough to have multiple separate vehicle assembly buildings designed to house OC2's and similar rockets, with a dedicated launch rail that can efficiently move them to the launch pad. Once assembled, a launch can be organised in under 36 hours, though that can affect the quality of the pre-launch checks.

MISSION CONTROL
The launch is monitored by the Base Control Centre, its reduced size from mission controls of old reflecting the more routine nature of space travel in the year 2013. From here, all aspects of the vehicle can be remotely controlled and monitored, and in extreme situations the vehicle can be self-destructed should it be heading for a sensitive or populated area.

OTC Stages Breakdown

Like multi-stage rockets before it, the OC2 uses multiple engines and fuel tanks on its way to orbit, shedding them as they become depleted. Unlike the Saturn V, only two stages are required to get the payload capsule into orbit. These detach and deploy parachutes allowing them to soft-land in the ocean for reuse.

- **A** Launch escape system
- **B** Crew capsule
- **C** Supply pod
- **D** Second stage
- **E** First stage

OC2-19 Supply Shuttle

A common sight mounted on top of the OC2 is the OC2-19 supply shuttle, an orbital vehicle that is designed to deliver supplies to space stations. This shuttle can be outfitted with a Cargo Transfer Pod that automatically delivers supplies to a station once docked, or a Maintenance Pod which features external cargo doors that can deploy a small satellite or allow a team of astronauts to work on an orbital structure.

DOCKING PROCEDURE

Upon arriving at an OTC station, the capsule can use the homing beacon to guide the vehicle into its docking berth where air, water and other supplies will be transferred automatically. Should there be a fault with the station's homing beacon, supplies can be connected manually using extra vehicular activity (EVA) suits.

CLASSIFIED

VEHICLE: OC2-19 Supply Shuttle

- RCS THRUSTERS
- STATION DOCKING SYSTEM
- CAPSULE DOCKING CLAMPS
- MAIN ENGINES
- ENTRY HATCH
- HEAT SHIELD
- CREW CAPSULE THRUSTERS
- CREW CAPSULE
- TELEMETRY ANTENNA
- COMMUNICATION ANTENNA

PROJECT 90

Complications with OTC's supply fleet and an accident at the Cornwall facility had left an OTC space station dangerously close to running out of air. With the station secretly forming the basis of WIN's new global positioning system, Shane Weston was forced to take over the rescue mission, and with no astronauts available, the position of pilot fell to Joe 90. Equipped with the brain pattern of Charles Drayton, Joe was able to successfully dock at the station and save the crew.

U59 WILDCAT

Keeping such a vast operation as the World Army running is a massive logistical effort with personnel, equipment and munitions needing to be transported all over the world in some of the roughest terrain imaginable. Most countries have a type of heavy transport vehicle, usually improvised from a civilian design with varying degrees of success. It quickly became apparent that a mass-produced "workhorse" should be commissioned in an effort to provide a standard of equipment across the entire force. This design became the U59.

Nicknamed "Wildcat" during production, the U59 entered service in the mid-1990s and quickly became a common sight across bases worldwide. A spacious and easily modifiable cargo bay, a powerful diesel engine that can achieve speeds of almost 100 kilometres per hour (unladen) and an impressive suspension that can clear incredibly uneven terrain while providing a reasonably smooth ride for the occupants. While unable to float, the vehicle can cross flooded areas with its intakes and exhausts mounted on the top of the vehicle. The hull of the craft has been hardened to repel most standard armour-piercing rounds, though it is generally not designed for combat, instead using its speed and manoeuvrability to repel fire.

The vehicle is designed to undergo varying levels of modification once in the field making it quite rare to find two in operation that are the same. Tow cables, weapons, snowploughs and many other peripherals can be mounted to the various hard points around the body.

INTERNATIONAL USAGE
While the main buyer for Wildcats has been the World Army, a few older models have been purchased for private or government use. Cherook Penitentiary located high in the Canadian Rockies, for example, utilises two of these vehicles for transporting prisoners.

WILDCAT VS. WARTHOG
While the Wildcats represent the workhorse of the World Army logistical fleet, the U87 "Warthog" is, in many ways, its newer, faster cousin reserved for transporting troops and command personnel across light to mid-terrain quickly. While much more stylish, it lacks a lot of the hard terrain capabilities of the Wildcat meaning it is most often reserved for transporting VIPs in the field.

TOW CABLE
The main aspect that makes the Wildcat so popular is how effortlessly it is able to traverse difficult terrain. The most extreme example is a variant of the vehicle with a forward mounted tow cable that it can use to climb near-vertical inclines. This manoeuvre is, however, considered extremely dangerous and must be carried out by an expert.

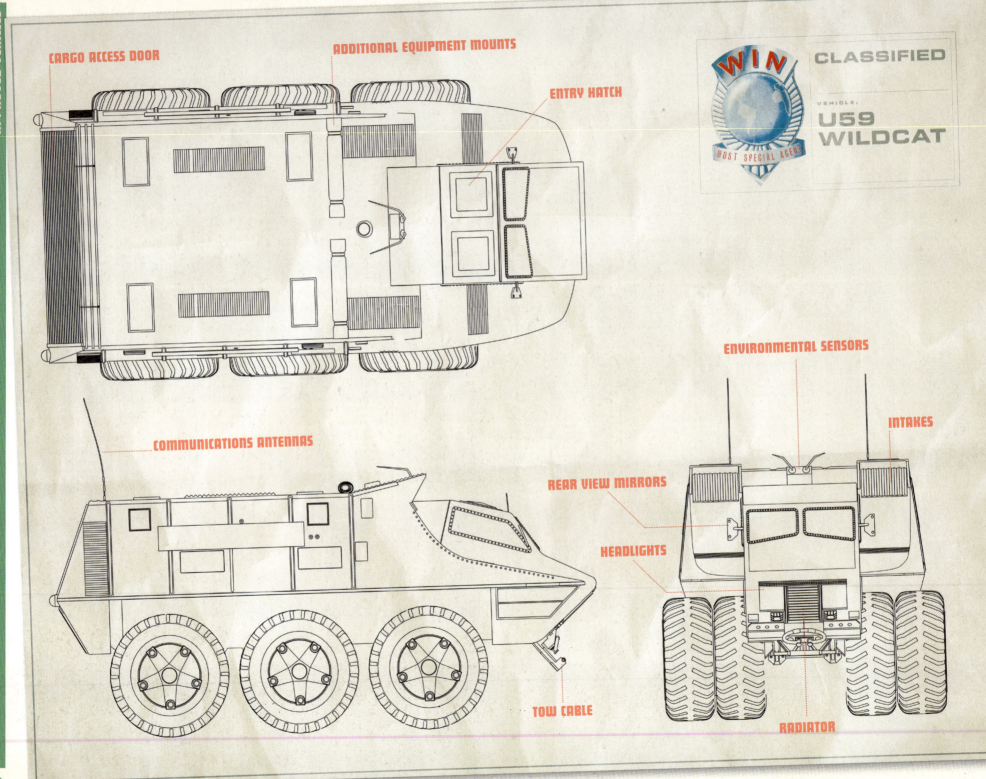

PROJECT 90

When corporate sabotage halted efforts to build a tunnel to link two African nations, WIN was brought in to ensure a highly volatile explosive material, U114, could be transported overland to the dig site. With the brain pattern of an expert driver overlaid on the brain pattern of an explosives expert, Joe took command of a convoy of Wildcats along with Sergeant Files and Private Johnson.

Sergeant Files was unfortunately killed when two of the Wildcats collided during a difficult hill climb in rough weather. Realising the explosives that they were carrying had been sabotaged by a corrupt Lieutenant and were rigged to detonate at 5,000 feet, Joe drove the one remaining vehicle down a sheer hillside. Once at a safe altitude, Joe disembarked and Private Johnson delivered its load of explosives to the destination, Katunga.

Accessories

The Wildcat outer shell is usually covered in hooks and hardpoints that are used to carry emergency equipment such as ropes, tents and medical items. Two tarpaulins are stowed on top of the vehicle which can be pulled out to shelter troops from the sun or rain when stopping for longer periods of time.

APPROVED VEHICLES | 3

VG 104

In keeping with the modern doctrine of the World Armed Forces, the VG 104 was designed as a "lone-wolf" surgical strike fighter. The design needed to be fast enough to evade fighter defences, manoeuvrable enough to fly low to the ground to avoid radar, and big enough to deliver a hefty payload which could eliminate a target with just one or two direct hits. The result of this brief is a surprisingly large but super sleek fighter aircraft that, for the moment, is unparalleled in its field.

The VG 104 is modular in design with the base frame of the aircraft featuring several hard points where additional fuel canisters and bombs can be fitted. Rather than attaching these to the wing, they are added via nacelles connected to the body itself; this way additional lift is generated that offsets the extra payload. Once spent, these can be jettisoned for the return trip.

The unmodified aircraft stores missiles and rockets within the body of the craft itself to prevent drag, while a small bomb bay is located on the underside.

As advanced as the aircraft is, its main weakness lies in its air-to-air capability. The larger size of the aircraft makes for a bigger target in the sky and a well-placed hit can easily penetrate its light armour. It is believed the Eastern Alliance are developing aircraft and missiles to counter the VG 104's speed. If successful, then the operational effectiveness of the craft will be greatly negated.

MID-AIR REFUELLING

The VG 104 has an operational range of 3,400 kilometres with additional fuel tanks in place, and either needs to be delivered within range of its target or refuelled in flight. The front of the craft contains a refuelling probe that allows it to dock with a tanker and hold position while its tanks are resupplied. This method is considered preferable as not all runways can accommodate the VG 104's requirements for take off.

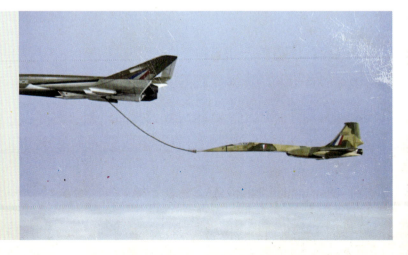

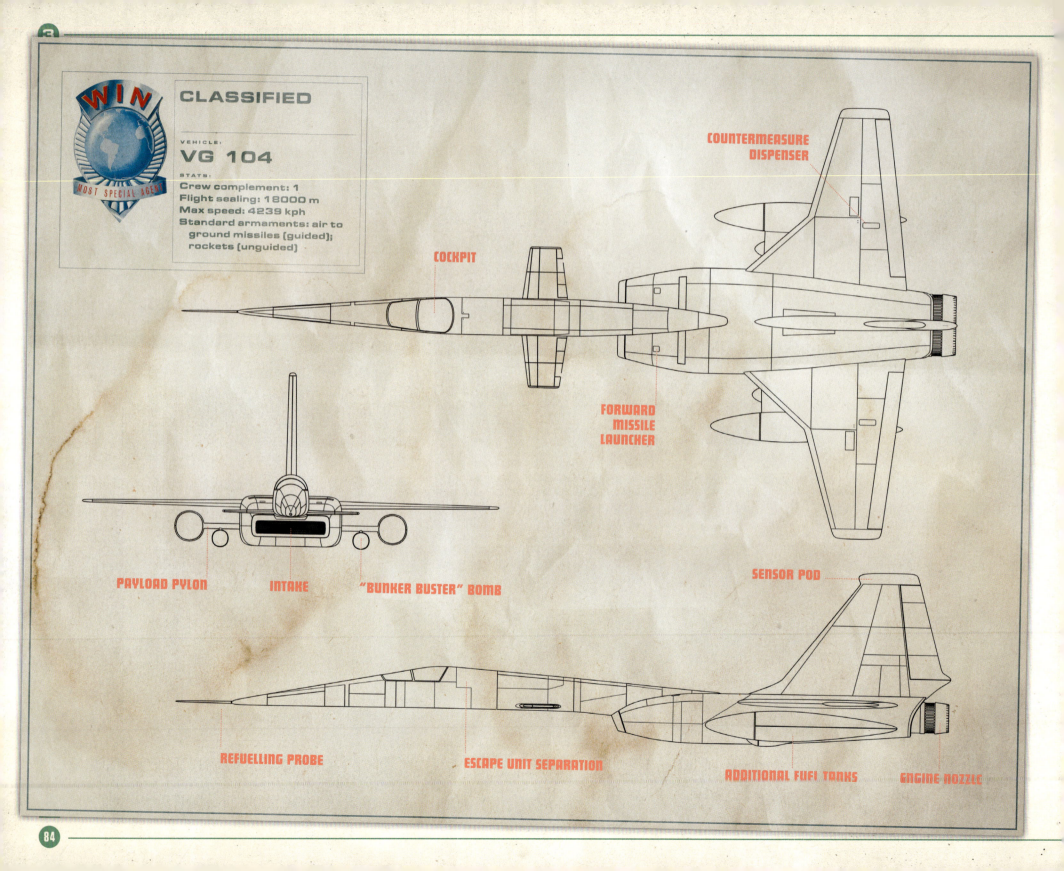

PROJECT 90

As part of a mission into Eastern Alliance territory to destroy a clandestine launch facility threatening to put a nuclear device into orbit, Joe flew a modified VG 104 under the callsign "Tiger" with additional fuel tanks added to extend the craft's operational range. Two "bunker buster" bombs were fitted to penetrate the protective shield the base was hidden under and the cockpit was also modified to fit Joe's significantly smaller frame. While Joe's initial attack run only did superficial damage to the base, his inherited skill allowed him to evade the aerial defences and then bullseye the rocket on the launchpad destroying the base through a chain reaction.

The modified components for the VG 104 have been retained in storage, enabling Joe to more easily use the aircraft in future missions.

MIG 242

The VG 104 was developed in direct competition with the Russian branch of the World Air Force and their design, the MIG 242. A slightly smaller and more robust craft, the MIG 242 displayed great potential to be a better option when confronting air-to-air targets but ultimately development was stalled in favour of the much more versatile VG 104. The MIG 242 has continued to live on as the subject of a hypothetical WIN training scenario for new recruits in London and Washington.

FEARLESS FOLEY

The VG 104 is an extremely complex aircraft that requires extensive simulator training before pilots are allowed to train on the aircraft in flight. Only five pilots in the World Air Force are considered adept enough to take the VG 104 out on missions including "Fearless Foley", the famous test pilot who worked on the aircraft through its experimental testing phase. Foley's knowledge and experience made him an ideal candidate for Project 90's purposes.

APPROVED VEHICLES

F116

While still one of the most advanced aircraft ever developed, there have been signs of the Eastern Alliance developing countermeasures to the VG 104 that, if implemented, run the risk of significantly reducing its effectiveness. In an attempt to stay ahead of the air superiority arms race, the Experimental Flight Division has been developing the F116 as a potential replacement.

The two main aspects of the F116 that make it truly special are the extremely sleek shape of the fuselage which allows the craft to maximise the amount of lift at high altitudes and the three high-power jet engines that hang from the reinforced tailplane.

Under the assumption that the Eastern Alliance have been able to beat the VG 104's speed, the F116 is designed to function at extreme altitude, beyond the range of enemy defences. The craft can then divebomb a target at Mach 4 and pull up to return to its flight ceiling, outrunning missiles and intercepting fighters.

Still in its flight-test phase, the F116 hasn't yet flown with armaments but the design has left room for multiple hardpoints to be fixed to the underside of the craft. In order to maximise the potential of the craft's sub-orbital bombing ability, a new kind of ordinance will need to be designed that can be deployed accurately at intense speed. At this stage, it is believed that the F116 would be too fast to be able to properly engage with other fighters so there are no plans to incorporate air-to-air missiles or machine cannons.

As the development of the craft is still at the experimental stage, WIN has shown an interest in the craft as a spy plane. Tests are being conducted by adding a series of cameras under the fuselage in order to take high-altitude photographs of sites and therefore reducing the need to redirect satellites.

F116 LANDING

The F116 is currently being tested at the Experimental Flight Division facility in Snowdonia, Wales. While the existence of the plane is known to the public, the World Armed Forces are keen for the vehicle to be kept out of view for the time being and a one-kilometre radius exclusion zone around the base is maintained at all times. While that may not prevent the vehicle from being seen in the air, its high-speed and agility would render any photographs taken by enemy agents or corporate spies while the F116 was in-flight effectively useless.

APPROVED VEHICLES

CLASSIFIED

VEHICLE:
F116

STATS:
Crew: 1, potentially 2 on future models
Top speed: Mach 4
Flight ceiling: 22000 m

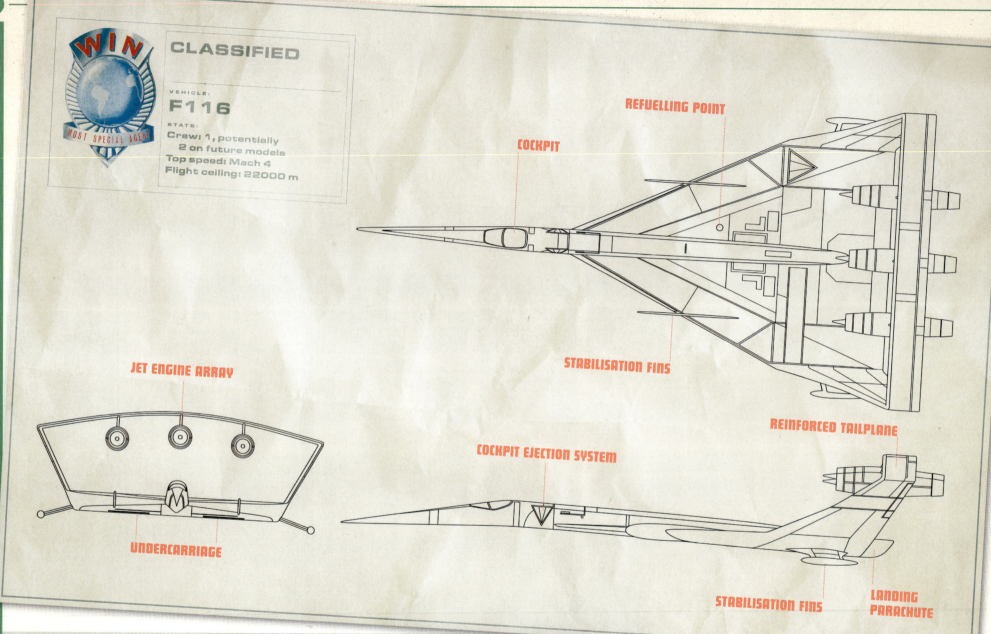

- REFUELLING POINT
- COCKPIT
- STABILISATION FINS
- REINFORCED TAILPLANE
- JET ENGINE ARRAY
- COCKPIT EJECTION SYSTEM
- UNDERCARRIAGE
- STABILISATION FINS
- LANDING PARACHUTE

COCKPIT

The cockpit of the F116 is the standardised cockpit used on multiple World Armed Forces designs. This allows equipment parts to be transferrable across multiple types of aircraft allowing for ease of maintenance and supply. The cockpit also incorporates two forms of escape unit: a standard ejector seat for slower speeds and lower altitudes; and a nose section that allows the entire cockpit to detach and be propelled to safety.

PROJECT 90

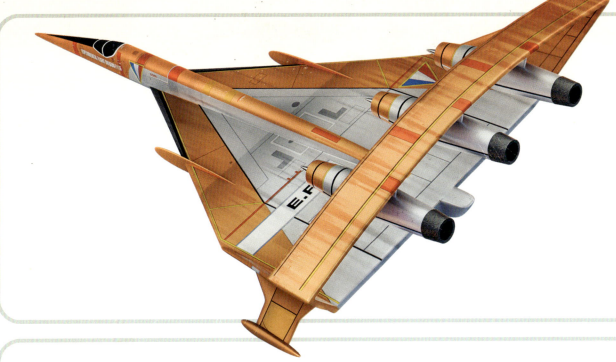

Wraparound Tail

The concept of using a wraparound tail on hypersonic vehicles has gained traction within the last few years with various bombers and transport craft adopting the design to take advantage of the stability it provides. The F116 is probably the most extreme version of this design to date with much of the craft essentially acting as a lifting body.

The F116 is currently painted in the iconic, high-visibility colours of the Experimental Flight Division. Once it enters service it will be repainted in line with other craft such as the VG 104. The version WIN is considering will most likely sport a semi-reflective livery in order to make it harder to see when in the air.

When one of the F116 prototypes mysteriously crashed during a test flight, WIN was brought in to ensure the security of the second prototype when it was being presented to the World Armed Forces top brass.

As an added precaution, Joe was given the brain pattern of Jim Grant, the programme's chief test pilot who was injured during the first crash. Once Joe was airborne and the presentation was underway, it came to light that Grant had suffered a memory blackout stemming from a fear of landing which had caused him to crash - a blackout that Joe was now experiencing. Professor McClaine and Sam Loover quickly forced Grant to face his fears and talk Joe through the landing procedure. Joe was able to bring the aircraft down safely and the F116 programme was ultimately deemed successful.

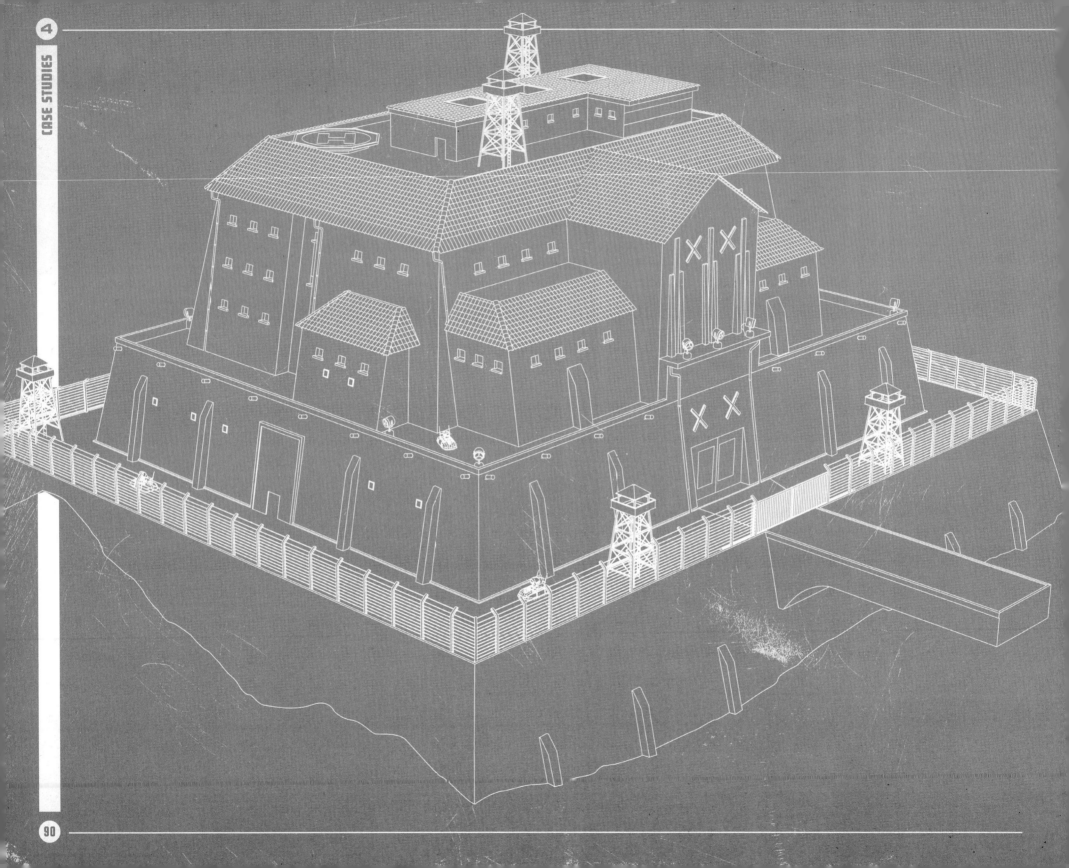

SECTION 4
CASE STUDIES

PROJECT 90

MISSION CASE FILES

AUTHORISED ACCESS ONLY

TOP SECRET

MIG-242 FACILITY OKOSK
CAPTURED JAN 07-2013

U85 - ATTACK SUBMARINE

SPIDER ANTI RIOT VEHICLE
CAPUTRED 17 AUGUST 2013
TESTING FACILITY LOCATED IN NORTH PELAGIA

VOSTULA BASE
CAPTURED SEPT 14-2011
AGENT CHRIS 21

FORWARD STABILISING FIN

AFT STABILISING FIN

SPECIAL AUTHORISATION BY SAMUEL LOOVER

The World Intelligence Network hereby authorises the declassification of specified case files related to the operations undertaken by Special Agent Joe 90 and the Project 90 team. The BIG RAT and Joe 90 have ensured the protection and security of the world against some of the most extraordinary criminal masterminds bent on spreading terror and destruction.

These files have been selected for release from the vault and added to this document in order to demonstrate the expansive opportunities that the BIG RAT offers for maintaining world peace. They also demonstrate the technological advances which have been developed or manipulated by the crime and terrorist organisations that WIN is committed to combating.

It is my hope that the information contained within these files will be used for the benefit of humankind so that others may learn from the groundbreaking work undertaken by Project 90 for the continued protection of global security.

Regards,

Sam Loover
Deputy Chief of WIN London

PROJECT 90

MISSION CODE NAME:
HIJACKED

CONFIDENTIAL

DATE: ███████████
APPROVAL: ███████████
CASE No: ███████████

SUPPORTING IMAGERY:

Mario Coletti, the ruthless gunrunner, had set up operations in England to illegally intercept and distribute armaments to politically unstable nations. Undercover WIN agent, Edward Johnson, obtained plans for an ambush to be carried out on a shipment of automatic rifles supplied by Hudson Armaments.

Joe 90, equipped with Johnson's brain pattern, was smuggled aboard the shipment in a packing case and the hijack was carried out as scheduled by Coletti's operatives. The packing case containing Joe was transported into the heart of Coletti's underground hideout where the stolen arms were to be processed for shipment. Having gained access to top security areas, Joe was eventually discovered by Coletti's gang and a gunfight resulted in Joe's capture. A car crash was staged to dispose of Joe but he escaped from the locked vehicle before it made impact.

Regaining access to Coletti's facility, Joe was engaged in an armed showdown with Coletti himself. The battle concluded when Joe launched a grenade attack, killing Coletti and devastating his supply of armaments in a blazing inferno. Joe was overcome by the fire and found unconscious in the complex's security office by Sam Loover and Professor McClaine, who traced Joe's location via a homing device. The surviving members of Coletti's gang were captured and later imprisoned. Joe quickly recovered from the ordeal which proved to be a successful first assignment for the young agent.

BRAIN PATTERN SPECIFICATIONS

NAME: Edward Johnson

CODE: E14

BORN: February 16, 1978

PLACE OF BIRTH: Liverpool, England

KNOWLEDGE & EXPERIENCE:
Undercover operations. Seven years of field work in Southern Europe infiltrating mafia groups including three months tracing Mario Coletti's gun running operations. Eight years of military service. Expert marksman. Explosives training.

HUDSON ARMAMENTS

Shipping Invoice

SHIPMENT: T120 - October 20, 2012 - 19:30

DEPART:
Hudson Armaments,
Warehouse 7A, Barwick Road,
Peterborough,
PE6 9NF

SHIP TO:
Gordon Barracks,
Bridge of Don,
Aberdeen,
AB23 8DB

QTY	DESCRIPTION	UNIT PRICE
300	MK.17 AUTOMATIC RIFLES	£2,200.99

TOTAL: £660,297

NOTE: MILITARY POLICE ESCORT REQUIRED

TRANSCRIPT:

Micro-recorder - property of WIN Agent: Edward Johnson

Recovered by Samuel Loover following an attack on Johnson in London on October 20, 2012.

_ REPORT A-17...
_ JOHNSON...
_ 1100 HOURS...
_ FOX WILL ATTACK...
_ 2100 HOURS TODAY...
_ TARGET: SHIPMENT TANGO 120...
_ WILL CONFIRM...
_ OUT.

PROJECT 90

CONFIDENTIAL

WORLD INTELLIGENCE NETWORK

::CRIMINAL FILE::

NAME:	KNOWN ASSOCIATES:
MARIO COLETTI	JOHN KREGSON, CARL DAVIS, HARRY CARTER (DECEASED)
BORN: JUNE 26, 1965	PLACE OF BIRTH: SICILY, ITALY

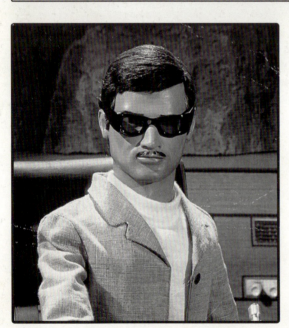

STATUS:

DECEASED

AFFILIATION:

"FOX" GANG

REPORT FILED BY: ▓▓▓▓▓▓▓▓▓
DATE: ▓▓▓▓▓▓▓
REF: ▓▓▓▓▓▓

Born into a Sicilian mafia clan, Mario Coletti was no stranger to murder and extortion. His father climbed the ladder of the family through fair means and foul, but Mario held little respect for his elders and their strict code of conduct. Mario watched his brother's unjust execution at the hands of family members. The young Coletti knew then that the institution was rotten to the core and he had to get out.

On his 16th birthday, the penniless Coletti fled and lived rough on the streets of London. Fearing the consequences of abandoning the mafia, Coletti formed a gang of like-minded criminals who served as his own private army. This group of talented con artists and thieves proved a great source of income for Coletti. Utilising his natural business acumen, he invested his employees' ill-gotten gains and was able to provide luxurious accommodation in exchange for services rendered.

The majority of the gang's fortune was invested in the arms trade. Coletti saw ample opportunity in selling weapons to the rebel factions across Europe and Asia, battling the global governments. He learned the security weaknesses of some of the world's biggest arms manufacturers, and soon realised that there was no need to purchase weapons legitimately when they could easily be taken by force. Coletti built an empire on capturing arms shipments and reselling them to the highest bidder fast enough to avoid detection. His ruthless attention to detail created a flawless operation which earned him millions. Payment was always in gold.

Coletti was committed to his work, and would only spend his wealth on assets and travel to further expand his gun running business, eluding world security forces for years with his increasingly sophisticated vehicles and technology. It was only when Coletti placed his trust in a new recruit, undercover WIN Agent Edward Johnson, that the operation was brought to its knees.

MISSION: HIJACKED

SUPPORT FILE:
0009+0010

COLETTI'S WAREHOUSE FACILITY

The base for Coletti's British operations housing armament shipments, vehicle depot, offices and living accommodation. Located in Buckinghamshire, England and damaged by fire, the remains of the underground complex have now been acquired by WIN for testing top secret equipment.

The main entrance to the complex was concealed inside an abandoned barn. The back wall could be raised remotely by a control inside the security office,

allowing access for vehicles. The barn, and many of the surrounding farm buildings, were destroyed when the armaments inside the base exploded.

HOVERVAN

Two cargo trucks from Coletti's fleet of vehicles were impounded by WIN during the operation. It is thought that the hovervan was chosen for intercepting armament shipments due to its range, its versatility in all terrains and its near-silent running. Such a vehicle would be ideal for transporting heavy goods over long distances in a covert manner. It is capable of speeds up to 110 miles per hour and is commonly used by military courier services.

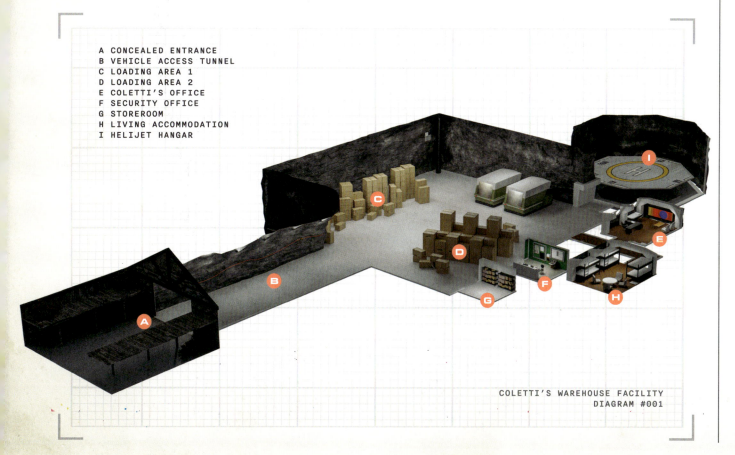

A CONCEALED ENTRANCE
B VEHICLE ACCESS TUNNEL
C LOADING AREA 1
D LOADING AREA 2
E COLETTI'S OFFICE
F SECURITY OFFICE
G STOREROOM
H LIVING ACCOMMODATION
I HELIJET HANGAR

COLETTI'S WAREHOUSE FACILITY
DIAGRAM #001

PROJECT 90

MISSION CODE NAME:
SPLASHDOWN

DATE: ▓▓▓▓▓▓▓▓▓▓
APPROVAL: ▓▓▓▓▓▓▓▓▓▓
CASE No: ▓▓▓▓▓▓▓▓▓▓

SUPPORTING IMAGERY:

When two Cyranian Airways flights crashed in the Mediterranean, the bodies of passengers Professor Maria Brynovsky and Dr Frank Adams were the only two not recovered. Both were electronics experts assigned to a government project in Istanbul, Turkey and were likely kidnapped prior to the incidents. Joe 90 and electronics expert Professor McClaine were instructed to board a third identical flight from London to Istanbul in a ruse to draw out the kidnappers.

Flight T-4063 was hijacked two minutes before beginning its descent into Athens by passenger Lucien Kramer and flight attendant Sofia Bauer. The pilots, William Barry and Frank Casper, were tranquilised and the AV21 jet locked into a crash-dive. Professor McClaine was forcefully ejected in an escape unit, with Kramer and Bauer also escaping, and met by gang leader Pierre Anton aboard his speedboat, *Dolphin*. Anton had ambitions to construct an orbital satellite armed with paralysing ultrasonic rays, and planned to rendezvous with a submarine which would smuggle the Professor back to their facility to commence work.

Joe regained control of the AV21 using the brain pattern of test pilot Major Charles Taylor. Sam Loover notified the navy of the intended rendezvous at area reference TU.74.1 but the nearest naval gunship was unlikely to intercept *Dolphin* in time. Joe therefore delayed *Dolphin* by using low altitude combat manoeuvres to steer the boat off course while dodging gunfire from the *Swordfish* submarine. Heading for a collision with the AV21, Bauer panicked and stopped the speedboat, forcing Anton to surrender to the naval patrol. WIN arrested the gang and tracked down the missing scientists successfully.

BOARDING PASS

CA

FLIGHT: T-4063
AIRCRAFT ID: AV21

PASSENGER:
PROFESSOR IAN MCCLAINE

SEAT NO. B12

LONDON INTERNATIONAL AIRPORT
DEPARTURE 0801

ATHENS AIRPORT
STOP 0915

ISTANBUL AIRPORT
ARRIVAL 0946

FLIGHT: T-4063
AIRCRAFT ID: AV21

BRAIN PATTERN SPECIFICATION

NAME: Major Charles Taylor
CODE: A5
BORN: April 19, 1976
PLACE OF BIRTH: Chicago, USA

KNOWLEDGE & EXPERIENCE:
Chief test pilot in the US Air Force with 18 years of combat flight experience. Led 30 successful low altitude bombing raids on Eastern Alliance military infrastructure. Small arms expert.

PROJECT 90

CASE STUDIES

CONFIDENTIAL

WORLD INTELLIGENCE NETWORK
::CRIMINAL FILE::

NAME:	KNOWN ASSOCIATES:
PIERRE ANTON	LUCIEN KRAMER, SOFIA BAUER

BORN:	PLACE OF BIRTH:
AUGUST 9, 1962	NICE, FRANCE

STATUS:
INCARCERATED

AFFILIATION:
"LA MER" GANG

REPORT FILED BY:
DATE:
REF:

Pierre Anton enjoyed a privileged upbringing as the only child of ruthless casino magnate, Julien Anton. As a boy, Anton witnessed the blackmail and violence behind his father's global gambling empire. Pierre appreciated such mediaeval methods but also yearned to modernise the casinos with profitable gadgets and luxuries. Upon his father's death, Anton inherited the corporation and employed top computer programmers to develop mind-altering gambling computer systems – eliminating anyone who questioned the ethics of the project.

One expert on Anton's payroll was Dr Lucas James, an untamed genius in the field of electronics but who was also troubled by psychological distress. Dr James never questioned the pursuit of profit over morality, thus making his ideas an invaluable resource to Anton. Dr James's unbeatable gambling programs quadrupled Anton's profits. The two men appeared to become friends, but Anton had a tight hold over James's private demons and could manipulate the scientist's brilliant mind in any direction.

Growing sicker and more tormented, Dr James presented his benefactor with a proposal for an entirely different device which could reap unthinkable rewards for Anton. James had drawn up blueprints for an orbital satellite which could destroy infrastructure and threaten life using powerful ultrasonic radiation emitters. While the idea sounded impossible, Anton theorised that a bluff could be played against the World Government. He could blackmail his many debtors in foreign space agencies to cooperate and offer the project credibility. Anton financed the project and relished the notion of holding his finger over the launch button while demanding an enormous ransom to ensure the protection of major cities.

Within weeks, however, Dr James had become intolerably unstable. He initiated a gunfight with his employer during a mindless rage and lost his life. Unperturbed, Anton planned to recruit more scientists by force to complete the satellite. He plotted with trusted conspirators to sabotage commercial airliners and kidnap the world's greatest minds. The operation was foiled by WIN during an attempt to capture leading electronics expert Professor McClaine. Anton is now fighting the charges against him in a lengthy legal battle.

MISSION: SPLASHDOWN

SUPPORT FILE:
0015

AV21 PASSENGER JET

The AV21 is a medium-range passenger jet aircraft capable of carrying up to 150 passengers and cargo in comfort and safety at a top speed of Mach 2. For use in an emergency situation, the first class cabin houses six ejectable escape units with a further nine in the rear of the plane.

AV21 PASSENGER JET
DIAGRAM #001

AV21 PASSENGER JET
DIAGRAM #002

MISSION: SPLASHDOWN

SUPPORT FILE:
0019

"SWORDFISH" SUBMARINE

Swordfish was a prototype Clam Class submarine used in the final phase of Anton's kidnapping operation to escape to safety. Armed with gun turrets and torpedoes, such a vessel is usually strictly reserved for military purposes. The submarine's owner was an associate of Anton's who was terminated by naval forces during the destruction of the craft.

"DOLPHIN" SPEEDBOAT

Dolphin is the hand-built, luxury yacht which belonged to Anton and was used in the operation to meet the gang members and kidnapped scientists from the plane's escape unit, and to transfer prisoners to the submarine. Adapted for high speed, the boat can reach up to 90 miles per hour, and sleep a crew of four.

PROJECT 90

MISSION CODE NAME:

BUSINESS HOLIDAY

CONFIDENTIAL

DATE: ~~~~~~~~~~~~~~
APPROVAL: ~~~~~~~~~~~~~~
CASE No: ~~~~~~~~~~~~~~

SUPPORTING IMAGERY:

The new government of Borova ordered the World Army to leave the country. The withdrawal was agreed on the understanding that Beneleta Base was dismantled. However, the Borovan military's Major Rossi reoccupied the base to threaten world peace. Agreement breached, the World Army planned a covert attack on the base to be spearheaded by General Henderson. Unfortunately, Henderson was injured and unable to take command.

WIN intended to recruit Joe 90 in Henderson's place, but Professor McClaine had planned a holiday that would prevent Joe's participation. Undeterred, Sam Loover convinced the McClaines to visit Borova for their vacation and suggested giving Joe the brain pattern of an aquanaut to improve his swimming. However, Joe unknowingly received a transfer of General Henderson's military vehicle expertise instead.

One consequence to Sam's deception was the need for WIN to surreptitiously prevent Joe from swimming in order to ensure his safety. Professor McClaine uncovered Shane Weston's plan, but with his agreement, the attack on Beneleta Base was set to proceed with Joe in command. The A14 Armoured Attack Vehicle was delivered by submarine and concealed 20 miles from the base. The attack run commenced at dawn through treacherous swamp and desert terrain. Beneleta's outer alarms were triggered and a counterattack was launched. However, the A14 was unharmed and destroyed all enemy vehicles easily. Joe ejected himself from the craft as the A14 stormed Beneleta under automatic control, demolishing the primary installation and triggering a chain reaction across the base's missile turrets.

NOTE:
Invoice for McClaine "business holiday" rejected by WIN Head Office. Weston ordered to cover expense personally. Signed, Thor Jurgens.

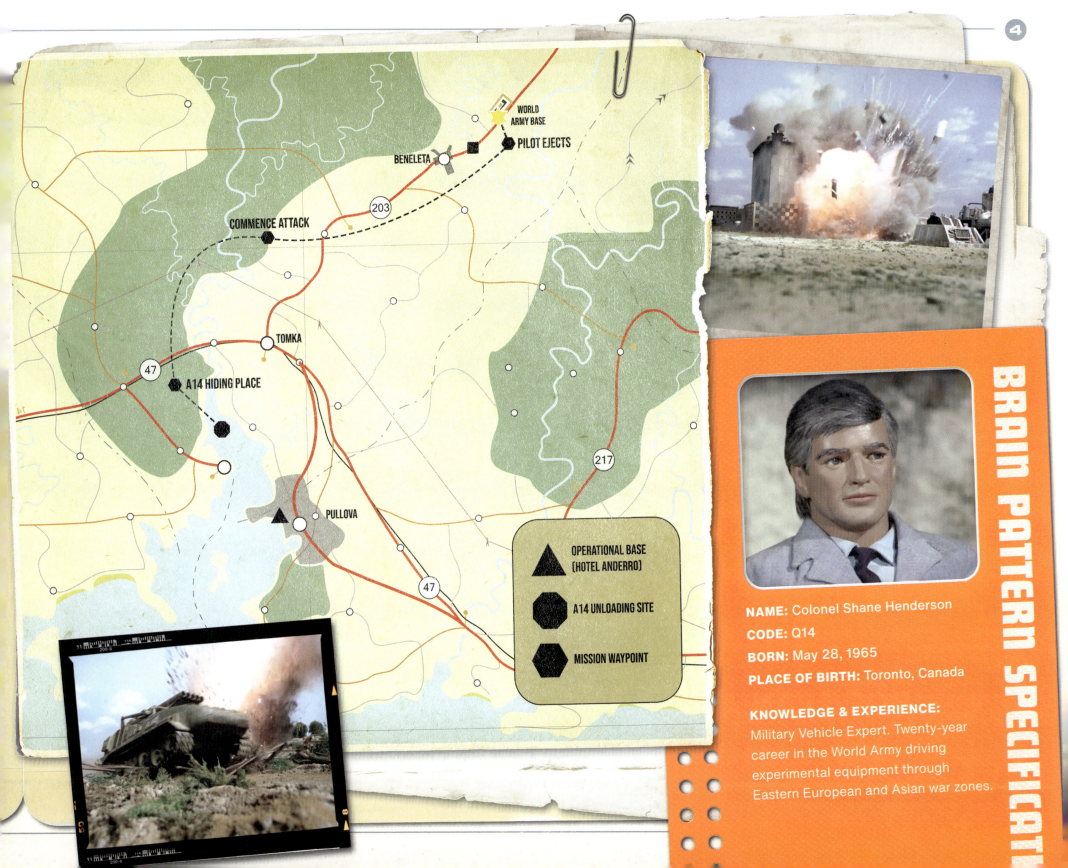

Map Legend

- ▲ OPERATIONAL BASE (HOTEL ANDERRO)
- ⬢ A14 UNLOADING SITE
- ⬡ MISSION WAYPOINT

Map locations: WORLD ARMY BASE, PILOT EJECTS, BENELETA, COMMENCE ATTACK, TOMKA, A14 HIDING PLACE, PULLOVA

BRAIN PATTERN SPECIFICATION

NAME: Colonel Shane Henderson
CODE: Q14
BORN: May 28, 1965
PLACE OF BIRTH: Toronto, Canada

KNOWLEDGE & EXPERIENCE:
Military Vehicle Expert. Twenty-year career in the World Army driving experimental equipment through Eastern European and Asian war zones.

PROJECT 90

CONFIDENTIAL

WORLD INTELLIGENCE NETWORK

::CRIMINAL FILE::

NAME:	KNOWN ASSOCIATES:
MAJOR GUERRIERO ROSSI	LIEUTENANT ALEXANDER SOKOLOV
BORN: AUGUST 24, 1970	PLACE OF BIRTH: RËTVA, BOROVA

STATUS:

AFFILIATION:
BOROVAN ARMY

REPORT FILED BY: ▮▮▮▮▮▮▮▮▮▮
DATE: ▮▮▮▮▮▮▮▮
REF: ▮▮▮▮▮

Major Guerriero Rossi was a career soldier in the Borovan Army with an impressive record of service spanning over two decades. However, he was also known for his arrogance, being desperately hungry for promotion, and was only respected by a few of his fellow soldiers.

Born and raised in a military family, Major Rossi believed that he was entitled to greatness despite his questionable reputation. At the age of 18, he enlisted in the Borovan Army where his brash personality quickly caught the attention of his superiors leading to rapid promotion through the ranks. Throughout his career, Major Rossi had been involved in several military campaigns, both at home and abroad. While he had earned several decorations and commendations for his bravery and strategic prowess, he had also made numerous enemies due to his inflated ego and abrasiveness.

His recent mission to reoccupy Beneleta Base had proven to be the most significant challenge of his career. Major Rossi saw the base as a crucial strategic asset and believed that its capture would further cement his reputation and secure his next promotion. Aware of the opposition from the World Army and protests from the international community, Major Rossi was determined to succeed and was unafraid of putting his team in danger. He had assembled a task force of soldiers including his obedient Lieutenant, Alexander Sokolov. Many were wary of Rossi's leadership and, despite sharing his passion for their country's security, did not respect him as a commander.

When Beneleta Base was on the receiving end of an attack from the World Army, Major Rossi unleashed the full fury of his military power. However, this proved ineffective against the incoming A14 and Rossi was killed alongside his colleagues when the base was devastated by a final attack charge which eliminated all installations. Major Rossi had no next of kin, dedicating his entire life to an egotistical desire for power which ultimately failed to serve him during his final moments.

MISSION: BUSINESS HOLIDAY

SUPPORT FILE:
0006

BENELETA BASE

Beneleta Base was strategically positioned 70 miles north of the Borovan capital of Pullova. It was originally built during the Cold War as one of many defensive showcases to ward off attacks along the borders of the Soviet Union. However, following the declaration of world peace and Borova's independence, the World Army occupied the base. Then, after years of peaceful democracy, a radical new government took hold which was only interested in defending Borova's own interests. As such, an order was issued to the World Army to abandon Beneleta Base, which the Borovan Army had every intention of reoccupying once again.

The arsenal of guided missiles would serve as the perfect deterrent to anyone opposing Borova's new regime. Under the command of Major Rossi, Beneleta Base remained fully operational until an attack was launched by the World Army in cooperation with WIN to ensure the base was dismantled once and for all. The full force of Beneleta Base's fleet of tanks and hovercraft were no match for the World Army's A14 attack vehicle. Beneleta Base and all its personnel were wiped out during the battle, and the site is due to be redeveloped to service Borova's growing tourism industry.

ANTI-AIRCRAFT MISSILE
DIAGRAM #001

Beneleta Base was known for its impressive array of anti-tank and anti-aircraft missile turrets. These defensive systems were a crucial part of the base's security infrastructure and were designed to repel any hostile military action. The turrets were strategically placed throughout the base and equipped with the latest in armour-piercing technology. The vast number of turrets should have ensured that the base was virtually impregnable to enemy attacks from the ground or air, but they proved ineffective against the A14.

BENELETA BASE
DIAGRAM #001

PROJECT 90

MISSION CODE NAME:
TEST FLIGHT

APPROVAL: DATE:
CASE No:

SUPPORTING IMAGERY:

At its disastrous inaugural flight, the Orbital Glide Transporter (OGT) crash-dived on top of its adjacent launch control building having failed to accelerate sufficiently at take off or respond to the shutdown procedure. The crew evacuated the building via the escape capsule while Dr Slade, the flight controller, remained inside to locate the fault-monitoring computer hidden by project lead, Brad Johnson. The collapsing roof, buried under the OGT wreckage, had blocked the reserve escape capsule and all means of access from outside.

Assigned to rescue Slade and the computer, Joe 90 was given Johnson's brain pattern, since he was both a computer programmer and a demolitions expert. A borehole too small for an adult to use was drilled for Joe to be lowered into launch control and search for Slade and the computer. The OGT wreckage above was being subjected to high winds and an electrical storm which risked unbalancing the craft or igniting its fuel and obliterating the building with Joe and Slade inside.

Joe located the computer but was threatened at gunpoint by Slade who revealed himself as the saboteur working on behalf of a rival operation. Slade had been attempting to find and eliminate Johnson's computer, thus setting back the OGT's development. Joe used explosive charges to clear the debris blocking the escape capsule, knocking Slade out in the process. Joe, Slade and the computer were ejected seconds before lightning detonated the OGT wreckage and destroyed the launch control building.

BRAIN PATTERN SPECIFICATIONS

NAME: Brad Johnson
CODE: P19
BORN: January 7, 1965
PLACE OF BIRTH: Vancouver, Canada

KNOWLEDGE & EXPERIENCE:
Studied electronics and aeronautical engineering at Stanford alongside Ian McClaine. Eight years of military service leading a covert demolitions squad. Ten years as managing director of Johnson Aerospace Ltd designing and manufacturing experimental aircraft. Five years as Senior Project Controller developing the Orbital Glide Transporter.

MISSION: TEST FLIGHT

SUPPORT FILE: **0096**

ORBITAL GLIDE TRANSPORTER 780

The OGT (or OGC) built by Johnson's team at Johnson Aerospace Ltd is the first machine of its kind and currently the most advanced concept in aerospace passenger transportation. The OGT can fly 500 passengers halfway around the world in 50 minutes from blast off to touch down.

The OGT achieves this record-breaking speed by building upon the achievements of aerospace pioneers such as the X-15 programme operated by the United States Air Force and NASA throughout the 1960s. Only recently has it become feasible to engineer components for hypersonic orbital craft at a scale associated with commercial airlines, while keeping passenger safety and comfort in mind. The 12 motors necessary to achieve adequate propulsion, where the X-15 only needed one, would be a typical example of the advancements required to make the OGT viable.

The precision required to pilot the OGT is beyond current human ability and so the flight is controlled entirely by an automatic pilot system supplemented with a secondary back-up. The OGT can therefore fly unmanned and is instead monitored remotely by three operatives plus one flight controller in the launch control building.

Only one other country has achieved the necessary investment to develop a similar method of mass orbital passenger transit, but this is still in its early stages. It is believed that this rival manufacturer has made numerous sabotage attempts on the OGT in order to create a monopoly.

PROJECT 90

CONFIDENTIAL

WORLD INTELLIGENCE NETWORK ::CRIMINAL FILE::	NAME: DR RYAN SLADE	KNOWN ASSOCIATES: ECLIPSE ENGINEERING LTD
	BORN: DECEMBER 21, 1983	PLACE OF BIRTH: WINCHESTER, ENGLAND

STATUS: **INCARCERATED**

Dr Ryan Slade stands out as an unlikely candidate for espionage. He was educated at Winchester College and the University of Oxford achieving a PhD in Aerospace Engineering. Slade received multiple job offers from the world's top aviation corporations upon graduating but the best opportunity came from Brad Johnson.

Slade had been invited to join Johnson's team developing the OGT. His enthusiasm for the project led to many of its early successes, such as solving the OGT's gargantuan propulsion requirements. Slade's responsibilities quickly grew to include overseeing the construction of the OGT's unique launch facilities. He initially baffled the engineers with his vision for a horizontal launch ramp which stood alongside the flight control facilities. However, the building work was completed within budget and ahead of schedule earning Slade the promotion to Flight Controller.

At this stage, Slade was approached by rival company, Eclipse Engineering Ltd, to shift allegiances and join their own mission to launch an orbital passenger craft in the East. Unwilling to leave England following his recent marriage, Slade declined, but he naively agreed to meet with Eclipse's representatives at a London hotel restaurant. The crooked corporation then revealed their ties to a global crime syndicate and told Slade that his new spouse would be assassinated unless the OGT's testing phase was sabotaged. If Slade notified the authorities, his parents' lives would also be threatened.

Left with no alternative, Slade embarked on a series of life-threatening operations to sabotage the OGT during ground trials. This ultimately resulted in the craft's destruction during the first test flight and his arrest soon after. WIN has placed Slade's family in a protection programme while he serves a prison sentence.

REPORT FILED BY: ▓▓▓▓▓▓▓▓
DATE: ▓▓▓▓▓▓▓▓
REF: ▓▓▓▓▓▓▓▓

MISSION: TEST FLIGHT

SUPPORT FILE:
0102

JOHNSON AEROSPACE TESTING FACILITY

This site covering five square miles of Cornwall was adapted from an old RAF testing base for the OGT project by Brad Johnson and Dr Slade. The layout of the base is unlike any other aerospace launch facility due to the OGT's unique requirements to achieve take off and landing. Upon the completion of the OGT's exhaustive test flight programme, Johnson Aerospace is committed to rolling out similar infrastructure to the world's leading airfields in order to realise the vision of mass transport within minutes on a global scale.

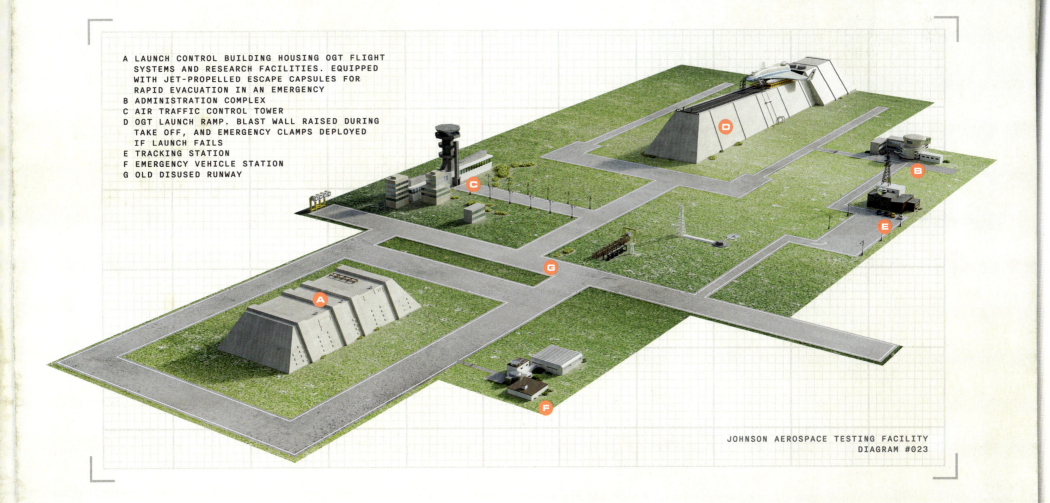

A LAUNCH CONTROL BUILDING HOUSING OGT FLIGHT SYSTEMS AND RESEARCH FACILITIES. EQUIPPED WITH JET-PROPELLED ESCAPE CAPSULES FOR RAPID EVACUATION IN AN EMERGENCY
B ADMINISTRATION COMPLEX
C AIR TRAFFIC CONTROL TOWER
D OGT LAUNCH RAMP. BLAST WALL RAISED DURING TAKE OFF, AND EMERGENCY CLAMPS DEPLOYED IF LAUNCH FAILS
E TRACKING STATION
F EMERGENCY VEHICLE STATION
G OLD DISUSED RUNWAY

JOHNSON AEROSPACE TESTING FACILITY
DIAGRAM #023

PROJECT 90

MISSION CODE NAME:

THE PROFESSIONAL

CONFIDENTIAL

DATE:
APPROVAL:
CASE No:

SUPPORTING IMAGERY:

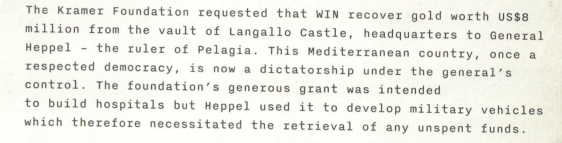

The Kramer Foundation requested that WIN recover gold worth US$8 million from the vault of Langallo Castle, headquarters to General Heppel - the ruler of Pelagia. This Mediterranean country, once a respected democracy, is now a dictatorship under the general's control. The foundation's generous grant was intended to build hospitals but Heppel used it to develop military vehicles which therefore necessitated the retrieval of any unspent funds.

Sam Loover and Professor Ian McClaine (Mac) recorded the brain pattern of expert safe-cracker Henry George Summerfield while overflying the yard of Maryfield Prison in Mac's car. Joe 90, assigned to the mission with Summerfield's capabilities, requisitioned a proton lance which can heat its point of impact to 6000 °F and disintegrate its target. The McClaines entered Pelagia under the cover of Mac attending a conference, but diverted to Langallo Castle to begin the heist at 0131 hours.

Joe bypassed the electrified perimeter, avoiding initial detection from the guard vehicles known as Spiders - a small, unmanned version of an attack vehicle Heppel's team had manufactured. Once inside, Joe used the proton lance to destroy an approaching Spider and break through the barriers into the vault. Because the damaged Spider missed its patrol checkpoint, however, Heppel sounded a red alert and scrambled all Spiders to eliminate Joe. With the cargo driven out of the castle on a transporter, Mac's car airlifted the gold using its mechanical grabs, while Joe defeated the remaining Spiders. The reclamation of the gold has successfully set back the development of General Heppel's military technology.

BRAIN PATTERN SPECIFICATIONS

NAME: Henry George Summerfield

CODE: X12

BORN: February 17th, 1959

PLACE OF BIRTH: Southend-on-Sea, UK

KNOWLEDGE & EXPERIENCE:
Safecracker who evaded capture for six separate high-profile heists totalling US$15 million before his arrest three months ago. Weapons expert, formerly a military armourer before joining a private security force responsible for the Bank of England among other key financial establishments.

MISSION: THE PROFESSIONAL

SUPPORT FILE: **0023**

LANGALLO CASTLE

Langallo Castle is located in the wild and rocky countryside, 15 miles south of Pelagia's capital city, Grand Fortuna.

Originally built in 1572 by the mysterious Knights of Langallo, the castle has been occupied by some of the richest families in Eastern Europe, eventually becoming the property of the Pelagian government's historical society for preservation. The society was violently disbanded when General Heppel took power, leaving him free to adapt the castle for his own purposes.

Today, it is heavily protected by an electrified perimeter fence, guard towers and roaming armed patrols on constant alert. The interior of the castle has been fully modernised with no shred of its former history remaining. This enormous structure now houses laboratories, barracks, training facilities, workshops, offices and prison cells to serve Heppel's desire to build a ruthless military force.

The crowning achievement of the facility is the new underground vault protecting Pelagia's ill-gotten wealth and Heppel's best kept secrets. The vault is enclosed behind reinforced concrete walls, steel grilles and an impregnable automatic alarm system. Securing the area is an 18-inch thick steel and ferrotungsten door.

LANGALLO CASTLE
DIAGRAM #001

PROJECT 90

CONFIDENTIAL

WORLD INTELLIGENCE NETWORK

::CRIMINAL FILE::

NAME:	KNOWN ASSOCIATES:
GENERAL VIKTOR HEPPEL	DR NICHOLAS KOVAČEVIĆ

BORN:	PLACE OF BIRTH:
JANUARY 3, 1963	DUBROVNIK, CROATIA

General Viktor Heppel, the tyrant ruler of Pelagia, was educated at a military academy in Dubrovnik and showed an early aptitude for strategy and tactics. Heppel rose through the ranks of the army, earning a reputation for his ruthlessness and cunning. Heppel showed heroism in the Pelagian War of Independence, leading successful guerrilla operations and key victories against occupying forces. His strong leadership quickly made him a hero among the Pelagian resistance fighters. He used his newfound popularity to consolidate power, eventually seizing control of the government and establishing a brutal dictatorship.

Heppel's regime is marked by corruption, brutality and a disregard for human rights. He views the people of Pelagia as weak and in need of a strong, ruthless leader like himself to maintain order. Dissent is not tolerated, and those who oppose his regime are imprisoned or executed. Heppel maintains control through a widespread security force, and a chokehold on the country's media. He has built a cult of personality around himself through propaganda and grand displays of military might. While many in Pelagia believe he is a necessary evil, those who have seen his regime up close know he is a ruthless tyrant who will do anything to maintain his grip on power.

The General's latest scheme to suppress the people of Pelagia involves the construction of advanced military vehicles, funded by charitable donations meant for a hospital building programme. Heppel assigned Dr Nicholas Kovačević to lead the research and development of these vehicles, with the end goal of constructing the greatest fighting force the world has ever seen, and to keep his people in their place.

Heppel rarely leaves the walls of his headquarters at Langallo Castle, seeing himself as a modern-day king, ruling with an iron fist and the support of a powerful military. He is feared but also respected by many in Pelagia. He has managed to maintain a fragile peace in the country, and deceived many into believing that his leadership is the only thing keeping Pelagia from descending into chaos.

STATUS:

REPORT FILED BY: ▓▓▓▓▓▓▓▓
DATE: ▓▓▓▓▓▓▓▓
REF: ▓▓▓▓▓▓▓▓

MISSION: THE PROFESSIONAL

SUPPORT FILE:
0038

SPIDER RIOT CONTROL AND GUARD VEHICLE

The Spider, developed by Dr Nicholas Kovačević under the instruction of General Heppel, is an advanced riot control vehicle built to be virtually indestructible. Its unique design makes it capable of withstanding guerilla attacks that would render any other machine inoperable. Its high manoeuvrability allows it to traverse rugged terrain and overcome obstacles that would immobilise any other vehicle.

Prototypes for a smaller, unmanned version of the Spider are currently being used to guard Langallo Castle. Each machine is equipped with a large electronic eye mounted on top and two antennas linked to the castle's main security system, allowing them to operate autonomously. They hover along flat surfaces, offering versatility and agility, and can patrol the perimeter of Langallo Castle in just four minutes. Furthermore, these automated guards can recognise security breaches and conduct investigations, making them an efficient tool for maintaining a secure environment. The Spider guards' power and weapon systems are electrically charged and each unit requires re-energising every three hours to guarantee peak operational effectiveness and firepower.

To ensure the integrity of the Spider guards, they are required to pass through an electronic checkpoint at the end of each patrol. This measure is in place to prevent sabotage from intruders going undetected. In the event that a Spider guard fails to pass through the checkpoint, a red alert is triggered, and security personnel are immediately notified to respond accordingly.

The Spider was developed using mismanaged charitable funds which have since been reclaimed by WIN, thus stalling further production. However, while General Heppel remains in power, it is inevitable that more military weapons will be built under his regime, requiring future intervention from World Government agencies to maintain peace.

SPIDER RIOT CONTROL
AND GUARD VEHICLE
DIAGRAM #001

PROJECT 90

MISSION CODE NAME:
BIG FISH

DATE: ███████████
APPROVAL: ███████████
CASE No: ███████████

SUPPORTING IMAGERY:

While on exercise, the missile outlet of World Navy Submarine U85 jammed and flooded the craft sinking it 145 feet down on the sea bed. The U85 then drifted 10 miles off course into forbidden Porto Guavan territorial waters. The crew evacuated during the cover of night but the submarine itself still required stealthy removal in order to prevent hostilities between the World Government and the ruthless police state in Porto Guava.

Joe 90 flew out with Professor McClaine to recover the U85. Upon the McClaines' arrival, the military police were suspicious of their cover story that they were holidaying in Porto Guava. Remaining undercover, Mac hired the fishing boat of one Miguel Umberto dos Passos Francesca so that Joe could investigate the U85 crash site. Joe's diving goggles were adapted with the electrodes necessary for the brain pattern of aquanaut Bill Fraser to take effect, and he dived down to find everything as reported.

Later that night, Joe, Professor McClaine and Francesca returned to commence the operation. While Joe worked on accessing the submarine, the Professor and Francesca were arrested by a police patrol boat and charged with murdering the boy. Meanwhile, Joe had been delayed when a giant clam trapped his leg and limited his remaining oxygen supply. Joe eventually broke free and successfully entered the U85 to pump out the floodwater and steer it back into World Navy territory. The next morning, Joe returned to Porto Guava proving the Professor and Francesca's innocence before they faced a firing squad. With a diplomatic incident prevented, free elections later took place in Porto Guava as scheduled ending President Chaves's dictatorship.

BRAIN PATTERN SPECIFICATIONS

NAME: Bill Fraser
CODE: Q8 (formerly Q14)
BORN: March 10, 1982
PLACE OF BIRTH: Goldcoast, Australia

KNOWLEDGE & EXPERIENCE:
Aquanaut, with 10 years experience undertaking covert operations for the World Navy. Accomplished submariner, surfer and scuba diver.

SUPPORT FILE: **0102**

MIGUEL UMBERTO DOS PASSOS FRANCESCA

Miguel Umberto dos Passos Francesca was born into a large family in the coastal town of Porto Guava. Growing up, he spent much of his childhood fishing with his uncle who taught him everything he knew about the trade. Francesca developed a great admiration for the ocean and the peacefulness it brought him, and decided to make fishing his career.

As he got older, Francesca met a kind and gentle woman named Rosa. They fell in love and were married shortly after. Together, Francesca and Rosa worked hard to build a life for themselves. Francesca purchased his own boat and began fishing independently, while Rosa managed the household and cared for their six children. Despite being part of a large family, Francesca valued his alone time and could often be found busking in the streets to earn some extra money and get out of the house.

Porto Guava was a police state and Francesca knew the importance of staying out of trouble in order to keep his livelihood. He learned the boundaries of the system and the best ways to operate within them to remain profitable. Francesca and Rosa were content with their simple life and had no desire to get involved in the political turmoil that often plagued their country.

Francesca unknowingly became involved in the recovery of the U85 by offering Professor McClaine and Joe 90 the hire of his boat at a suitable price. When history eventually reveals the truth of those events, Francesca may be remembered as instrumental to Porto Guava's return to a peaceful democracy by aiding the prevention of a diplomatic breakdown. Today, however, Francesca continues to run his fishing and tourism operation, and enjoys the care-free lifestyle he had always dreamt of.

PROJECT 90

CONFIDENTIAL

NAME:	KNOWN ASSOCIATES:
PRESIDENT GENERAL JUAN CHAVES	POLICE CHIEF NATALIA GOMEZ, POLICE CAPTAIN ANTONIO GARCÍA
BORN:	PLACE OF BIRTH:
DECEMBER 23, 1952	SÃO PAULO, BRAZIL

STATUS: RETIRED

REPORT FILED BY: ███████████
DATE: ███████████
REF: ███████

President General Juan Chaves is the former ruler of Porto Guava. His parents were immigrants from Argentina who had moved to Brazil to start a new life. As a child, Chaves was known to be rebellious and often got into trouble with the law. However, his life took a turn when he joined the military at the age of 18. In the army, Chaves excelled in his training and became known for his brutal tactics and willingness to do whatever it takes to get the job done. Chaves quickly became a favourite of the military higher-ups and was given command of his own unit.

During his career, Chaves became obsessed with power and control. He saw how easy it was to use fear and violence to manipulate people, and became determined to do the same on a larger scale. Porto Guava was a small, impoverished island nation off the coast of South America and Chaves saw an opportunity to take control. With the help of his loyal associates such as mob boss Natalia Gomez who became his chief of police, Chaves staged a coup and overthrew the government.

Once in power, Chaves quickly turned Porto Guava into a police state. He used his militaristic mindset to suppress any opposition and crush dissent. Chaves plastered his image on every building in Porto Guava, serving as a constant reminder of his power and control. He was known from the propaganda posters for his distinctive moustache and his large, overweight frame but was actually rarely seen in person. Chaves's reign lasted for over a decade during which time he enriched himself and his obedient public servants at the expense of the people of Porto Guava. However, his rule came to an end when the World Government enforced democratic elections in the country, and he was outvoted in favour of libertarian Christina Sánchez.

Today, Chaves lives in luxury in an undisclosed foreign country surrounded by his wealth and the remnants of his former power. He is despised by the people of Porto Guava who remember his brutal rule and the atrocities committed in his name.

MISSION: BIG FISH

SUPPORT FILE:
0076

WORLD NAVY SUBMARINE U85

The U85 was developed by the World Navy for covert attack operations in hostile waters. Requiring a crew of just two, the submarine is equipped with a simplified control layout, allowing operators quick and efficient access to a full range of capabilities.

The tour-de-force in the U85's development is undoubtedly its missile ejector units which occupy the bow of the craft. Four missile outlets are positioned strategically around the submarine's nose, with concealed hatchways allowing the U85's hydrodynamics to remain optimised during both combat and cruising.

The agile machine can reach speeds of up to 50 knots underwater and can, on paper at least, outmanoeuvre all known enemy craft. It is expected that the U85 will become a staple of World Navy operations once it has completed final testing exercises. However, one such exercise demonstrated that the missile hatches were prone to jamming, nearly causing a major disaster. Engineers have consulted with the crew and damage reports to ensure the fault is thoroughly rectified before the next trials commence.

WORLD NAVY
SUBMARINE U85
DIAGRAM #001

PROJECT 90

MISSION CODE NAME:
RUNAWAY

DATE: ▇▇▇▇▇▇▇▇▇▇
APPROVAL: ▇▇▇▇▇▇▇▇▇▇
CASE No: ▇▇▇▇▇▇

SUPPORTING IMAGERY:

Enroute to a conference planning the expansion of global intelligence services, Sam Loover was protecting World Justice Minister, Dana Louise Hart, on a monorail journey to Glasgow, Scotland. Loover noticed the train accelerating unexpectedly and bypassing scheduled stops, so proceeded to investigate the driver's cab. There, he discovered the crew unconscious and a gang led by Harper Weyland hijacking the train. Weyland wanted to assassinate Hart and prevent the security conference from taking place. Loover attempted to overpower the three hijackers but was unable to defeat them in hand-to-hand combat.

Fearing further interference, Weyland threw Loover from the train and sabotaged the automatic controls so that the locomotive could not be stopped. While the gang escaped the doomed train, Loover transmitted an emergency signal to WIN before passing out from his injuries. Joe 90 and Professor McClaine rushed towards Glasgow in their jet-powered car with orders from Shane Weston to meet the runaway train minutes before it was due to crash at the terminus. Joe had received the brain pattern of a leading mono-train engineer and was going to board the train to try and stop it.

After a daring leap at high speed from Professor McClaine's flying car onto a train carriage roof, Joe accessed the cab and began to work on the damaged control system. As passengers panicked upon seeing the fast-approaching Glasgow Central station, Joe successfully engaged the brakes and brought the mono-train to a standstill. With the conference able to go ahead as planned, all that remained was for the Professor to get Loover to hospital and for Weston to arrest Weyland's gang.

BRAIN PATTERN SPECIFICATIONS

NAME: Lionel Jones
CODE: T3
BORN: November 1, 1973
PLACE OF BIRTH: Manchester, England

KNOWLEDGE & EXPERIENCE:
Twenty years working with the National Monorail Service as the lead engineer on projects to automate and improve the safety of key areas on the rail network. Responsible for designing the Class 31 mono-train automatic control system rolled out across the fleet.

13:59 NMS SERVICE TO GLASGOW

DATE:
February 26, 2012

DEPARTURE:
13:59 London Euston (EUS)

SCHEDULED ARRIVAL:
15:03 Glasgow Central (GLC)

CALLING AT:
London Euston, Wigan, Lancaster, Carlisle, Glasgow Central

DRIVERS:
Keith Wilkin and Jeremy Alexander

NOTES:
Service hijacked approximately five miles from Lancaster. WIN intervened to bring the train back under control before arriving at Glasgow Central ahead of schedule at 14:46. National Monorail Service ordered by the World Government to conduct a network-wide security inquiry. Details to follow.

PROJECT 90

CONFIDENTIAL

NAME:	KNOWN ASSOCIATES:
HARPER WEYLAND	CHASE NEWLEY, ANGUS CRANE
BORN: DECEMBER 15, 1976	PLACE OF BIRTH: HOUSTON, USA

Harper Weyland was born and raised in the heart of Houston, Texas. Growing up in a tough neighbourhood, she quickly learned to defend herself and stand up for what she believed in. She had a sharp wit and a quick mind which she used to her advantage throughout her life. As a young woman, Weyland became involved with a local gang and her natural leadership skills soon made her one of the most respected members. She quickly rose through the ranks, and before long she was running her own crew.

Under Weyland's leadership, the gang became one of the most powerful criminal organisations in the United States, and later the world. They operated in the shadows using their underground network to smuggle goods, launder money and commit all manner of other crimes. Weyland was ruthless when it came to defending her operations and would stop at nothing to protect her gang's interests. Despite her criminal activities, Weyland was respected by many in her community. She was known for her intelligence and charisma, and had a way of inspiring loyalty in those around her.

Her career was cut short, however, when WIN uncovered her plot to hijack a train carrying World Justice Minister, Dana Louise Hart. After an encounter with Sam Loover and her eventual arrest by Shane Weston, Weyland was sentenced to a life in prison alongside associates Chase Newley and Angus Crane. She now resides in a maximum security establishment but proudly boasts her plots for revenge against the World Intelligence Network.

The remaining members of the gang have tried and failed to break Weyland out of the facility, lacking the charisma and expertise of their leader to develop a viable escape plan. Nevertheless, Weyland's bold spirit has earned her heroic status in the criminal underworld, and enemies of world peace are determined to get her back - an eventuality that WIN must be prepared for.

STATUS: INCARCERATED

AFFILIATION: "THREE BEARS" GANG

REPORT FILED BY: ███████████
DATE: ███████████
REF: ███████████

MISSION: BIG FISH

SUPPORT FILE:
0076

NATIONAL MONORAIL SERVICE CLASS 31 TRAIN

The Class 31 is the pride of the National Monorail Service (NMS) in the United Kingdom and is fast becoming a symbol for the public transit industry as a whole. Taking inspiration from Canada National Railway's remarkable Ontario class hovertrains, the NMS commissioned 100 units equipped with a brand new automatic control system designed by lead NMS engineer, Lionel Jones. The new stock joined the network in March 2012 and has already become a hit with passengers.

The locomotive typically pulls two additional compartments carrying a total of 600 passengers at speeds in excess of 200 metres per hour. A round-the-clock service travels between London and Glasgow, Scotland within two hours, in addition to a number of new, high-speed routes up and down the country.

The NMS plans to continue rolling out the Class 31 across the network as soon as monorail infrastructure in local areas has been updated to the necessary safety standards.

The Ontario class, on which the Class 31 is based, is widely used across the world; most famously as the Canadian Prime Minister's personal rail transport. The Class 31 streamlines this design as well as improving the efficiency of the vehicle.

NATIONAL MONORAIL
SERVICE CLASS 31 TRAIN
DIAGRAM #013

PROJECT 90

MISSION CODE NAME:

MURDER ON THE MOON

DATE:
APPROVAL:
CASE No:

SUPPORTING IMAGERY:

Niel Acree, a WIN whistleblower, had been killed under mysterious circumstances whilst serving on the International Moonbase (IMB). The decision was made to smuggle in Joe 90 to undertake a covert investigation. Joe was equipped with the brain pattern of Colonel Joe Steeler, Moonbase's original designer, and then hidden in a secret compartment built into a new scientific module bound for the Moon, with his exit route aboard a supply shuttle set for eight days later.

Joe's first task was to survey the outer hull wearing an EVA suit, and he quickly realised that the exterior of Acree's module had been tampered with: there was indeed a murder on the IMB. Acree's information had pointed towards the theft and disruption of top secret research from several laboratories across Moonbase. Morale was clearly low, and the Moonbase Commander, Gwen Akamore, had been put under pressure to resolve the issue.

Under the cover of night, Joe began to investigate the base using the incredibly small ventilation ducts to get around. After several days observing the activities of various personnel, Joe stumbled upon the lead researcher, Dr Philip Oliver - representing the Bluegrave Pharmaceutical wing of the base - breaking into a rival department. An unlucky fall from the ventilation shaft revealed Joe's presence, leading to a frenzied shoot-out. Oliver made a run for it and set an escape plan into motion, shutting off the Moonbase reactor, stealing a power core and making a break in a lunar rover for a foreign settlement.

CONTINUED ON NEXT PAGE

MISSION: MURDER ON THE MOON

SUPPORT FILE: **0096**

INTERNATIONAL MOONBASE

The IMB is a remote outpost on the lunar surface, built in the early 2000s as a scientific research station for use by members of the World Government. Modular in design, the Moonbase is designed to be expanded over time with new facilities and laboratories, and currently houses 37 scientists and eight operational crew.

The IMB is designed to be reasonably self-sufficient with a hydroponics garden supplementing air and food, and a reactor generating power. While water can be recycled to a degree, it requires replenishment from Earth. A tour of duty on the IMB is about three months.

A shuttle services the Moonbase every four weeks for supplies and crew rotation while new modules can be specifically scheduled to land at any time. The landing pad is located away from the Moonbase for safety reasons and requires a lunar rover to access it.

The IMB is the largest structure on the lunar surface but not the only one, with several satellite laboratories and mining complexes in the surrounding area.

With the Moonbase air supply starting to fail, Joe took a second rover and made chase using his superior knowledge of the base's geography to catch up with Oliver, forcing his rover to crash and rendering him unconscious. Joe retrieved the stolen power core and returned to Moonbase in time to restore the oxygen supply, before taking refuge once again in his undiscovered hiding spot.

Under interrogation, Oliver revealed that he had been paid by a foreign government to steal information and eventually set back the research being conducted on the IMB. When questioned regarding his captor, he kept babbling about a child who was hiding in the air ducts. Joe smuggled himself onto the supply shuttle as planned, and returned to Earth.

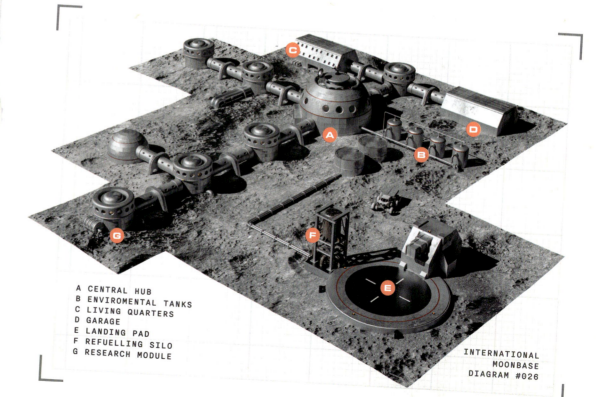

A CENTRAL HUB
B ENVIROMENTAL TANKS
C LIVING QUARTERS
D GARAGE
E LANDING PAD
F REFUELLING SILO
G RESEARCH MODULE

INTERNATIONAL MOONBASE DIAGRAM #026

PROJECT 90

CONFIDENTIAL

WORLD INTELLIGENCE NETWORK
::CRIMINAL FILE::

NAME:	KNOWN ASSOCIATES:
DR PHILIP OLIVER	ALEXIS GRAVES, COMMANDER OF EASTERN ALLIANCE MINING FACILITY AT NAV POINT BRAVO
BORN:	PLACE OF BIRTH:
APRIL 9, 1979	ARMAGH, NORTHERN IRELAND

STATUS:

INCARCERATED

REPORT FILED BY: ▓▓▓▓▓▓▓▓
DATE: ▓▓▓▓▓▓▓▓
REF: ▓▓▓▓▓▓▓▓

Dr Phillip Oliver was a child prodigy. He excelled academically and earned a PhD in Molecular Biology at Queen's University Belfast. Oliver's passion for medicine was beyond question but his boisterous bravado started to hold back his career in the heavily regulated pharmaceutical industry. The biologist's extravagant proposals for costly, experimental projects were met with red tape and quiet intolerance. The frustrated young scientist bounced between pharmaceutical companies across the globe making his brash and exasperated hypotheses.

In Japan, Oliver married and settled for a junior research position far below his abilities at a Tokyo medical laboratory. The honeymoon period of the relationship had initially eased Oliver's fiery ambition, but he soon became bitter about the career sacrifices he had made for his partner. Within 18 months, Oliver declared the marriage a failure and sought freedom to continue his quest for scientific excellence away from the restrictions of family life.

Returning to Europe after a swift separation, Oliver's work consumed him. His irritation with the field's inflexibility fuelled more outlandish research proposals which faced continual rejection. Oliver's new employers at Bluegrave Pharmaceuticals voiced complaints about his disruptive attitude, but they also couldn't fault his supreme knowledge of molecular science. They resolved to invite the pushy scientist to lead Bluegrave's modest research wing at the International Moonbase. Oliver was acutely aware that his appointment to the IMB served to keep him away from Bluegrave's more critical projects taking place on Earth. It was relegation disguised as promotion.

Ultimately, Oliver adapted to life on the IMB and began to play politics with the base's hierarchy, earning him access to confidential information and corporate secrets. Commander Graves of the Eastern Alliance had learned of Oliver's dissatisfaction in tandem with his resourcefulness, and employed the scientist to sabotage all IMB research projects. Oliver accepted a generous payment and set to work, but was captured by Joe 90 and charged by WIN for his acts of murder and espionage.

MISSION: MURDER ON THE MOON

SUPPORT FILE:
0122

LUNAR ROVER

An integral part of life on the Moon, the lunar rover is an extremely versatile form of transport designed to act as a shuttle, scout and mobile laboratory all in one. Using a gimballed suspension system the vehicle is able to clear most forms of terrain with ease while four vertical jets enable it to "jump" over impassable objects.

In order to transport them to the Moon, the rovers are designed to be extremely lightweight and easy to disassemble. This makes them difficult to drive by anyone who is not a trained expert as a miscalculated boost or jump can send them out of control at lethal speeds.

The IMB operates eight of these craft, all with various levels of modification to tailor them to specific responsibilities.

LUNAR ROVER
DIAGRAM #015

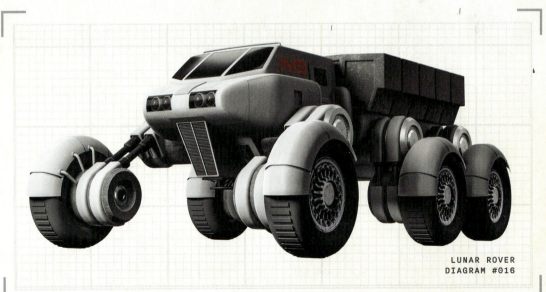

LUNAR ROVER
DIAGRAM #016

BRAIN PATTERN SPECIFI

NAME: Colonel Joe Steeler

CODE: G17

BORN: February 16, 1962

PLACE OF BIRTH: Liverpool, England

KNOWLEDGE & EXPERIENCE:
Aerospace construction, EVA expert, engineer, Moonbase designer.

EXIT STRATEGY

CONCLUSION

As with any ongoing operation, an exit strategy is required in order to allow the participants to disband and return to normal life safely. As an "off-books" operation, Project 90 has to take a slightly different approach to this strategy to meet certain scenarios.

THE MCCLAINES DECIDE TO END THEIR PARTICIPATION

In every way, this operation hinges on two civilians who have offered their time, talents and lives in the pursuit of world peace. Sam Loover and I have both sworn that both Professor McClaine and Joe can rescind their participation in this project at any point should they feel their safety has been compromised or if the operation has pushed any moral boundaries. In this situation, WIN may still have access to the BIG RAT on an occasional basis for use by Sam and myself.

JOE IS KILLED ON A MISSION

Unfortunately in this line of business, we are forced to consider the unthinkable. Should Joe 90 be killed on assignment a prepared cover-up will move into place to explain Joe's death to the local authorities. Understandably, Professor McClaine's association with WIN will likely cease.

THE OPERATION IS EXPOSED TO WIN AND THE PUBLIC

Given the questionable ethics around Project 90, exposing the operation will almost certainly lead to an enquiry and serious consequences for those involved. In the event our cover is blown, a plan has been put together to lean into the plausible deniability of the operation. Professor McClaine has made arrangements that the BIG RAT can be hastily buried, hiding the existence of the underground laboratory while one of the many decoy File 90s will be taken from the WIN archive to explain the situation. Should this succeed and the authorities be placated it would theoretically be possible to resume the operation, but in the light of increased scrutiny, this may be impossible.

BRAIN PATTERN SPECIFICATIONS

NAME: Lionel Jones
CODE: T3
BORN: November 1, 1973
PLACE OF BIRTH: Manchester, England

KNOWLEDGE & EXPERIENCE:
Twenty years working with the National Monorail Service as the lead engineer on projects to automate and improve the safety of key areas on the rail network. Responsible for designing the Class 31 mono-train automatic control system rolled out across the fleet.

13:59 NMS SERVICE TO GLASGOW

DATE:
February 26, 2012

DEPARTURE:
13:59 London Euston (EUS)

SCHEDULED ARRIVAL:
15:03 Glasgow Central (GLC)

CALLING AT:
London Euston, Wigan, Lancaster, Carlisle, Glasgow Central

DRIVERS:
Keith Wilkin and Jeremy Alexander

NOTES:
Service hijacked approximately five miles from Lancaster. WIN intervened to bring the train back under control before arriving at Glasgow Central ahead of schedule at 14:46. National Monorail Service ordered by the World Government to conduct a network-wide security inquiry. Details to follow.

CONCLUSION

For now, Project 90 continues, every day fixing one small part of the world in an attempt to slide the scales closer to true peace in our time. Whenever it will end, who will know. It is my hope that one day the services of young Joe will be used less and less until eventually, WIN will be once again able to handle these cases on its own. I fear though that those days are a long way off.

One thing I do know, however, is that every morning when people wake up and open their newspapers, they will read stories of arms dealers being thwarted, dictators being toppled, lives being saved. And some things they will never know – the wars being averted, all the triggers not being pulled – and the fact that all this is due to a 10-year-old boy in Dorset and his father, who together, just for a moment, make the world a safer place...

Shane Weston

Shane Weston

5 CONCLUSION

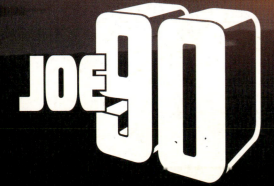

Written by
CHRIS THOMPSON
JACK KNOLL

Illustrated by
CHRIS THOMPSON

Edited by
STEPHANIE BRIGGS

Design by
AMAZING15

Produced by
JAMIE ANDERSON

Joe 90 created by
GERRY AND SYLVIA ANDERSON

WITH THANKS TO:
Tim Collins, Sophia Halliday, David Hirsch

Joe 90 ™ and © ITC Entertainment Group 1968 and 2023. Licensed by ITV Studios Limited. All Rights Reserved.

Additional imagery © Shutterstock

First published in the UK in 2023 by Anderson Entertainment

Hardback 978-1-914522-60-4

Printed at Interak Printing House, Poland

Apart from any use permitted under UK copyright law, this publication may only be reproduced, stored, or transmitted, in any form, or by any means, with prior permission in writing by the publishers or, in the case of reprographic production, in accordance with the terms of licences issued by the Copyright Licensing Agency. Cataloguing in Publication Data is available from the British Library.

Every effort has been made to fulfil requirements with regard to reproducing copyright material. The author and publisher will be glad to rectify any omissions at the earliest opportunity.

GERRYANDERSON.COM